The Effects of Housing and Management on the Behaviour and Welfare of Hens and Broilers

The Effects of Housing and Management on the Behaviour and Welfare of Hens and Broilers

Editors

Victoria Sandilands
Tina Widowski

Basel • Beijing • Wuhan • Barcelona • Belgrade • Novi Sad • Cluj • Manchester

Editors
Victoria Sandilands
Department of Agriculture
and Land-Based Engineering
Scotland's Rural College
(SRUC)
Edinburgh
United Kingdom

Tina Widowski
Department of Animal
Biosciences
University of Guelph
Guelph
Canada

Editorial Office
MDPI
St. Alban-Anlage 66
4052 Basel, Switzerland

This is a reprint of articles from the Special Issue published online in the open access journal *Animals* (ISSN 2076-2615) (available at: https://www.mdpi.com/journal/animals/special_issues/ The_Effects_of_Housing_and_Management_on_the_Behaviour_and_Welfare_of_Hens_and_Broilers).

For citation purposes, cite each article independently as indicated on the article page online and as indicated below:

Lastname, A.A.; Lastname, B.B. Article Title. *Journal Name* **Year**, *Volume Number*, Page Range.

ISBN 978-3-7258-1104-5 (Hbk)
ISBN 978-3-7258-1103-8 (PDF)
doi.org/10.3390/books978-3-7258-1103-8

Contents

animals

Impacts of Rearing Enrichments on Pullets' and Free-Range Hens' Positive Behaviors across the Flock Cycle

Dana L. M. Campbell *, Sue Belson, Tim R. Dyall, Jim M. Lea and Caroline Lee

Agriculture and Food, Commonwealth Scientific and Industrial Research Organisation (CSIRO), Armidale, NSW 2350, Australia; sue.belson@csiro.au (S.B.); tim.dyall@csiro.au (T.R.D.); jim.lea@csiro.au (J.M.L.); caroline.lee@csiro.au (C.L.)
* Correspondence: dana.campbell@csiro.au

Simple Summary: Enrichment during the indoor rearing of young laying hens (pullets) destined for free-range systems may improve pullet development and increase motivated natural behaviors (termed 'positive behaviors') such as foraging, dust bathing and chick play. Hy-Line Brown®chicks ($n = 1700$) were floor-reared indoors across 16 weeks with three enrichment treatments ($n = 3$ pens/treatment): (1) standard control, (2) weekly novel objects—'novelty', (3) perching/navigation structures—'structural'. Pullets (16 weeks old: $n = 1386$) were then transferred to nine identical pens within rearing treatments, with outdoor range access from 25 to 65 weeks. Video cameras recorded the pullet pens, adult indoor pens, and outside range. During rearing, observations of play behavior in chicks at 2, 4 and 6 weeks showed no overall effect of rearing treatment. At 11 and 14 weeks only the novelty hens were observed to increase their foraging across age with no differences between treatments in dust bathing. Observations of adult hens at 26, 31, 41, 50, 60 and 64 weeks showed that the structural hens exhibited more dust bathing and foraging overall than the control hens, but that both novelty and/or structural hens showed small increases relative to control hens depending on the behavior and location. Across age, adult hens differed in the degree of dust bathing performed inside or outside and foraging outside but not inside. For litter-reared pullets, additional enrichments may result in some long-term increases in positive behaviors.

Citation: Campbell, D.L.M.; Belson, S.; Dyall, T.R.; Lea, J.M.; Lee, C. Impacts of Rearing Enrichments on Pullets and Free-Range Hens' Positive Behaviors across the Flock Cycle. *Animals* **2022**, *12*, 280. https://doi.org/10.3390/ani12030280

Academic Editors: Victoria Sandilands and Tina Widowski

Received: 23 November 2021
Accepted: 20 January 2022
Published: 23 January 2022

Abstract: Enrichment during the indoor rearing of pullets destined for free-range systems may optimize pullet development including increasing motivated natural behaviors (termed 'positive behaviors') including foraging, dust bathing and chick play. Hy-Line Brown®chicks ($n = 1700$) were floor-reared indoors across 16 weeks with three enrichment treatments ($n = 3$ pens/treatment): (1) standard control, (2) weekly novel objects—'novelty', (3) perching/navigation structures—'structural'. At 16 weeks, pullets ($n = 1386$) were transferred to nine identical pens within rearing treatments with outdoor range access from 25 to 65 weeks. Video cameras recorded the pullet pens, adult indoor pens, and outside range. During rearing, observations of play behavior (running, frolicking, wing-flapping, sparring) in chicks at 2, 4 and 6 weeks (total of 432 thirty-second scans: 16 observations × 3 days × 9 pens) showed no overall effect of rearing treatment ($p = 0.16$). At 11 and 14 weeks only the 'novelty' hens were observed to increase their foraging across age ($p = 0.009$; dust bathing: $p = 0.40$) (total of 612 thirty-second scans per behavior: 17 observations × 2 days × 2 age points × 9 pens). Observations of adult hens at 26, 31, 41, 50, 60 and 64 weeks showed that the structural hens exhibited overall more dust bathing and foraging than the control hens (both $p < 0.04$) but both novelty and/or structural hens showed small increases depending on the behavior and location (total of 4104 scans per behavior: 17 observations × 2 days × 6 age points × 9 pens × 2 locations = 3672 + an additional 432 observations following daylight saving). Across age, adult hens differed in the degree of dust bathing performed inside or outside (both $p \leq 0.001$) and foraging outside ($p < 0.001$) but not inside ($p = 0.15$). For litter-reared pullets, additional enrichments may result in some long-term increases in positive behaviors.

Keywords: dust bathing; foraging; play; laying hen; novel objects; perching structures; navigation

1. Introduction

Free-range laying hen production systems are prevalent within Australia due to their popularity with consumers [1]. Across Australia and internationally, free-range hens are perceived to have improved welfare and the eggs are preferred by consumers for perceived better quality and health benefits [2,3]. However, the welfare status of hens in free-range systems can be complex as there are both benefits and challenges to providing birds with outdoor access [4]. Free-range systems can provide hens with increased space, a more natural outdoor environment and greater ranging can improve plumage quality [5–7]. Conversely, the outdoor environment is more unpredictable than the controlled indoor setting and may place greater physical stressors on hens which could increase mortality [8,9].

One of the potential benefits of free-range systems is the increased space and outdoor environment that provides more freedom to exhibit behaviors that are viewed as part of a positive behavioral repertoire for hens [10]. These positive species-specific behaviors are typical natural behaviors that are not abnormal or negative (e.g., severe feather pecking, smothering) and hens show motivation to perform [11]. These include dust bathing and foraging (scratching and pecking at the ground) where environments that facilitate these (and other) natural behaviors are believed to provide positive welfare experiences for commercial hens [10]. These behaviors are thwarted in conventional cage systems but are facilitated in indoor litter-based systems and may be greater in systems with outdoor access. Previous observations in an experimental free-range setting showed behavioral repertoires differed between hens located inside the shed versus out on the range with hens located outside showing greater foraging and dust bathing relative to what was exhibited by hens inside on the litter [12]. On a commercial free-range farm, hens were primarily observed to be foraging when outside on the range [13], with foraging likely to occur more frequently than dust bathing [14], although behaviors can vary depending on the vegetation and topography outside [15].

While behaviors of dust bathing and foraging are innate and are performed even in the absence of suitable substrates [16,17], appropriate behavioral development can be affected by the environment pullets are reared in. The rearing environment overall is critical for physical, physiological, and behavioral development of the pullets (young, developing laying hens). There can be long-term effects on bird welfare if rearing environments are sub-optimal, or not best matched for the laying environment the pullets are transferring to later in life [18,19]. For example, rearing with access to ramps can improve use of the elevated areas for hens housed in aviaries and decrease keel bone damage [20]. Access to litter in the first four weeks of life can have long-term impacts on the development of feather pecking behaviors [21,22], which may be related to litter stimulating natural foraging behavior versus undesirable pecking of conspecifics. Adult hens may readily utilize an available dust bathing substrate even if they were reared without substrate exposure [23,24]; however, early experience without a suitable substrate may explain why some hens still sham dust bathe in the presence of litter [25]. Evidence to date demonstrates specific types of environmental enrichment can have effects on specific related behaviors as birds mature (e.g., older pullets used ramps to a greater extent when they had access to them as chicks; [26]) but can also have more generalized effects, such as rearing complexity reducing fear responses in young adult hens [27]. More generalized effects of optimizing bird development could include impacts on positive species-specific behaviors such as dust bathing and foraging. While the presence of litter can affect these, additional enrichments to a litter substrate may have even greater impacts. Peat and hay enrichments increased ground pecking in broilers, even when the birds were not in proximity of the enrichments [28]. If enrichments have generalized impacts on increasing positive behaviors the effects could be first apparent in the early weeks of life through changes in chick play behavior, which are expressed early in life and generally viewed as being positive [29,30].

Play behavior is performed across many animal species and may be associated with better preparing animals for the unexpected later in life [29]. Play may also be indicative of improved welfare, although not in every case, with some animals showing increased play following stressful experiences [30]. While there is limited research conducted on play behavior in chickens, spontaneous play has been observed in broiler chickens [28,31,32]. Play typically occurs in the early weeks of a chicken's life and will decrease across age [28,31,32]. Play has previously been categorized to include behaviors such as running, worm/food running (running with an object in the beak) wing-flapping, frolicking (running with wings flapping) and sparring/play fighting, all of which are performed spontaneously in young chicks or can be stimulated in experimental contexts [28,31,32]. In the limited research available, the effects of enrichment on play are unclear. No effects have been found on spontaneous play [28,31,32] but non-enriched birds have shown more play during specific play tests [32].

Pullets destined for free-range systems in Australia (and elsewhere) are typically reared indoors before being transferred to a laying facility with outdoor access. The discrepancies between the rearing and laying housing systems may impact how the adult hens adapt to their new housing. With outdoor access for pullets being logistically difficult, enriching the rearing environment may be a strategy to optimize pullet development, improving welfare and adaptability. A rearing enrichment trial was designed to measure the impacts of different types of rearing environments on the behavior, health, production, and welfare of a flock of free-range hens in an experimental setting and showed long-term effects of the type of environment the pullets were reared in across multiple measures [5,33,34]. The aim of this study was to assess how these different rearing enrichments affected chick play behavior as well as foraging and dust bathing in the pullets and adult hens across the flock cycle. It was predicted that both types of rearing enrichments would increase these positive species-specific behaviors through generalized impacts on optimizing pullet behavioral development. While the enrichment types were distinct, there were no clear predictions on how they may differentially affect the birds, as there is limited literature available and neither of the enrichment types specifically target litter-related behaviors.

2. Materials and Methods

2.1. Ethical Statement

All research was approved by the University of New England Animal Ethics Committee (AEC17-092).

2.2. Animals and Housing

2.2.1. Rearing (0–16 Weeks of Age)

This study used 1700 Hy-Line® Brown layers that were first reared indoors for 16 weeks in the Rob Cumming Poultry Innovation Centre of the University of New England, Armidale, Australia, before transfer to the Laureldale free-range facility of the University of New England, where they remained until the conclusion of the trial at 65 weeks of age. The housing set-up for the birds has been described previously (e.g., [5,33,34]). In November 2017, day-old, beak-trimmed chicks from a commercial hatchery were placed into nine floor-litter pens (6.2 m L × 3.2 m W) within three rooms. Chicks arrived in multiple boxes with boxes randomly allocated to treatment pens (approximately two boxes/pen). All pens within a room were visually separated with shade cloth attached to the wire pen dividers. Rice hulls covered the ground as floor litter, four round feeders per pen provided ad libitum access to commercially formulated mash and drinking water was provided via nipples (20 nipples/pen). Resources either met or exceeded the current Australian Model Code of Practice for the Welfare of Animals—Domestic Poultry [35]. Three separate rearing enrichment treatments were applied with one treatment replicate per room, balanced for location. (1) a 'control' group with just the floor litter, (2) a 'novelty' group where novel objects were changed at weekly intervals (e.g., balls, bottles, bricks, brooms, brushes, buckets, containers, pet toys, plastic pipes, strings, water bottles) and (3) a 'structural' group where five custom-designed H-shaped perching/navigation structures (L, W, H: all 0.60 m) with two solid panels and an open-framed side were provided in different orientations within each pen. Pullets could perch on these structures and the solid side in some orientations created a visual/physical barrier requiring the birds to navigate around it, thus the structures were designed to add complexity and stimulate both perching and navigation in the pen. There was approximately 6 cm perching space/bird during rearing provided by these structures. By 16 weeks of age, bird density was approximately 15 kg/m^2 (average 174–190 pullets/pen resulting from both chick mortality and chick placement error). Temperature and lighting schedules followed the Hy-Line® Brown alternative management guidelines [36] except the LED lighting was maintained at 100 Lux as the pullets were being reared for a free-range system. Rooms were mechanically ventilated but there was no cooling system present. Litter was visually assessed during daily routine bird health checks and was deemed dry and friable throughout the rearing period across all pens. Chicks and pullets were vaccinated as per regulatory requirements and standard recommendations.

2.2.2. Free-Range Facility—Indoor Pens (16–65 Weeks of Age)

At 16 weeks of age, 1386 pullets were transferred to nine identical, visually isolated pens within a single shed at the Laureldale free-range facility of the University of New England with rearing treatments balanced for location across the shed. Additional pullets, surplus to the space restrictions of the layer shed, were rehomed including those birds that were either heavier or lighter than the mean body weight, and then some additional randomly selected pullets to reach the desired quota. Pullets were socially remixed within pen replicates of their rearing treatments (three pen replicates per rearing treatment) to simulate the social remixing that occurs commercially. However, pullets were placed into three pens containing only birds from the same rearing treatment. Bird density was approximately 9 birds/m^2 (n = 154 hens/pen; 3.6 m W × 4.8 m L). Each pen contained nest boxes, perches, feeders, and water nipples to meet or exceed requirements of the *Australian Model Code of Practice for the Welfare of Animals-Domestic Poultry* [35]. Pen space logistics restricted perching space to 10 cm per bird, but hens also perched on the tops of the waterline and feeders. Rice hulls formed the floor litter substrate with regular raking management and one complete mid-lay litter replacement to ensure dryness and friability of the litter. By 30 weeks of age, the LED lighting schedule had gradually reached 16 hours light and 8 hours dark with an average pen intensity of 10.0 (±0.84 SE) Lux (Lutron Light Meter, LX-112850; Lutron Electronic Enterprise CO., Ltd., Taipei, Taiwan) as measured at birds' eye height from three pen locations (front, middle, back) when the pop-holes were closed. This lux was the highest that could be achieved with the shed lighting system. The shed was fan-ventilated with no temperature or humidity control.

2.2.3. Free-Range Facility—Outdoor Range (16–65 Weeks of Age)

Each indoor pen was connected to an outdoor range area accessible via two pop-hole openings (18 cm W × 36 cm H). The nine range areas were visually isolated from each other via shade cloth on the wire fences. The pop-holes first opened at 25 weeks of age (May 2018) and provided daytime range access on an automated schedule from 09:15 until after sunset resulting in approximately 9 h of available ranging time across winter and approximately 11 h of available ranging time daylight saving time onward (October 2018). The range area comprised of a 1.2 m length concrete path, followed by 3.6 m length of river rock and then a 26.2 m length of grassed area with no trees or artificial shelters. Monthly photos of the range areas allowed visual estimation of vegetation coverage. Initially the range areas were 90% covered in grass, which was destroyed by hens or seasonal die-off after 8 weeks of range access. Six months after first range access (hen age: 48 weeks), there was some spring grass regrowth with up to 40% coverage in some pens (3 pens 0%, 4 pens 20%, 2 pens 40%) but by summer (8 months after first range access: hen age: 56 weeks) the ranges were only bare dirt with some scattered hen-resistant weeds.

2.3. Video Recording and Data Collection

Hikvision Network cameras (Model DS-2CD2232-I5 4 mm, Hikvision, Hangzhou, China) were installed to capture the indoor rearing pens during light hours at 2, 3, 4, 6, 8, 11 and 14 weeks of age. The same cameras were installed to capture the indoor pens and range area of each pen at the layer facility at 26, 31, 41, 50, 60 and 64 weeks of age. Across the ranging period there was typically little rain due to severe drought in the region. Due to camera angles, across all pens equally, approximately 0.5 m in front of the pop-holes inside and approximately 1.2 m in front of the pop holes outside was unavoidably excluded from video capture. Video recordings were later decoded by observers who were blind to rearing treatment or blind to the aims of the trial where rearing enrichments were visible in the video. The observers were all trained by a single researcher who simultaneously did sections of video with the trainee to ensure the correct behaviors were being identified. For instances where two observers were collecting data on the same behavior, both observers watched one identical section of video first independently (one day of observations across one pen). If inter-observer reliability was initially below 90% as assessed by correlation in Microsoft Excel (agreement values ranged from 76–89%), the two observers then discussed the section of video to reach 100% agreement in the identification of birds performing each behavior at each time point the observers previously showed discrepancies on before proceeding with their independent observation days. Where two observers watched one behavior, the allocation of pens and treatments to observe were balanced to minimize any potential observer bias per a specific rearing treatment.

The observers collected data as follows:

1. Rearing—Enrichment interactions: counts of birds using/interacting with enrichments in the rearing pens at 3, 6, 8, 11 and 14 weeks of age. These data were collected by one observer to document the use of the enrichments during rearing with no intended comparison between treatment groups. At each age point a single day of video per enriched pen (novelty and structural pens, not control pens) was observed with point counts made every 30 min from 08:00 until 17:30 (total 20 counts per day × 5 days × 6 enriched pens = total dataset of 600 counts). Days were selected to be at least 2 days after the new novel objects had been added but also at least 2 days after other/disturbances interventions such as body weight assessment as part of a separate dataset [37] and vaccinations. Interaction with the enrichments was defined as a bird perched on, pecking at, or standing/sitting directly next to an enrichment (i.e., less than a bird body width away). The structural enrichment remained the same throughout the rearing period, but the novel objects changed weekly and thus were variable across the 5 days assessed. The video was played at each time point for up to 10 s as needed to confirm bird behavior at the specified time point.

2. Rearing—Play behavior of chicks: counts of chicks exhibiting play behavior were observed in each pen across one day at 2, 4, and 6 weeks of age. Ages were selected based on previous literature on broiler chickens [31,32] of when play behavior may be most prevalent. While the match between broiler chickens and laying hens is limited given broiler chickens reach maturity (and hence slaughter) at 6 weeks of age, this selected age period provided a starting reference point for documenting (potentially peak) play in laying hen chicks. Observations by a single observer of running, frolicking, wing-flapping and sparring were made based on the ethogram as described in Table 2 of Liu et al. (2020) [32]. Wing-flapping was included, as although it is often classified as a comfort behavior in older birds, it has been observed to be associated with play or aggressive interactions in chicks [31,32,38]. Data were collected across a 30 s period every 30 min throughout the day totalling 432 observations (16 observations per day × 3 days × 9 pens). Only a single day was chosen at each age point as it was uncertain how much play behavior would be observed (if any) and there were shorter time intervals between the observation ages relative to the dust bathing and foraging observations in the pullets and adults.

3. Rearing—Pullet foraging and dust bathing: at 11 and 14 weeks of age, all pullet pens were observed by a single observer across two days per age point to count the number of birds dust bathing or foraging (defined as feet scratching backwards in the litter typically

followed by pecking in the litter) across 30 s every 30 min from 09:30 until 17:30 (total 17 observations × 2 days × 2 age points × 9 pens = 612 observations for each behavior). This definition of foraging has been used in previous studies [12,39] although some authors include other exploratory behaviors within their foraging definition [40].

4. Free-range facility—Hen foraging and dust bathing: at each age, approximately one week of video was recorded with the specific days of observation within the week selected based on a full set of recordings with no missing video due to technical issues, and predominantly dry weather. Across two days each at 26, 31, 41, 50, 60 and 64 weeks of age, the number of hens dust bathing or foraging inside were counted by two observers across a 30 s period every 30 min from 09:30 (pop-holes opened at 09:15) until 17:30 (just before sunset) or until 19:30 from 50 weeks onwards following daylight saving time change (total 17 observations × 2 days × 6 age points × 9 pens = 1836 + an additional 216 observations following daylight saving: total 2052 observations each of dust bathing or foraging across the flock cycle). At each observation point, the corresponding observations of hens' dust bathing or foraging were made outside on the range (n = 2052 observations each for dust bathing and foraging outside) by a different two observers. Selected days within age points had one full day between them that was not observed (i.e., the selected days per age week were not consecutive).

5. Free-range facility—Time budgets of hens: across two days at 50 weeks of age, a 10 m length portion of the range area for each pen was selected for time budget observations by a single observer (the same area was selected for each range, in mid-view of the video capture). This age point was selected as foraging and dust bathing on the range were observed at higher levels and daylight hours were extended for more observations. Scan sampling was applied every 30 min from 09:30 until 19:00 with hens in the designated area first counted and then a behavior allocated per hen based on the ethogram in Table 1. At each time point the video was played for a few seconds to confirm the behavior the hen was exhibiting (total 20 observations points × 9 pens × 2 days = 360 observations points).

Table 1. Ethogram of the behaviors observed for each hen whilst out on the range at 50 weeks of age.

Behavior	Description
Body shaking	Hen completes a full shake of her body ruffling her feathers
Dust bathing	Hen is lying on the ground, kicking dirt onto her feathers and tossing it over her body with her wings and full body movement
Fighting	Two hens are jumping up and pecking at each other with force
Jumping in air/flying	Hen jumps into air, flaps wings, and travels a short distance
Foraging	Hen scratches her feet backwards in the dirt and then pecks the ground
Pecking	Hen is using her beak to touch the ground or surrounding environment Hen may pick up something (e.g., dirt) with her beak
Pecking other chickens	Hen is using her beak to touch another hen
Piling	Hens are in a group tightly clustered together
Preening	Hen is using her beak on her feathers to align them or pull off debris (e.g., dirt)
Standing	Hen is upright and remaining in one location.
Sunbathing	Hen is lying in the dirt with wings spread out and is motionless (i.e., not moving around as per dust bathing activity)
Running	Hen is upright and moving forward at a fast pace
Tail shaking	Hen shakes tail feathers whilst walking or standing
Walking	Hen is upright and moving forward at a slower pace than when classified as running
Wing flapping	Hen's wings are outstretched and rapidly flapped while hen remains on the ground (i.e., not airborne)

2.4. Data and Statistical Analyses

All analyses were conducted in JMP 14.0.0 (SAS Institute, Cary, NC, USA) with $\alpha = 0.05$. All proportions were calculated taking cumulative mortality into account. All data were checked for normality and transformed where necessary for parametric tests. The studentized residuals were visually inspected to ensure homoscedascity. Non-parametric tests were conducted where transformations could not make the data normally distributed.

The count data for interaction with enrichments were converted to proportion of birds within the pen at each time point and visually displayed. No statistical analyses were conducted on these data as there were no specific comparisons to be made among treatment groups. The counts of chicks exhibiting play behaviors (running, frolicking, wing-flapping and sparring) were converted to proportions of chicks within each pen performing each behavior at each age point. Observations across the day were summed into a daily mean per pen per age point ($n = 27$: 3 × daily means × 9 pens) and were logit transformed. A constant of 0.001 was added to the sparring proportions only prior to transformation to account for zero values. A General Linear Mixed Model (GLMM) was applied to each behavior (and all play behaviors summed together) with rearing treatment, age and their interaction as fixed effects, including pen nested within treatment as a random effect. Where significant differences were present, post-hoc Tukey's tests were applied to the least squares means.

The counts of pullets dust bathing and foraging were converted to proportions of pullets performing the behaviors, logit transformed and mean daily values were calculated per pen for each behavior ($n = 36$: 4 daily means × 9 pens). A General Linear Mixed Model (GLMM) was applied with rearing treatment, age, and their interaction as fixed effects including pen nested within treatment and observation day as random effects. Where significant differences were present, post-hoc Tukey's tests were applied to the least squares means.

The counts of hens dust bathing or foraging inside and outside were converted to proportions of all hens in the pen and summed across the two locations. The conversion to proportions of all hens in the pen rather than proportions of hens specifically inside or outside was to display the proportions of the total group that were exhibiting each behavior in each location (i.e., conversions based on hens present inside or outside would inflate the proportions of hens exhibiting the behavior). The original dataset (2052 observations per behavior) was summarized to include one mean value per pen per day each for dust bathing and foraging ($n = 108$ per behavior: 2 days × 6 age points × 9 pens). The proportions were logit transformed but were not normally distributed and were analyzed for an effect of rearing treatment using separate Kruskal-Wallis tests, including blocking for the effect of age (only one blocking factor was permitted in the analyses). Where significant differences were present, post-hoc tests were conducted between all pairs using the Steel-Dwass method. The proportions of hens dust bathing or foraging inside the shed or outside on the range were then analyzed separately for an effect of rearing treatment using separate Kruskal-Wallis tests that included blocking for the effect of age. A constant of 0.001 was added to these data prior to logit transformation to account for values of zero. Finally, effect of age for dust bathing or foraging in indoor and outdoor locations was analyzed using separate Kruskal-Wallis tests blocking for effects of rearing treatment.

The counts of hens performing specific behaviors at 50 weeks of age were converted into proportions of hens in the observation area exhibiting each behavior. Due to low incidences of some behaviors, observations of body shaking, preening, sunbathing, tail shaking and wing flapping were combined into a single category of 'comfort behaviors'. The original dataset was summarized into one mean value per behavior per pen per day (summarized dataset: 2 days × 9 pens = 18 datapoints per 11 behaviors). The behaviors of jumping/flying, piling, pecking other chickens and fighting occurred too infrequently (~ 1% of the hens' time budget combined) and were not included in any further analyses. The proportions of comfort behaviors, dust bathing, foraging, pecking, running, standing and walking were analyzed for an effect of rearing treatment using separate non-parametric Kruskal-Wallis tests.

3. Results

3.1. Rearing

Observations of the proportions of pullets utilizing enrichment in the two enriched rearing treatments showed that the birds were interacting with the provided enrichment across the rearing period with approximately 10% of pullets using them at any single point in time (overall mean $\pm$ SEM; novelty pullets: 9.22% $\pm$ 0.22; structural pullets: 11.47% $\pm$ 0.25, Figure 1).

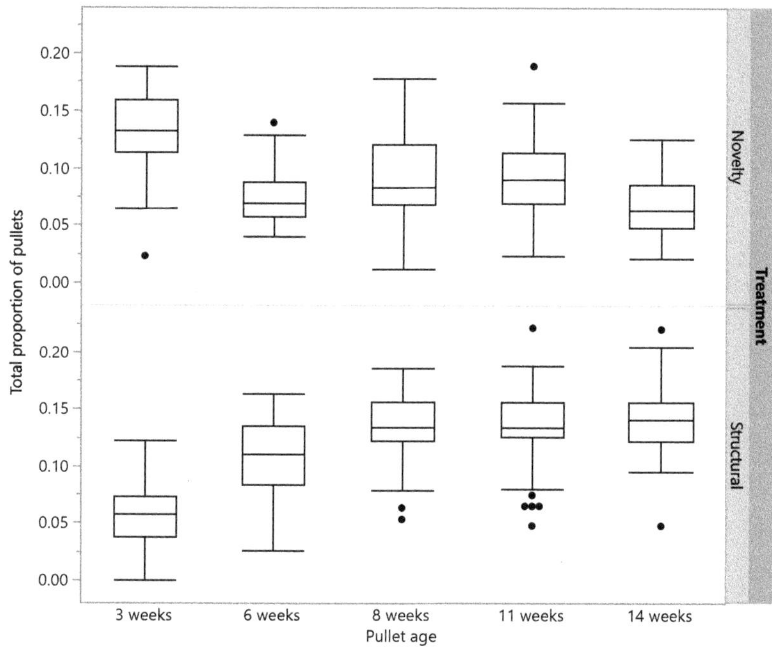

Figure 1. Box plots indicating the proportion of pullets in the pen that were interacting with enrichments in the novelty and structural treatment pens across five age points (3 to 14 weeks of age). The horizontal line within each box indicates the median value with the box ends representing the 1st and 3rd quartiles. The whiskers extend to the outer datapoints that fall within a distance 1.5 $\times$ outside the 1st or 3rd quartiles. If datapoints do not reach these computed ranges, the whiskers represent the upper and lower data points (excluding outliers).

For play behaviors, there was an interaction between rearing treatment and age for the proportion of chicks running ($F_{(4,12)}$ = 5.28, p = 0.02), with the enriched chicks showing less running at four weeks of age compared with the control chicks (Figure 2). There was no overall effect of rearing treatment ($F_{(2,6)}$ = 2.57, p = 0.16), but running decreased linearly across age ($F_{(2,12)}$ = 59.38, p < 0.0001, Figure 2). There was only an effect of age on the proportion of chicks frolicking ($F_{(2,12)}$ = 5.76, p = 0.02) with less frolicking at six weeks (Table 2). There was no effect of rearing treatment ($F_{(2,6)}$ = 2.81, p = 0.14), or interaction between age and rearing treatment ($F_{(4,12)}$ = 0.44, p = 0.78). There was a significant effect of rearing treatment on the proportion of chicks showing wing-flapping ($F_{(2,6)}$ = 9.50, p = 0.01) with the structural chicks showing less than control and novelty chicks (Table 2). There was also a significant effect of age ($F_{(2,12)}$ = 4.67, p = 0.03) with less wing-flapping at two weeks of age compared with six weeks of age (Table 2). There was no interaction between rearing treatment and age ($F_{(4,12)}$ = 0.57, p = 0.69). There was a significant interaction between age and rearing treatment on the proportion of chicks sparring ($F_{(2,12)}$ = 3.29, p = 0.048), with the control chicks only showing more sparring at six weeks relative to two weeks of age. There was a significant effect of age ($F_{(2,12)}$ = 5.44, p = 0.02) with more sparring at four weeks than at two weeks, but no overall effect of rearing treatment ($F_{(2,6)}$ = 0.25, p = 0.79). When all play behaviors were combined, there was only a significant effect of age ($F_{(2,12)}$ = 29.05, p < 0.0001) with play linearly decreasing across age. There was no effect of rearing treatment ($F_{(2,6)}$ = 2.94, p = 0.13) and no interaction between rearing treatment and age ($F_{(4,12)}$ = 1.22, p = 0.35). Running was the most frequently observed play behavior but was still observed in less than 10% of the chicks during observations, with sparring only observed in a few chicks (Table 2).

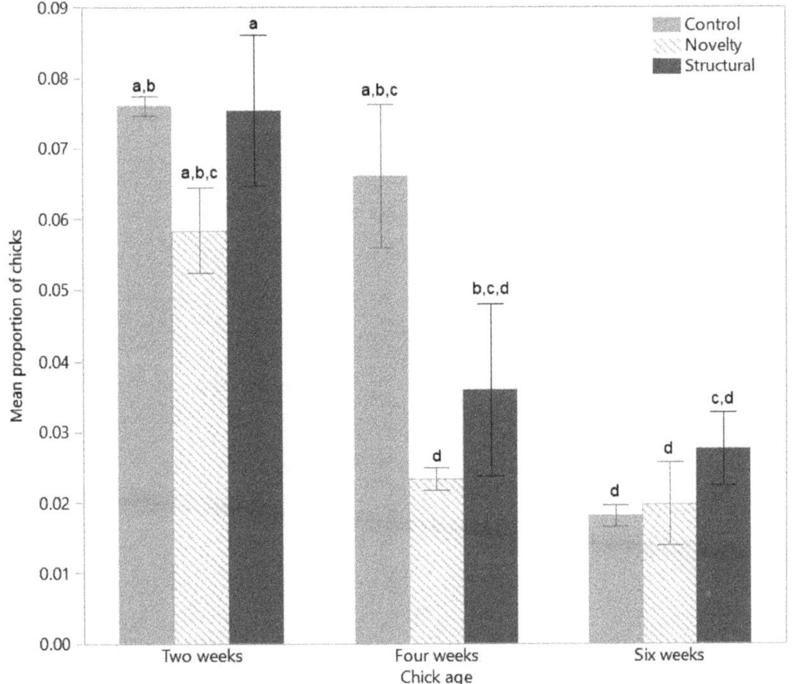

Figure 2. The mean (±SEM) proportion of chicks from three rearing treatments (control, novelty, structural) running within the pens across three age points (2, 4, 6 weeks). [a–d] Dissimilar superscript letters indicate significant differences across rearing treatments and age. Raw data are presented with analyses conducted on transformed means.

The proportions of pullets dust bathing in their rearing pens were similar across rearing treatments ($F_{(2,6)}$ = 1.06, p = 0.40), but the proportions decreased from 11 to 14 weeks ($F_{(1,2)}$ = 332.18, p = 0.003) with no interaction between treatment and age ($F_{(2,22)}$ = 2.40, p = 0.11, Figure 3). There was a significant interaction between age and rearing treatment for the proportion of pullets foraging ($F_{(2,22)}$ = 5.67, p = 0.01) with the pullets from the novelty treatment increasing their foraging with age, but the control and structural pullets remained at similar levels between 11 and 14 weeks (Figure 3). Visually, all groups showed similar patterns of dust bathing and foraging across the day (Figure 3).

Table 2. The mean (±SEM) percentages of chicks that performed each play behavior (frolicking, wing-flapping and sparring) across three rearing treatments (control, novelty, structural) at three different observation age points (2, 4, 6 weeks). [a,b] Dissimilar superscript letters indicate differences across rearing treatments or across age. Raw means are presented with analyses conducted on transformed data.

	Behavior (Mean % ± SEM)	Frolicking	Wing-Flapping	Sparring
Treatment	Control	2.27 ± 0.51	1.12 ± 0.10 [a]	0.42 ± 0.09
	Novelty	1.75 ± 0.26	1.09 ± 0.07 [a]	0.29 ± 0.07
	Structural	2.27 ± 0.34	1.0 ± 0.13 [b]	0.35 ± 0.05
Age	Two weeks	2.43 ± 0.18 [a]	0.87 ± 0.07 [b]	0.22 ± 0.04 [b]
	Four weeks	2.42 ± 0.30 [a]	1.07 ± 0.06 [a,b]	0.41 ± 0.06 [a]
	Six weeks	1.44 ± 0.17 [b]	1.32 ± 0.10 [a]	0.43 ± 0.08 [a,b]

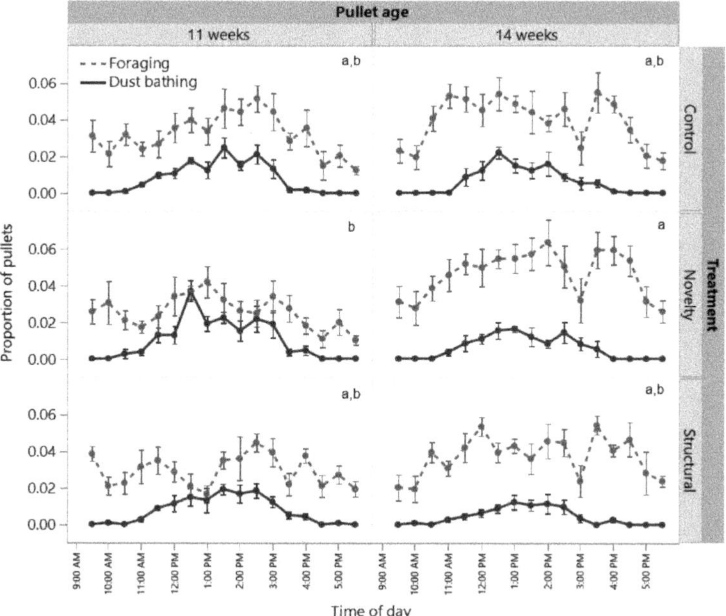

Figure 3. The proportion of pullets from three rearing treatments (control, novelty, structural) dust bathing or foraging across the day as assessed at 11 and 14 weeks of age. [a,b] Dissimilar letters indicate significant differences between treatments across age for foraging behavior. The raw mean (±SEM) values are presented across the day with statistical tests conducted on transformed daily total means.

3.2. Free-Range Facility

Across all age points there was a significant effect of rearing treatment on the total proportions of adult hens dust bathing (χ^2 = 13.81, df = 2, p = 0.001) and foraging (χ^2 = 6.53, df = 2, p = 0.04) with the structural hens showing more dust bathing and foraging than the control hens only (Figure 4).

There was a significant effect of rearing treatment on the proportion of hens dust bathing inside (χ^2 = 22.54, df = 2, p < 0.001) with both the novelty and structural groups showing more dust bathing than the control hens (both $p \leq$ 0.0007, Figure 5). There was a significant effect of rearing treatment on the proportion of hens dust bathing outside (χ^2 = 11.78, df = 2, p < 0.003) with the structural hens showing more dust bathing than the novelty hens only (p = 0.002), Figure 5). There was a significant effect of rearing treatment on the proportion of hens foraging inside (χ^2 = 8.39, df = 2, p = 0.02) with the novelty hens showing more foraging than the control hens only (p = 0.01, Figure 5). There was also a significant effect of rearing treatment on the proportion of hens foraging outside (χ^2 = 9.79, df = 2, p < 0.008) with the structural hens showing more foraging than the control hens only (p = 0.006, Figure 5).

There were significant differences across age for hens dust bathing inside (χ^2 = 22.40, df = 5, p = 0.0004) and outside (χ^2 = 64.59, df = 5, p < 0.0001) and significant differences across age for hens foraging outside (χ^2 = 64.55, df = 5, p < 0.0001), but not across age for hens foraging inside (χ^2 = 8.12, df = 5, p = 0.15, Figure 6). There was more variation across age for dust bathing and foraging behaviors observed outside on the range than inside the shed (Figure 6).

Analyses of the time budgets of hens on the range at 50 weeks of age showed no treatment differences in the proportion of hens performing comfort behaviors, dust bathing, foraging, pecking, running, standing and walking (χ^2 = 0.46–5.10, df = 2, $p \geq$ 0.08; Figure 7). The most frequent behaviors observed were walking, pecking and then standing (Figure 7).

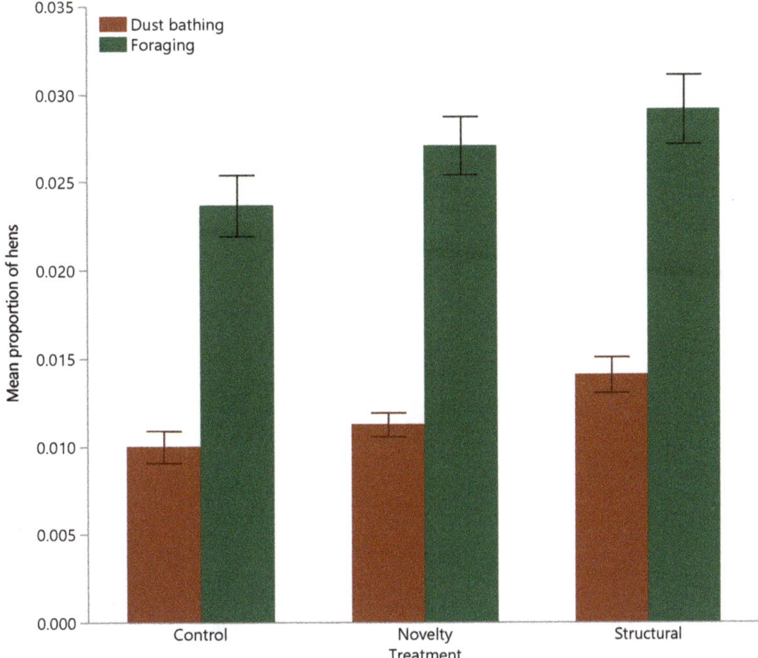

Figure 4. The mean (±SEM) proportion of hens summed for both inside the shed and outside on the range exhibiting dust bathing or foraging behavior across the flock cycle from three rearing treatments (control, novelty, structural). Raw data are presented.

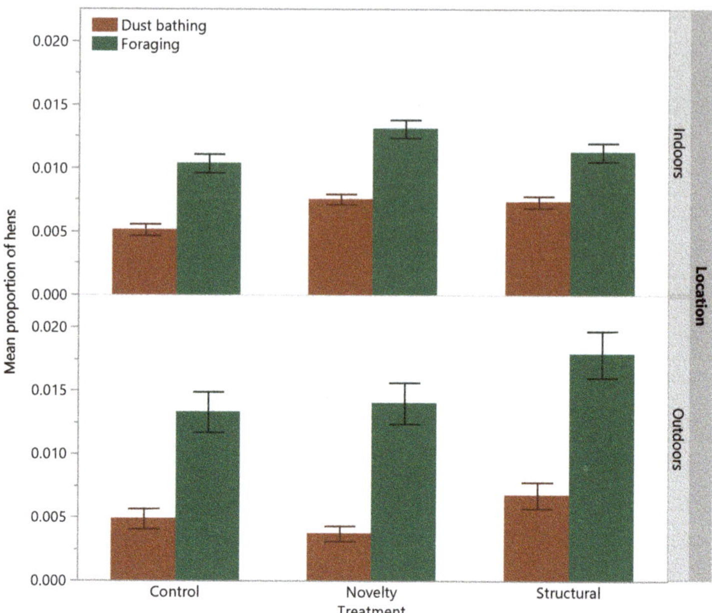

Figure 5. The mean (±SEM) proportion of hens inside the shed (indoors) or outside on the range (outdoors) exhibiting dust bathing or foraging behavior across the flock cycle from three rearing treatments (control, novelty, structural). Raw data are presented.

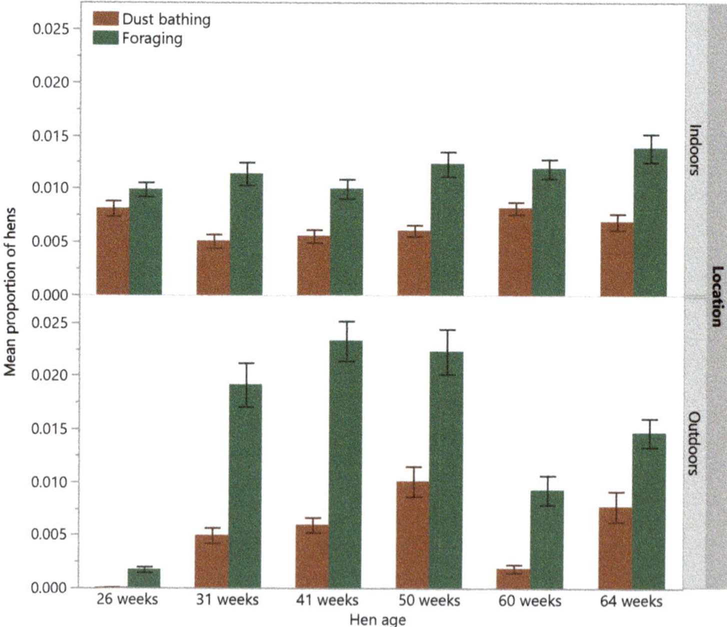

Figure 6. The mean (±SEM) proportion of hens inside the shed (indoors) or outside on the range (outdoors) exhibiting dust bathing or foraging behavior across hen ages (26, 31, 41, 50, 60, 64 weeks). Raw data are presented from all rearing treatments combined.

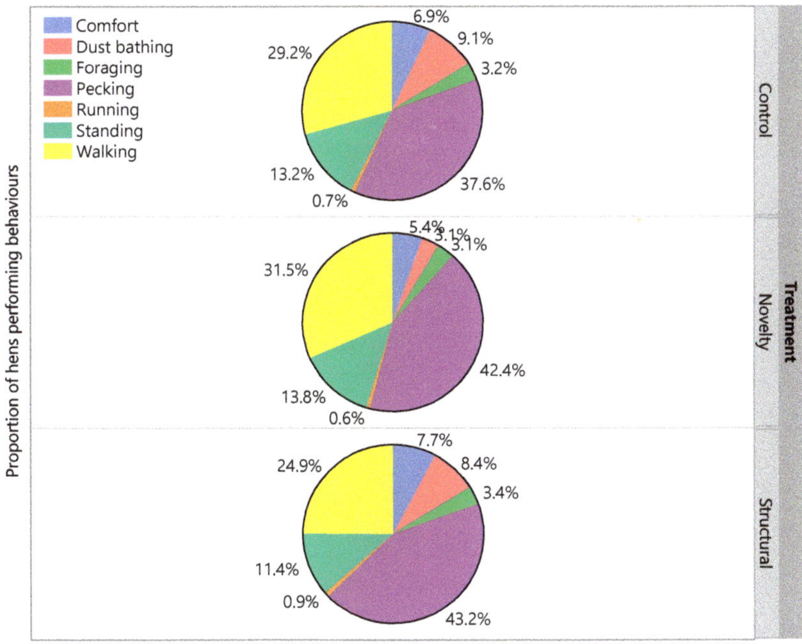

Figure 7. The percentages of hens on a portion of the range performing comfort behaviors, dust bathing, foraging, pecking, running, standing and walking across three different rearing treatments (control, novelty, structural). Observations were made when hens were 50 weeks of age across two days.

4. Discussion

Enrichments provided during the indoor rearing phase for free-range laying hens may have beneficial effects on multiple aspects of behavior and physical health. The aim of this study was to assess how different types of rearing enrichments may affect species-specific, motivated, natural laying-hen behaviors (termed positive behaviors) including foraging and dust bathing for pullets and adult hens across their production cycle as well as play behavior in chicks. Providing novel objects or perching/navigation structures to pullets raised on floor litter had some effects on play behavior but most play behaviors were observed in low frequencies equally among treatments. The novel objects increased the foraging behavior of pullets across age during the rearing treatment phase to a greater degree relative to the control and structural treatment groups. In the adult hens, the perching/navigation structure enrichments increased overall foraging and dust bathing relative to the control hens but there were treatment effects of both enrichment types, dependent on the location and behavior observed. Hens differed in the amount of foraging and dust bathing performed inside the shed versus outside on the range across age. These results demonstrate that for pullets reared on litter, additional enrichments can still result in some increases in these species-specific positive behaviors in the adult hens. These beneficial effects of rearing enrichments may become more apparent in the longer term as the adult birds are exposed to a new environment, come into lay and likely encounter various stressors across the production cycle.

Rearing enrichments resulted in a greater drop in running behavior relative to the control chicks from 2 to 4 weeks of age but when all play behaviors were combined, there was no effect of the rearing treatments indicating enrichments resulted in minimal impacts on play behaviors. It is possible that a different observation method such as extended continuous observations across fewer periods [32] may have increased observations of play occurrences. All play behaviors did decrease across age which is consistent with other studies [31,32]. The results on enrichment effects are similar to observations of play behavior in broiler chicks provided with enrichment. Broiler results showed no treatment differences in spontaneous play [32] or in play stimulated by personnel walking through and creating an open space [31], but when specific tests of play were conducted, the non-enriched birds were more responsive [32]. The higher proportions of running in the control chicks and greater increase in sparring behavior relative to enriched groups across age may have been a result of few other stimulatory objects in their environment to engage the chicks [32], or the greater open litter space in their comparatively empty pens. Currently the literature on play behavior in chickens is limited and while this study adds knowledge, there is still much to be understood about what stimulates play in young chickens and how this may affect their welfare, both as developing chicks and longer term.

All birds in this study were raised on litter, which is important for the development of foraging behavior and reductions of feather pecking [21,22], although the effects on dust bathing development are less clear [23–25]. The relationship between foraging and feather pecking has been hypothesized to be a redirection of food-related pecking at feathers when a substrate is not present [41], although conflicting results across a multitude of studies highlight the complexities around this relationship [42]. Despite the presence of the litter, there were still some impacts of the rearing enrichments on these behaviors. The perching/navigation structures and novel objects were not intended to specifically increase litter-related behaviors, although there were pecking strings provided as some of the novel objects within some weeks across the 16-week rearing period. These increases suggest that the enrichments had more generalized impacts in optimizing the behavioral development of the birds, resulting in more performance of behaviors that are believed to be positive for laying hens to engage in [43], and that, when thwarted, have been shown to increase the occurrence of abnormal behaviors [11].

Increases in foraging only were shown in the novelty pullets across age which may have been a result of changes in the degree of engagement with the varying enrichments in their pens across time. Once hens moved into the laying system and had a choice of engaging in these behaviors both inside the shed in the floor litter, or outside the shed in the dirt, there were differences between rearing treatments for both dust bathing and foraging, but these differences were in part dependent on the location being observed. Overall, the structural hens did show the most foraging and dust bathing, and this in part may have been related to differences in ranging behavior. Through individual range-use tracking using radio-frequency identification technology, the structural hens spent the longest daily time outside on the range, with the novelty and structural hens showing the longest times for individual visits relative to the control birds [34]. This increased time outside with more space may have led to more observations of dust bathing and foraging as previous research with a separate flock in the same experimental setting showed that hens exhibited more of these behaviors outside relative to what was observed inside the shed [12]. However, there were still some treatment effects when comparing just the behavior exhibited inside the shed in the floor litter. It is difficult to conclude from this study whether that was indirectly related to range-use differences among treatment groups (i.e., increased space available inside per hen with more hens outside) or a separate effect of the rearing enrichment that increased the motivation to perform these behaviors. The variation between rearing treatments and behaviors performed inside or outside does highlight how free-range hens have a choice of locations within this type of system and different locations may be preferred for certain behaviors. This can extend to different locations out on the range as well, where open range areas may elicit different behaviors to sheltered areas [14,15].

The increase in foraging behavior may have had other welfare benefits across the trial, although confirmation of a causal relationship in this study is limited. The control hens overall showed less foraging relative to the structural hens, and they also exhibited the most plumage damage across time [5]. Foraging is proposed to function as both food searching as well as environmental exploration [11]. In this study, foraging was defined as scratching followed by pecking, but foraging in other studies has encompassed walking, pecking and scratching (e.g., [14]). It is uncertain if the discrepancies in these definitions would also correspond with different motivations behind the behaviors, where walking while pecking may be a greater representation of explorative foraging. In terms of food searching, all adult hens had equal feed available indoors which should have met their nutritional requirements. Thus, it is possible that the structural hens were performing more foraging under increased motivation to explore their environment. This may have resulted from the structures provided during rearing that were intended to improve physical development as well as improve spatial navigation around their pens (each structure included opaque sides designed to provide a visual block for development of navigation abilities) [44]. This would be consistent with the research by Rudkin (2021, [42]), who found no direct correlation between foraging and feather pecking in hens provided with a range of foraging enrichments in their cages, indicating that the foraging substrates enabled development of exploration foraging. If control hens were less engaged in exploring their environments, then this could have increased conspecific pecking behavior and/or stress resulting in this negative pattern of behavior [45]. However, the hens exposed to multiple different novel objects during rearing did not show more foraging overall, only more foraging than the control hens inside the shed. Thus, the relationship between rearing enrichments, exploration, foraging and feather pecking in this study is uncertain and requires further investigation.

5. Conclusions

This research demonstrates some long-term benefits of rearing enrichments in the form of novel objects or perching/navigation structures for pullets destined for a free-range environment, where additional complexity in litter-based environments may optimize behavioral development of the pullets and increase performance of positive species-specific behaviors. The effects were most prominent in the pullets reared with the perching/navigation structures throughout development, although the increases were small. Future research should seek to further understand mechanisms behind these effects to design rearing environments that will facilitate desirable behaviors across the laying cycle. Benefits may be seen for laying hens in loose-housed systems with or without range access.

Author Contributions: Conceptualization, D.L.M.C. and C.L.; methodology, D.L.M.C.; formal analysis, D.L.M.C.; investigation, D.L.M.C., S.B., T.R.D. and J.M.L.; resources, D.L.M.C. and C.L.; data curation, S.B., T.R.D. and J.M.L.; writing—original draft preparation, D.L.M.C.; writing—review and editing, D.L.M.C. and C.L.; visualization, D.L.M.C.; supervision, D.L.M.C.; project administration, D.L.M.C.; funding acquisition, D.L.M.C. and C.L. All authors have read and agreed to the published version of the manuscript.

Funding: This research and APC was funded by Poultry Hub Australia, grant number 2017-20.

Institutional Review Board Statement: The study was conducted according to the guidelines of the Declaration of Helsinki and approved by the Animal Ethics Committee of the University of New England, Armidale, NSW, Australia (protocol code 17-092, approved 17 October 2017).

Informed Consent Statement: Not applicable.

Data Availability Statement: Data will be made available upon any reasonable request to the corresponding author.

Acknowledgments: Thank you to Jannala McNally, Belinda Niemeyer, Celia Lacomme, and Baptiste Le Polodec (CSIRO) for data collection and husbandry assistance. Thank you to Andrew M. Cohen-Barnhouse (University of New England) for technical expertise.

Conflicts of Interest: The authors declare no conflict of interest. The funders had no role in the design of the study; in the collection, analyses, or interpretation of data; in the writing of the manuscript, or in the decision to publish the results.

References

1. Scrinis, G.; Parker, C.; Carey, R. The Caged Chicken or the Free-Range Egg? The Regulatory and Market Dynamics of Layer-Hen Welfare in the UK, Australia and the USA. *J. Agric. Environ. Ethics* **2017**, *30*, 783–808. [CrossRef]
2. Bray, H.J.; Ankeny, R. Happy Chickens Lay Tastier Eggs: Motivations for Buying Free-range Eggs in Australia. *Anthrozoös* **2017**, *30*, 213–226. [CrossRef]
3. Pettersson, I.C.; Weeks, C.; Wilson, L.R.M.; Nicol, C. Consumer perceptions of free-range laying hen welfare. *Br. Food J.* **2016**, *118*, 1999–2013. [CrossRef]
4. Campbell, D.L.M.; Bari, M.S.; Rault, J.-L. Free-range egg production: Its implications for hen welfare. *Anim. Prod. Sci.* **2020**, *61*, 848. [CrossRef]
5. Bari, S.; Downing, J.A.; Dyall, T.R.; Lee, C.; Campbell, D.L.M. Relationships Between Rearing Enrichments, Range Use, and an Environmental Stressor for Free-Range Laying Hen Welfare. *Front. Vet. Sci.* **2020**, *7*, 480. [CrossRef]
6. Lambton, S.; Knowles, T.; Yorke, C.; Nicol, C. The risk factors affecting the development of gentle and severe feather pecking in loose housed laying hens. *Appl. Anim. Behav. Sci.* **2010**, *123*, 32–42. [CrossRef]
7. Rodriguez-Aurrekoetxea, A.; Estevez, I. Use of space and its impact on the welfare of laying hens in a commercial free-range system. *Poult. Sci.* **2016**, *95*, 2503–2513. [CrossRef]
8. Singh, M.; Ruhnke, I.; De Koning, C.; Drake, K.; Skerman, A.G.; Hinch, G.N.; Glatz, P.C. Demographics and practices of semi-intensive free-range farming systems in Australia with an outdoor stocking density of ≤1500 hens/hectare. *PLoS ONE* **2017**, *12*, e0187057. [CrossRef]
9. Weeks, C.A.; Lambton, S.L.; Williams, A.G. Implications for Welfare, Productivity and Sustainability of the Variation in Reported Levels of Mortality for Laying Hen Flocks Kept in Different Housing Systems: A Meta-Analysis of Ten Studies. *PLoS ONE* **2016**, *11*, e0146394. [CrossRef]
10. Edgar, J.L.; Mullan, S.; Pritchard, J.C.; McFarlane, U.J.C.; Main, D.C.J. Towards a 'Good Life' for Farm Animals: Development of a Resource Tier Framework to Achieve Positive Welfare for Laying Hens. *Animals* **2013**, *3*, 584–605. [CrossRef]
11. Hemsworth, P.H.; Edwards, L.E. Natural behaviours, their drivers and their implications for laying hen welfare. *Anim. Prod. Sci.* **2020**, *61*, 915. [CrossRef]
12. Campbell, D.L.M.; Hinch, G.; Downing, J.A.; Lee, C. Outdoor stocking density in free-range laying hens: Effects on behaviour and welfare. *Animal* **2017**, *11*, 1036–1045. [CrossRef] [PubMed]
13. Diep, A.T.; Larsen, H.; Rault, J.-L. Behavioural repertoire of free-range laying hens indoors and outdoors, and in relation to distance from the shed. *Aust. Vet. J.* **2018**, *96*, 127–131. [CrossRef] [PubMed]
14. Chielo, L.I.; Pike, T.; Cooper, J. Ranging Behaviour of Commercial Free-Range Laying Hens. *Animals* **2016**, *6*, 28. [CrossRef]
15. Larsen, H.; Cronin, G.; Smith, C.; Hemsworth, P.; Rault, J.-L. Behaviour of free-range laying hens in distinct outdoor environments. *Anim. Welf.* **2017**, *26*, 255–264. [CrossRef]
16. Merrill, R.J.N.; Nicol, C.J. The effects of novel floorings on dust bathing, pecking and scratching behaviour of caged hens. *Anim. Welf.* **2005**, *14*, 179–186.
17. Moroki, Y. Impact of flooring type on the sham dustbathing behaviour of caged laying hens. *Appl. Anim. Behav. Sci.* **2020**, *230*, 105066. [CrossRef]
18. Campbell, D.L.M.; de Haas, E.N.; Lee, C. A review of environmental enrichment for laying hens during rearing in relation to their behavioral and physiological development. *Poult. Sci.* **2019**, *98*, 9–28. [CrossRef]
19. Janczak, A.M.; Riber, A.B. Review of rearing-related factors affecting the welfare of laying hens. *Poult. Sci.* **2015**, *94*, 1454–1469. [CrossRef]
20. Norman, K.I.; Weeks, C.A.; Tarlton, J.F.; Nicol, C.J. Rearing experience with ramps improves specific learning and behaviour and welfare on a commercial laying farm. *Sci. Rep.* **2021**, *11*, 8860. [CrossRef]
21. Bestman, M.; Koene, P.; Wagenaar, J.-P. Influence of farm factors on the occurrence of feather pecking in organic reared hens and their predictability for feather pecking in the laying period. *Appl. Anim. Behav. Sci.* **2009**, *121*, 120–125. [CrossRef]
22. Johnsen, P.F.; Vestergaard, K.S.; Nørgaard-Nielsen, G. Influence of early rearing conditions on the development of feather pecking and cannibalism in domestic fowl. *Appl. Anim. Behav. Sci.* **1998**, *60*, 25–41. [CrossRef]
23. Nicol, C.; Lindberg, A.; Phillips, A.; Pope, S.; Wilkins, L.; Green, L. Influence of prior exposure to wood shavings on feather pecking, dustbathing and foraging in adult laying hens. *Appl. Anim. Behav. Sci.* **2001**, *73*, 141–155. [CrossRef]
24. Wichman, A.; Keeling, L. Hens are motivated to dustbathe in peat irrespective of being reared with or without a suitable dustbathing substrate. *Anim. Behav.* **2008**, *75*, 1525–1533. [CrossRef]
25. Olsson, I.S.; Keeling, L.J.; Duncan, I.J. Why do hens sham dustbathe when they have litter? *Appl. Anim. Behav. Sci.* **2002**, *76*, 53–64. [CrossRef]
26. Norman, K.I.; Weeks, C.A.; Pettersson, I.C.; Nicol, C. The effect of experience of ramps at rear on the subsequent ability of layer pullets to negotiate a ramp transition. *Appl. Anim. Behav. Sci.* **2018**, *208*, 92–99. [CrossRef]
27. Brantsæter, M.; Tahamtani, F.M.; Moe, R.O.; Hansen, T.B.; Orritt, R.; Nicol, C.; Janczak, A.M. Rearing Laying Hens in Aviaries Reduces Fearfulness following Transfer to Furnished Cages. *Front. Vet. Sci.* **2016**, *3*, 13. [CrossRef]
28. Vasdal, G.; Vas, J.; Newberry, R.; Moe, R.O. Effects of environmental enrichment on activity and lameness in commercial broiler production. *J. Appl. Anim. Welf. Sci.* **2019**, *22*, 197–205. [CrossRef]
29. Spinka, M.; Newberry, R.; Bekoff, M. Mammalian Play: Training for the Unexpected. *Q. Rev. Biol.* **2001**, *76*, 141–168. [CrossRef]

30. Ahloy-Dallaire, J.; Espinosa, J.; Mason, G. Play and optimal welfare: Does play indicate the presence of positive affective states? *Behav. Process.* **2018**, *156*, 3–15. [CrossRef]
31. Baxter, M.; Bailie, C.L.; O'Connell, N.E. Play behaviour, fear responses and activity levels in commercial broiler chickens provided with preferred environmental enrichments. *Animal* **2019**, *13*, 171–179. [CrossRef] [PubMed]
32. Liu, Z.; Torrey, S.; Newberry, R.; Widowski, T. Play behaviour reduced by environmental enrichment in fast-growing broiler chickens. *Appl. Anim. Behav. Sci.* **2020**, *232*, 105098. [CrossRef]
33. Bari, S.; Cohen-Barnhouse, A.M.; Campbell, D.L.M. Early rearing enrichments influenced nest use and egg quality in free-range laying hens. *Animal* **2020**, *14*, 1249–1257. [CrossRef] [PubMed]
34. Campbell, D.L.M.; Dyall, T.R.; Downing, J.A.; Cohen-Barnhouse, A.M.; Lee, C. Rearing Enrichments Affected Ranging Behavior in Free-Range Laying Hens. *Front. Vet. Sci.* **2020**, *7*, 446. [CrossRef]
35. Primary Industries Standing Committee. *Model Code of Practice for the Welfare of Animals: Domestic Poultry*, 4th ed.; CSIRO PUBLISHING: Collingwood, Australia, 2002; Volume 83.
36. Hy-Line. Management Guide for Hy-Line Brown Laying Hen in Alternative Systems. 2016. Available online: https://www.hyline.com/userdocs/pages/B_ALT_COM_ENG.pdf (accessed on 15 May 2017).
37. Campbell, D.L.M.; Gerber, P.F.; Downing, J.A.; Lee, C. Minimal Effects of Rearing Enrichments on Pullet Behaviour and Welfare. *Animals* **2020**, *10*, 314. [CrossRef]
38. Fält, B. Differences in aggressiveness between brooded and non-brooded domestic chicks. *Appl. Anim. Ethol.* **1978**, *4*, 211–221. [CrossRef]
39. Campbell, D.L.M.; Ali, A.B.A.; Karcher, D.M.; Siegford, J.M. Laying hens in aviaries with different litter substrates: Behavior across the flock cycle and feather lipid content. *Poult. Sci.* **2017**, *96*, 3824–3835. [CrossRef]
40. Schutz, K.E.; Jensen, P. Effects of Resource Allocation on Behavioural Strategies: A Comparison of Red Junglefowl (*Gallus gallus*) and Two Domesticated Breeds of Poultry. *Ethology* **2001**, *107*, 753–765. [CrossRef]
41. Huber-Eicher, B.; Wechsler, B. The effect of quality and availability of foraging materials on feather pecking in laying hen chicks. *Anim. Behav.* **1998**, *55*, 861–873. [CrossRef]
42. Rudkin, C. Feather pecking and foraging uncorrelated—The redirection hypothesis revisited. *Br. Poult. Sci.* **2021**, in press. [CrossRef]
43. Widowski, T.M.; Duncan, I.J. Working for a dustbath: Are hens increasing pleasure rather than reducing suffering? *Appl. Anim. Behav. Sci.* **2000**, *68*, 39–53. [CrossRef]
44. Freire, R.; Cheng, H.-W.; Nicol, C.J. Development of spatial memory in occlusion-experienced domestic chicks. *Anim. Behav.* **2004**, *67*, 141–150. [CrossRef]
45. El-Lethey, H.; Aerni, V.; Jungi, T.; Wechsler, B. Stress and feather pecking in laying hens in relation to housing conditions. *Br. Poult. Sci.* **2000**, *41*, 22–28. [CrossRef] [PubMed]

Article

The Effect of Light Intensity, Strain, and Age on the Behavior, Jumping Frequency and Success, and Welfare of Egg-Strain Pullets Reared in Perchery Systems

Jo Ann Chew [1], Tina Widowski [2], Eugenia Herwig [1], Tory Shynkaruk [1] and Karen Schwean-Lardner [1,*]

[1] Department of Animal and Poultry Science, University of Saskatchewan, Saskatoon, SK S7N 5A8, Canada; joann.chew@usask.ca (J.A.C.); herwig.eugenia@usask.ca (E.H.); tory.shynkaruk@usask.ca (T.S.)
[2] Department of Animal Biosciences, University of Guelph, Guelph, ON N1G 2W1, Canada; twidowsk@uoguelph.ca
* Correspondence: karen.schwean@usask.ca

Citation: Chew, J.A.; Widowski, T.; Herwig, E.; Shynkaruk, T.; Schwean-Lardner, K. The Effect of Light Intensity, Strain, and Age on the Behavior, Jumping Frequency and Success, and Welfare of Egg-Strain Pullets Reared in Perchery Systems. *Animals* **2021**, *11*, 3353. https://doi.org/10.3390/ani11123353

Academic Editor: Marian Stamp Dawkins

Received: 8 October 2021
Accepted: 21 November 2021
Published: 24 November 2021

Publisher's Note: MDPI stays neutral with regard to jurisdictional claims in published maps and institutional affiliations.

Simple Summary: The effect of light intensity on pullet behavior and welfare is not well studied. In this study, two strains (Lohmann LSL-Lite and Lohmann Brown-Lite) of pullets reared in floor pens containing a perchery system were tested under one of three light intensities (10, 30, or 50 lux). Behavior, jumping frequency and success, fear, and stress levels were assessed throughout the study. Pullets reared at 50 lux spent more time preening (comfort behavior) than pullets reared at 10 lux, while pullets reared at 10 lux spent more time pecking at walls (exploratory behavior). All pullets increased their time spent preening with age. The number and accuracy of jumps also increased with age. Light intensity did not affect landing success, nor did it affect pullet fear or stress levels. Lohmann-LSL-Lite pullets performed more jumps than Lohmann Brown-Lite pullets, while Lohmann Brown-Lite pullets spent more time performing exploratory behaviors. Lohmann Brown-Lite pullets also scored higher on the fear and stress assessments, which might suggest genetic differences between the two strains. Overall, the results suggest that light intensity does not affect pullet behavior, although higher light intensity at 50 lux may slightly increase preening in the pullets, which may indicate positive welfare attributes.

Abstract: The effects of light intensity (L) are not well studied in pullets. Our research objective was to study the effect of L on navigational success, behavior, and welfare of two pullet strains (S). In two repeated trials, a 3 × 2 × 4 factorial arrangement tested three L (10, 30, 50 lux) and two S (Lohmann Brown-Lite (LB), LSL-Lite (LW)) at four ages. One thousand eight hundred pullets/S (0–16 wk) were randomly assigned to floor pens within light-tight rooms (three pens/S/room, four rooms/L) containing four parallel perches and a ramp. Data collection included jumping frequency and success (24h continuous sampling), novel object tests (fear), heterophil to lymphocyte (H/L) ratios (stress), and behavior (instantaneous scan sampling) during photoperiods. L did not affect injurious behavior, fear, or H/L. Pullets reared at 50 lux spent more time preening than at 10 lux. Pullets reared at 10 lux spent more time wall pecking than at 50 lux. Time spent standing and preening and total number and accuracy of jumping increased with age. Pullets reared at 30 lux had higher jumping frequency than at 10 lux; accuracy was not affected. LW jumped more than LB, but with similar success. LB spent more time exploring and scored higher in the fear and stress assessments, suggesting S differences.

Keywords: Lohmann Brown-Lite; Lohmann LSL-Lite; novel object test; heterophil/lymphocyte ratio; environmental navigation

1. Introduction

Studies showed that laying hens are more successful at using complex housing environments when reared in a similar type of housing, as it allows learning to occur early in life [1]. Light intensity (L) may play a role in helping pullets navigate these complex environments by increasing visual acuity. Light intensity may also have an impact on bird behavior and welfare. Many studies have been conducted on L on broilers, while only a few looked at hen behavior and jumping ability [2,3]. Further, information on the impact of L levels on pullets is not well known.

Several variables might affect a pullet's ability to successfully navigate an environment. Increasing L may result in better navigational ability, as a brighter environment may improve poultry vision [2,4]. Previous studies on laying hens reported increased bird activity and jumping success in a brighter environment [5,6]. Age could be another factor, as Kozak et al. [7] reported that pullets at 10–16 wk performed more aerial ascents than hens at 17–24 and 25–37 wk. It is possible that jumping behavior may decrease with age as younger animals have higher levels of energetic capacity and movements than older animals [8]. Therefore, results from hen studies cannot be directly applied to pullets. Bird strain (S) may also play a role in pullet behavioral and navigational qualities [9–11], and should be considered when evaluating the effect of L on pullets.

While increasing L may have positive impacts on birds, it may also result in increased injurious pecking and cannibalism [3,12,13]. Increased injurious behavior within the flock can impact fear and stress levels. Heterophil/lymphocyte (H/L) ratios are considered to be a reliable measure for chronic stress [14]; however, the effect of L on hen or pullet H/L ratios has not been well-studied, as most research focused on hen jumping ability at lower L settings. Information on fear levels in hens in relation to L is also not well-known; however, one study by Hughes and Black [15] reported that hens were more fearful in 17–22 lux than in 55–80 lux.

The present study was conducted to determine whether increasing L can aid pullets in navigating a complex environment without negative behavioral and welfare consequences. The objectives of this study were to examine the effects of L on the behavior, jumping frequency and success, fear, and stress of Lohmann Brown-Lite (LB) and Lohmann LSL-Lite (LW) pullets reared to 16 wk. Three light intensities were tested: 10 (current industry recommended value), 30, and 50 lux. The following hypotheses were tested: (1) L higher than 10 lux would increase active behavior, as well as jumping frequency and success, due to increased visual acuity and bird activity; (2) L higher than 10 lux would increase stress levels, which will increase injurious behavior, fear levels, and H/L ratio; (3) differences between S would result in different measured outputs; and (4) pullet behavior and jumping frequency and success would increase with age as pullets learned to navigate their surroundings.

2. Materials and Methods

2.1. Animal Housing and Husbandry

The effect of L during pullet rearing on use of space, behavior, fear, and stress were evaluated over two 16 wk blocked trials. Three L, 10, 30, and 50 lux, were evaluated on six individually controlled, light-tight rooms. In each trial, LB and LW pullets ($n = 900$ per S), obtained from a commercial hatchery (Clark's Poultry, Brandon, MB, Canada), were reared from 0 to 16 wk of age. Pullets were randomly assigned to a pen within a room (50 pullets per pen, six pens per room), and each room was randomly assigned to one of the three L treatments ($n = 300$ pullets per L × S). The average stocking density achieved was 6.5 birds/m^2, in accordance with the recommendations in the Lohmann Management Guide [16]. Each pen (4.0 m × 2.3 m) was bedded with 7–10 cm depth of wheat straw and furnished with a perching system, ramp, two pan feeders, and a drinker line with six nipples (Lubing Systems LP, Cleveland, TN, USA). The perching system (height 0.56 m × width 1.16 m × length 2.18 m) consisted of four wooden rectangle perches (length 3.8 cm × height 3.5 cm) spaced 30 cm apart with the top corners angled to allow for easy grasping. The ramp (length 81.3 cm × width 48.3 cm at an angle of 38°) was made of 14-gauge wire with 2.54 cm × 2.54 cm dimensions. The ramps were added to the perching system at 14 d to prevent pullets' toes from becoming trapped in the ramp wires prior to that time. The pan feeders used initially had a 36 cm diameter and 113 cm circumference and were replaced with larger pans (44 cm diameter and 138 cm circumference) at 6 wk of age. All birds had ad libitum access to water and commercial feed appropriate for their stage of development [16] and also had access to supplemental feeders and waterers in the first week. The pullets were vaccinated for Marek's Rispens, HVT-IBD, and Poulvac ST at the hatchery. They were also vaccinated for Newcastle bronchitis at the ages of 2, 6, and 10 wk, *Salmonella typhimurium* at 9 and 11 wk, and Newcastle bronchitis and *Salmonella enteriditis* at 15 wk. Birds were checked a minimum of twice daily throughout the trial.

Lighting was provided via eight 11-watt white-light-emitting diode (LED) light bulbs (2821 Kelvin, Greengage Agritech Limited, Roslin Innovation Centre, Midlothian, UK) per room. The light bulbs were positioned so L was similar in all pens when measured at bird level in the center of the pen. The pullets were provided with a photoperiod of 23L:1D for the first week, which gradually decreased until 7 wk of age, where the photoperiod remained at 8L:16D until the end of trial [17]. For the first week, L was set at 50 lux for all rooms to ensure all chicks were able to easily locate feed and water. The L settings were adjusted according to the assigned room-appropriate intensity treatment after the first week. Light intensity was measured with a lux meter every 2 wk (Extech LT300, Extech Instruments, Montreal, QC, Canada), and any variances corrected back to planned intensity. Dawn and dusk periods were simulated over two 15 min periods daily. Room temperature was set at 33 °C on the first day; heat was provided via hot water pipes running along the walls of the rooms. Room temperature was gradually decreased daily until 20 °C at week 5, where it was maintained in accordance with industry recommendations [16]. All rooms were ventilated via a negative pressure inlet-fan system.

2.2. Data Collection

2.2.1. Behavior

Pullet behavior was recorded on a pen basis (per S per room) with infrared cameras (Panasonic WV-CF224FX, Panasonic Corporation of North America, Secaucus, NJ, USA) for 24 h periods at 4, 8, 13, and 16 wk of age. The cameras captured the entire area of the pen. Videos were analyzed with the Genetec Omnicast software (Genetec Inc., Montreal, QC, Canada). Instantaneous scan sampling was conducted at 20 min intervals during the light period (the length of the photoperiod was 12 h at 4 wk, 8 h at 8, 13, and 16 wk), according to the ethogram presented in Table 1.

To determine bird navigation between pen furnishings, the 24 h recordings were also used to conduct continuous behavior sampling. Jumps and flights of both takeoff and landing locations were recorded, as well as whether the landing was a success or failure. Success was determined when a pullet jumped from one part of the pen environment or equipment to another and reached or landed in its target location without incident (falling or crashing), while failure was classified as when the pullet did not reach its targeted location, and instead crashed into it or fell. Pen jumping and landing locations included perches, ramps, drinker lines, top of feeder bins, and the floor. Jumping and landing success were also determined as a percentage based on the number of successful and failed jumps over the total number of jumps performed. Post observations, these jumps were categorized into jumps upward, downward, or across. Several jumps were too infrequent to justify analysis, including jumps from the perch to the top of the feeder bin, from the perch to the ramp and drinker line, from the ramp to the perch, floor, drinker line, or top of feeder bin, from the floor to the top of the feeder bin, from the drinker to the perch and top of feeder bin, and from the top of the feeder bin to the perch, floor, or drinker line.

One observer conducted the observations for both instantaneous scan sampling and continuous behavior sampling. The observer was blind to L treatments, but not to S treatments, since it was easy to tell between S in the video recordings. However, it was not possible to tell between L treatments; therefore, the observer was blind to L treatments. Prior to beginning observations, inter-observer reliability was tested by having a second observer watch the same footage (10 footages per strain at 4 wk), calculating the percent agreement for each behavior, and obtaining an average minimum of 80% consistency across data.

Table 1. Ethogram for pullets, adapted from [18–23].

Behavior	Definition
Active behavior	
Standing	Body in upright and idle position [20]
Walking	Taking at least two successive steps [18]
Jumping or flying	Both feet in the air with wings flapping [21]
Resting behavior	Lying down or crouching with breast on floor or head tucked under wing, otherwise inactive [22]
Comfort behavior	
Preening	Manipulating own feathers with beak while standing or laying [20]
Wing or leg stretching	Extending wing or leg out to side or behind body and returning wing or leg back under body without taking a step forward [20]
Tail wagging	Moving tail side to side without moving rest of body [20]
Head shaking	Head moving side to side or up and down rapidly, body immobile [20]
Head scratching	Extending leg forward and upward to scratch head or neck [20]
Feather ruffling	Raising or shaking out feathers of wings and body [22]
Dustbathing	Rubbing body against floor and performing full body shake [22]
Wing flapping	Extending wings away from body and flapping up and down rapidly but without flight [20]
Nutritive behavior	
At the feeder	Standing or sitting with head extended into feeder [18]
At the drinker	Pecking at nipple drinker [22]
Exploratory behavior	
Gentle feather pecking	Pecking at other birds that does not cause harm or damage to plumage [20]
Wall pecking	Pecking at pen walls [20]
Object pecking	Pecking at perch, ramp, feeder tube (not feed pan), drinker (not nipples) [20]
Litter pecking	Pecking at straw or litter [20]
Ground scratching	Scratching movements on ground while crouching slightly [20]
Head sweeping	Rubbing beak from side to side [20]
Injurious behavior	
Injurious pecking	Pecking at other birds directed at head and neck but may include feet, causes recipient to flinch or escape environment [23]
Fighting	Sparring, leaping, wing flapping toward opponent and can include pecking [19]
Unidentified	Behavior unidentifiable, action of bird cannot be seen

2.2.2. Novel Object Test

At 15 wk of age, pullet fear responses were assessed using a novel object test. A foil tie-dyed balloon weight (Unique 4927, Fancy Dress Worldwide, Worcester, UK) was placed on the pen floor, approximately 0.6 m from the pen entrance. Pullets housed in two pens per S per room were evaluated by recording the latency for three separate birds to peck at the novel object with a maximum allotted time of 900 s (15 min per observation). All pens in a room were tested by live observation at the same time with four different testers randomly assigned to each pen and with each pen observed individually. Tests began at 8 a.m. and were concluded at 9:30 a.m. An average latency to peck at the object for all three pecking times was recorded in seconds and used for analysis.

2.2.3. H/L Ratio

To assess chronic stress, blood was collected from two birds per pen per room at 15 wk of age for analysis of H/L ratio. Using a 22-gauge needle, 2 ml of blood were collected from the brachial vein in an ethylenediaminetetraacetic acid (EDTA) anti-coagulation vacutainer. Within 30 min of collection, the blood from each bird was used to create two duplicate smear stains. After drying for 24 h, the slides were stained using PROTOCOL™ Hema 3™ (Fisher Scientific, Ottawa, ON, Canada). The slides were then read using a light microscope (Optika© B-290TB, Bergamo, Italy) fitted with 100× field of view with oil magnification. Up to 100 heterophil or lymphocyte cells were counted, and the H/L ratio was determined by dividing the number of heterophils by the number of lymphocytes. One observer conducted the observation and was blind to both L and S treatments. Prior to beginning observations, inter-observer reliability was tested by having a second observer watch the same field of view, calculating the percent agreement for each field of view, and obtaining a minimum of 80% consistency across data.

2.3. Statistical Analyses

The experiment was designed as a 3 L × 2 S × 4 wk factorial arrangement within a randomized complete block design. Trial was treated as a block. Room was nested within L and was also the replicate unit for L (two repetitions per L treatments per trial). Pen was the replicate unit for S (three replicates per S per room per trial). All data were checked for normality using the UNIVARIATE Procedure in SAS 9.4® (SAS® 9.4, Cary, NC, USA), and any data not meeting normality assumptions were log transformed (data log+1) prior to analyses. An analysis of variance (ANOVA) test was done using the MIXED Procedure (SAS® 9.4, Cary, NC, USA) to determine differences among group means. Behavior and jumping frequency and success were analyzed as a two-way repeated measure ANOVA. For all data, a Tukey's range test was used to separate means. For all statistical analyses, significance was declared when $p < 0.05$ and trends noted at $0.05 \leq p < 0.10$.

3. Results

3.1. Behavior

The effects of L, S, and week on pullet behavior is reported in Table 2. There was no interaction between L, S, and week. There was an interaction between L and week on time spent jumping or flying. At 4 wk, pullets reared at 10 or 30 lux spent more time jumping or flying (0.11% and 0.12%) than pullets reared at 50 lux (0.04%, $p = 0.02$, Table 3). Time spent jumping or flying decreased with increasing week for pullets reared at 30 lux, while it remained constant for pullets reared at 10 lux. Pullets reared at 50 lux spent the most time jumping and flying at 8 wk (0.42%, Table 3).

There was also an interaction between L and week for time spent pecking at objects in the environment (Table 2). These included perches, ramps, feeder bins (without consumption), and drinker lines (without consumption). Pullets reared at 50 lux spent the least amount of time pecking at objects in the environment at 4 wk (Table 3). Overall, in all L treatments, time spent object pecking increased with week.

An interaction between S and week was also observed for object pecking. In fact, the two S expressed exploratory behaviors differently with week. Exploratory behaviors included gentle pecking, litter-directed pecking, wall pecking, and object pecking. Time spent gentle pecking increased in LB pullets with age (0.20% at 4 wk, 0.18% at 8wk, 0.37% at 13 wk, 0.77% at 16 wk), while for LW pullets, time spent gentle pecking peaked at 8 and 13 wk (0.65% and 0.47% vs. 0.22% at 4 wk and 0.16% at 16 wk, $p < 0.05$, Table 3). Time spent litter-directed pecking was higher in LB pullets than LW pullets at 4 (23.04% vs. 16.31%, $p < 0.05$), 8 (20.97% vs. 13.83%, $p < 0.05$), and 13 wk (18.58% vs. 15.06%, $p < 0.05$). Within strain, time spent litter-directed pecking decreased with age for LB pullets, and there was no S effect on time spent litter-directed pecking for LW pullets. For wall pecking, LB pullets spent more time wall pecking at 13 and 16 wk (4.43% and 5.61% vs. 1.41% at 4 wk and 3.11% at 8 wk), while LW pullets spent similar amounts of time performing this behavior throughout all recorded observations (1.99%, 2.32%, 2.22%, and 2.34% at 4, 8, 13, and 16 wk, respectively, $p > 0.05$). Both S increased the time spent object pecking with week. LW pullets spent more time object pecking at 8 (0.46%), 13 (0.60%), and 16 wk (0.66%) than 4 wk (0.16%), while LB pullets spent the most time object pecking at 16 wk (1.11% vs. 0.26%, 0.26%, and 0.39% at 4, 8, and 13 wk, respectively, Table 3).

There was an interaction between S and week for time spent at the feeder. Both S decreased time spent at the feeder at 16 wk (7.22% vs. 10.60% at 4 wk for LB, $p < 0.05$, 9.20% vs. 12.24% at 4 wk for LW, $p < 0.05$). There was also an interaction between L and S for the percentage of unidentified behaviors, where LB pullets' behaviors were consistently more difficult to identify than LW pullets at all L treatments (Table 3).

Table 2. Average percentage of time (%) spent on each behavior by Lohmann Brown-Lite (LB) and Lohmann Selected Leghorn Lite (LW) pullets reared in floor pens under light intensities of 10, 30, or 50 lux over 12 h of light at 4 wk, and 8 h of light at 8, 13, and 16 wk of age.

	Light Intensity (L)				Strain (S)			Week of Age (wk)				
	10	30	50	*p*	LB	LW	*p*	4	8	13	16	*p*
Standing	23.6	23.5	23.3	0.94	23.2	23.7	0.68	20.6 [c]	15.2 [d]	25.5 [b]	32.5 [a]	<0.01
Walking	4.5	5.0	5.1	0.09	4.7	5.0	0.39	5.3 [a]	5.4 [a]	4.0 [b]	4.9 [a]	<0.01
Jumping or flying	0.1	0.1	0.2	0.34	0.1	0.1	0.67	0.1	0.2	0.1	0.1	0.10
Resting	13.5	14.2	13.9	0.46	10.8	17.0	<0.01	11.4 [c]	21.0 [a]	14.5 [b]	8.7 [d]	<0.01
Preening	10.1 [b]	11.3 [a]	12.1 [a]	<0.01	9.8	12.6	<0.01	5.8 [c]	11.9 [b]	13.1 [ab]	13.8 [a]	<0.01
Comfort [1]	1.0 [b]	1.1 [ab]	1.5 [a]	0.01	1.1	1.2	0.17	0.9 [b]	1.6 [a]	1.5 [a]	0.7 [b]	<0.01
At the feeder	9.8	9.6	9.6	0.71	9.4	9.9	0.19	11.4 [a]	10.6 [a]	8.9 [b]	7.7 [c]	<0.01
At the drinker	3.4	3.3	3.4	0.46	3.3	3.4	0.12	3.0 [b]	4.0 [a]	3.4 [b]	3.0 [b]	<0.01
Gentle pecking	0.4	0.3	0.4	0.26	0.4	0.4	0.94	0.2 [b]	0.4 [a]	0.4 [a]	0.5 [a]	<0.01
Litter directed [2]	17.4	17.3	17.5	0.95	19.7	15.2	<0.01	19.7 [a]	17.4 [b]	16.8 [b]	15.9 [b]	0.02
Wall pecking	3.8 [a]	2.7 [b]	2.3 [b]	<0.01	3.6	2.2	<0.01	1.7 [c]	2.7 [bc]	3.3 [ab]	4.0 [a]	<0.01
Object pecking [3]	0.5	0.4	0.5	0.62	0.5	0.5	1.00	0.2 [c]	0.4 [b]	0.5 [ab]	0.9 [a]	<0.01
Injurious [4]	<0.1	<0.1	<0.1	0.06	<0.1	<0.1	0.70	<0.1	<0.1	<0.1	<0.1	0.82
Unidentified	11.8	11.3	11.0	0.66	13.6	9.1	<0.01	19.7 [a]	10.1 [b]	8.3 [c]	7.5 [c]	<0.01

p for Interactions	L × S	L × wk	S × wk	L × S × wk	SEM [5]
Standing	0.17	0.60	0.15	0.97	0.74
Walking	0.52	0.59	0.77	0.62	0.12
Jumping or flying	0.21	0.02 *	0.42	0.08	0.02
Resting	0.39	0.77	0.47	0.98	0.65
Preening	0.90	0.16	0.61	0.88	0.41
Comfort [1]	0.31	0.36	0.34	0.86	0.09
At the feeder	0.92	0.49	0.03 *	0.98	0.29
At the drinker	0.91	0.06	0.37	0.53	0.08
Gentle pecking	0.41	0.39	<0.01 *	0.34	0.03
Litter directed [2]	0.97	0.44	<0.01 *	0.82	0.38
Wall pecking	0.57	0.92	<0.01 *	0.58	0.21
Object pecking [3]	0.38	0.02 *	0.03 *	0.95	0.05
Injurious [4]	0.32	0.79	0.86	0.48	0.01
Unidentified	0.02 *	0.29	0.31	0.33	0.63

[a–d] Means within rows with different letters indicate a significant difference ($p < 0.05$). * Indicates a significant difference within interactions ($p < 0.05$). [1] Wing or leg stretching, tail wagging, head shaking, head scratching, feather ruffling, dustbathing, and wing flapping. [2] Behavior directed toward ground, including litter pecking, ground scratching, and head sweeping. [3] Pecking at perch, ramp, drinker, or feeder bin. [4] Injurious pecking and fighting. [5] SEM—Standard error of mean.

Table 3. Interactions between light intensity, strain, and week of age for behavioral expression of Lohmann Brown-Lite (LB) or Lohmann Selected Leghorn Lite (LW) pullets reared in floor pens under light intensities of 10, 30, or 50 lux.

		Week of Age (wk)			
		4	8	13	16
Percentage of Time (%) Spent on Each Behavior	Light Intensity (lux)				
Jumping or flying	10	0.11 [ab]	0.11 [abc]	0.02 [bc]	0.07 [bc]
	30	0.12 [ab]	0.10 [abc]	0.05 [bc]	0.05 [c]
	50	0.04 [c]	0.42 [a]	0.12 [abc]	0.04 [c]
Object pecking	10	0.29 [e]	0.33 [de]	0.61 [abc]	0.85 [ab]
	30	0.25 [e]	0.41 [bcde]	0.38 [cde]	0.75 [abcd]
	50	0.08 [f]	0.34 [de]	0.51 [abcd]	1.06 [a]

Table 3. *Cont.*

Percentage of Time (%) Spent on Each Behavior	Light Intensity (lux)	Week of Age (wk)			
		4	8	13	16
	Strain				
At the feeder	LB	10.60 [ab]	10.62 [a]	9.34 [ab]	7.22 [c]
	LW	12.24 [a]	10.65 [a]	8.43 [bc]	8.20 [bc]
Gentle pecking	LB	0.20 [d]	0.18 [cd]	0.37 [bcd]	0.77 [a]
	LW	0.22 [d]	0.65 [ab]	0.47 [abc]	0.16 [d]
Litter directed	LB	23.04 [a]	20.97 [ab]	18.58 [bc]	16.10 [d]
	LW	16.31 [cd]	13.83 [d]	15.06 [d]	15.62 [d]
Wall pecking	LB	1.41 [b]	3.11 [b]	4.43 [a]	5.61 [a]
	LW	1.99 [b]	2.32 [b]	2.22 [b]	2.34 [b]
Object pecking	LB	0.26 [bc]	0.26 [bc]	0.39 [abc]	1.11 [a]
	LW	0.16 [c]	0.46 [ab]	0.60 [ab]	0.66 [ab]
		Light intensity (lux)			
		10	30	50	
Unidentified	LB	14.77 [a]	13.67 [a]	12.34 [a]	
	LW	8.83 [b]	8.87 [b]	9.71 [b]	

[a–f] Means within a behavior with different letters indicate a significant difference ($p < 0.05$).

There was an effect of L on preening and wall-pecking behavior. Pullets reared at 30 (11.3%) and 50 lux (12.1%) spent more time preening than pullets reared at 10 lux (10.1%, $p < 0.05$, Table 2), while pullets reared at 10 lux spent more time pecking at walls (3.8%) than pullets reared at 30 (2.7%) or 50 lux (2.3%, $p < 0.05$, Table 2).

For the effect of S on pullet behavior, LW pullets spent more time resting (17.0%) and preening (12.6%) than LB pullets (10.8% and 9.8%), while LB pullets spent more time performing litter-directed pecking (19.7%) and wall pecking (3.6%) than LW pullets (15.2% and 2.2%).

All pullets increased the time spent preening with age. Standing was highest at 16 wk (32.5%), followed by 13 wk (25.5%), 4 wk (20.6%), and 8 wk (15.2%). Time spent walking was lowest at 13 wk (4.0%) compared to all other recorded periods (5.3%, 5.4%, and 4.9% at 4, 8, and 16 wk, respectively, $p < 0.05$, Table 2). Time spent resting was highest at 8 wk (21.0%), followed by 13 (14.5%), 4 (11.4%), and 16 wk (8.7%). Time spent performing other comfort behaviors was higher at 8 (1.6%) and 13 wk (1.5% vs. 0.9% and 0.7% at 4 and 16 wk, respectively, $p < 0.05$, Table 2). Time spent at the drinker was highest at 8 wk (4.0% vs. 3.0%, 3.4%, and 3.0% at 4, 13, and 16 wk, respectively, $p < 0.05$, Table 2). Altogether, pullets' behaviors were easier to identify at 13 and 16 wk (7.5% and 8.3% unidentified behavior) than at 4 (19.7%) or 8 (10.1%) wk. Finally, neither L, S, nor week affected injurious behavior in pullets.

3.2. Jumping Frequency and Success Rate

There were several interactions between S and week (Table 4). For jumps directed upward from the floor to the drinker, LW pullets had the highest number of successful jumps at 4 wk (average 7.75 jumps per pullet over 24 h vs. 1.62 jumps, Table 5). LW pullets also had the highest number of failed landings from the floor to the drinker at 4 wk (0.17 jumps vs. 0.03, 0.02, 0.01, and <0.01, $p < 0.05$, Table 5). There was no interaction between S and week on percent success of jumps from the floor to the drinker (Table 4). This same relationship was also observed for jumps from the floor to the ramp; LW pullets performed both more successful and failed jumps at 4 wk (1.77 and 0.03) than LB pullets (0.78 and 0.01) (Table 5). However, percent success of landing from the floor to the ramp was unaffected by the interaction between S and week (Table 4). Another type of jump directed upward was from the floor to the perch. At all recorded weeks, LW pullets performed more successful jumps from the floor to the perch than LB pullets (8.59, 15.22, 15.00, and 15.16 vs. 1.15, 5.49, 6.78, and 6.25 at 4, 8, 13, and 16 wk, respectively, $p < 0.05$, Table 5). Within S, overall, the number of successful jumps and jumping accuracy increased with week (Table 5).

For jumps directed downward, LW pullets successfully jumped from the drinker to the floor the most at 4 wk (7.62 vs. 3.01, 3.46, 3.84 at 8, 13, and 16 wk, respectively), while that specific behavior in LB pullets peaked at 8 wk (2.94 vs. 1.57, 2.37, and 2.55 at 4, 13, and 16 wk, respectively, Table 5). LB pullets had a higher number of failed landings from the drinker to the floor at 4 wk; however, the difference was numerically minute (0.01 at 4 wk vs. 0.00 at 8, 13, and 16 wk), and total jumping accuracy was unaffected. On the other hand, jumping accuracy was affected for jumps from the perch to the floor; at 4 wk, LB pullets had a lower percent success than the other weeks. Despite this, the difference in percent success was numerically minute (99.77% vs. 100% at 8, 13, and 16 wk, $p < 0.05$, Table 5). Percent success for LW pullets from the perch to the floor was similar across ages.

There was also an interaction between L and S for failed jumps from the perch to the floor. LW pullets reared at 10 lux had a higher number of failed landings (0.01) than LB pullets (0.00). However, differences were negligible.

For jumps between perches, LW pullets performed the most successful jumps at 4 (8.02) and 8 wk (9.97). LW pullets performed fewer successful jumps at 13 (5.92) and 16 wk (5.77) compared to LB pullets (7.79 and 8.04 at 13 and 16 wk, respectively), who peaked in jumps between perches at 8 wk (8.98).

Overall, for differences between S jumps across weeks, LW pullets had more total successful jumps than LB pullets (43.94 vs. 14.40 at 4 wk, 43.66 vs. 24.29 at 8 wk, 41.03 vs. 23.46 at 13 wk, and 42.41 vs. 23.86 at 16 wk), while within S, LB pullets had more total successful jumps at 8, 13, and 16 wk than 4 wk ($p < 0.05$, Table 5). LW pullets also had the most failed landings between S and across all weeks (0.33 vs. 0.12 at 4 wk, 0.10 vs. 0.05 at 8 wk, 0.04 vs. 0.02 at 13 wk, and 0.04 vs. 0.02 at 16 wk). Both S performed fewer failed landings with increasing week; however, total percent success was not affected between S and week.

L impacted pullets jumping between perches. Pullets reared at 30 lux had a higher number of successful jumps (8.3) than pullets reared at 10 lux (6.9), with pullets reared at 50 lux showing an intermediate response (7.6, Table 4). Despite the higher jumping frequency, jumping accuracy was not affected by L.

S impacted the frequency and success of several different jumps. For jumps directed upward from the floor to the perch, LW pullets had higher failed landings than LB pullets (<0.1 vs. <0.1, respectively, $p < 0.01$, Table 4). For jumps directed downward from the drinker to the floor, LB pullets had a higher jumping accuracy (100.00%) than LW pullets (99.97%). LW pullets had a higher number of successful jumps from the perch to the floor (11.8) than LB pullets (3.5, $p < 0.01$).

Finally, age had several effects on jumping frequency and success of pullets. Jumping accuracy increased with age for jumps from the floor to the drinker (97.83%, 98.77%, 99.66%, 99.85% at 4, 8, 13, and 16 wk, respectively, Table 4). Failed landings from the floor to the perch decreased with age (0.1 at 4 wk, <0.1 at 8, 13, and 16 wk). The number of jumps from the ramp to the floor decreased with age (0.4 at 4 wk vs. 0.1 at 8, 13, and 16 wk), while successful jumps from the perch to the floor increased with age (6.1, 7.5, 8.2, 8.9 at 4, 8, 13, and 16 wk, respectively). Failed landings decreased with age (<0.1 at 4, 8, and 16 wk, 0.0 at 13 wk). Additionally, failed jumps between perches decreased with age (0.1 at 4 wk, <0.1 at 8, 13, and 16 wk), and percent success increased with age (99.29%, 99.89%, 99.95%, 99.95% at 4, 8, 13, and 16 wk, respectively, Table 4). Overall, despite being numerically similar, total jumping accuracy increased with age (99.16%, 99.77%, 99.92%, 99.92% at 4, 8, 13, and 16 wk, respectively, Table 4).

Table 4. Average number of successful, failed, and percent success of jumps per bird directed upward, downward, and across by Lohmann Brown-Lite (LB) or Lohmann Selected Leghorn Lite (LW) pullets reared in floor pens under light intensities of 10, 30, or 50 lux over 24 h at 4, 8, 13, and 16 wk of age.

From	To		Light Intensity (L)				Strain (S)			Week of Age (wk)				
			10	30	50	p	LB	LW	p	4	8	13	16	p
Jumps upward														
Floor	Drinker	S	3.1	3.7	3.9	0.33	2.4	4.7	<0.01	4.7	3.2	3.1	3.2	0.19
		F	<0.1	<0.1	<0.1	0.71	<0.1	0.1	<0.01	0.1 [a]	<0.1 [b]	<0.1 [bc]	<0.1 [c]	<0.01
		%	98.44	99.22	99.42	0.19	99.20	98.85	0.47	97.83 [b]	98.77 [ab]	99.66 [a]	99.85 [a]	0.01
Floor	Ramp	S	0.4	0.4	0.6	0.22	0.3	0.6	0.04	1.3 [a]	0.4 [b]	0.2 [c]	0.1 [d]	<0.01
		F	<0.1	<0.1	<0.1	0.78	<0.1 [b]	<0.1 [a]	0.03	<0.1 [a]	<0.1 [b]	<0.1 [b]	0.0 [b]	<0.01
		%	98.71	99.50	99.82	0.35	99.12	99.59	0.40	98.74	99.74	98.96	100.00	0.44
Floor	Perch	S	9.0	9.3	9.3	0.70	4.9	13.5	<0.01	4.9 [b]	10.4 [a]	10.9 [a]	10.7 [a]	<0.01
		F	<0.1	<0.1	<0.1	0.24	<0.1 [b]	<0.1 [a]	<0.01	0.1 [a]	<0.1 [a]	<0.1 [b]	<0.1 [b]	<0.01
		%	99.07	99.12	99.20	0.95	98.64	99.61	0.01	97.15 [c]	99.64 [b]	99.89 [ab]	99.83 [a]	<0.01
Jumps downward														
Drinker	Floor	S	3.0	3.5	3.7	0.36	2.4	4.5	<0.01	4.6	3.0	2.9	3.2	0.06
		F	0.0	<0.1	<0.1	0.30	0.0	<0.1	0.03	<0.1	0.0	0.0	<0.1	0.07
		%	100.00	99.98	99.98	0.33	100.00	99.97	0.04	99.96	100.00	100.00	99.98	0.14
Ramp	Floor	S	0.2	0.2	0.2	0.42	0.2	0.2	0.09	0.4 [a]	0.1 [b]	0.1 [b]	0.1 [b]	<0.01
		F	0.0	0.0	0.0	-	0.0	0.0	-	0.0	0.0	0.0	0.0	-
		%	100.00	100.00	100.00	-	100.00	100.00	-	100.00	100.00	100.00	100.00	-
Perch	Floor	S	7.5	7.6	7.9	0.71	3.5	11.8	<0.01	6.1 [c]	7.5 [b]	8.2 [b]	8.9 [a]	<0.01
		F	<0.1	<0.1	<0.1	0.47	<0.1	<0.1	0.95	<0.1 [a]	<0.1 [ab]	0.0 [b]	<0.1 [ab]	0.04
		%	99.98	99.96	99.96	0.80	99.94	99.98	0.13	99.87 [b]	99.99 [a]	100.00 [a]	99.99 [a]	<0.01
Jumps across														
Perch	Perch	S	6.9 [b]	8.3 [a]	7.6 [ab]	0.04	7.8	7.4	0.35	7.2 [b]	9.5 [a]	6.9 [b]	6.9 [b]	<0.01
		F	<0.1	<0.1	<0.1	0.33	<0.1	<0.1	0.56	0.1 [a]	<0.1 [b]	<0.1 [c]	<0.1 [bc]	<0.01
		%	99.71	99.75	99.85	0.38	99.79	99.75	0.71	99.29 [b]	99.89 [a]	99.95 [a]	99.95 [a]	<0.01
Total														
		S	30.1	33.1	33.2	0.31	21.5	42.8	<0.01	29.2 [b]	34.0 [a]	32.2 [a]	33.1 [a]	<0.01
		F	0.1	0.1	0.1	0.42	0.1 [b]	0.1 [a]	<0.01	0.2 [a]	0.1 [b]	<0.1 [c]	<0.1 [c]	<0.01
		%	99.66	99.69	99.73	0.59	99.68	99.70	0.65	99.16 [c]	99.77 [b]	99.92 [a]	99.92 [a]	<0.01

From	To		p for Interactions				
			L × S	L × wk	S × wk	L × S × wk	SEM [1]
Jumps upward							
Floor	Drinker	S	0.30	0.99	<0.01 *	0.73	0.23
		F	0.75	0.64	0.01 *	0.93	0.01
		%	0.78	0.23	0.80	0.58	0.22
Floor	Ramp	S	0.91	0.30	0.01 *	0.52	0.07
		F	0.50	0.84	<0.01 *	0.93	<0.01
		%	0.15	0.58	0.46	0.71	0.30
Floor	Perch	S	0.71	0.54	<0.01 *	0.64	0.55
		F	0.51	0.77	0.28	0.84	<0.01
		%	0.70	0.79	0.04 *	0.91	0.21
Jumps downward							
Drinker	Floor	S	0.31	0.99	<0.01 *	0.78	0.22
		F	0.30	0.29	0.07	0.29	<0.01
		%	0.33	0.16	0.14	0.16	0.01
Ramp	Floor	S	0.33	0.53	0.89	0.18	0.02
		F	-	-	-	-	0.00
		%	-	-	-	-	0.00
Perch	Floor	S	0.95	0.97	0.39	0.84	0.46
		F	<0.01 *	0.37	0.03 *	0.38	<0.01
		%	0.07	0.43	0.01 *	0.24	0.02
Jumps across							
Perch	Perch	S	0.74	0.79	<0.01 *	0.97	0.22
		F	0.73	0.49	0.86	0.94	<0.01
		%	0.73	0.50	0.73	0.97	0.05
Total							
		S	0.76	0.97	<0.01 *	0.99	1.25
		F	0.45	0.71	0.01 *	0.97	0.01
		%	0.44	0.44	0.76	0.85	0.04

S—Success. F—Failure. %—Percent success. [a–c] Means within a row with different letters indicate a significant difference ($p < 0.05$). * Indicates a significant difference within interactions ($p < 0.05$). [1] SEM—Standard error of mean.

Table 5. Interactions between light intensity, strain, and week of age for jumping frequency and success of Lohmann Brown-Lite (LB) or Lohmann Selected Leghorn Lite (LW) pullets reared in floor pens under light intensities of 10, 30, or 50 lux.

Average Jumps Per Pullet over 24 h				Week of Age (wk)			
				4	**8**	**13**	**16**
	Upward		**Strain**				
Floor	Drinker	S	LB	1.62 [c]	3.04 [b]	2.43 [bc]	2.57 [bc]
			LW	7.75 [a]	3.39 [b]	3.68 [b]	3.71 [b]
		F	LB	0.01 [bc]	0.02 [bc]	0.01 [bc]	<0.01 [c]
			LW	0.17 [a]	0.03 [b]	0.01 [bc]	0.01 [bc]
Floor	Ramp	S	LB	0.78 [ab]	0.36 [bc]	0.14 [d]	0.05 [e]
			LW	1.77 [a]	0.36 [b]	0.17 [cd]	0.12 [de]
		F	LB	0.01 [b]	0.00 [b]	<0.01 [b]	0.00 [b]
			LW	0.03 [a]	<0.01 [b]	0.00 [b]	0.00 [b]
Floor	Perch	S	LB	1.15 [e]	5.49 [d]	6.78 [bc]	6.25 [cd]
			LW	8.59 [b]	15.22 [a]	15.00 [a]	15.16 [a]
		%	LB	95.18 [b]	99.68 [a]	99.90 [a]	99.81 [a]
			LW	99.12 [ab]	99.61 [a]	99.88 [a]	99.85 [a]
	Downward						
Drinker	Floor	S	LB	1.57 [c]	2.94 [b]	2.37 [bc]	2.55 [bc]
			LW	7.62 [a]	3.01 [b]	3.46 [b]	3.84 [b]
		F	LB	0.01 [a]	0.00 [b]	0.00 [b]	0.00 [b]
			LW	<0.01 [ab]	<0.01 [ab]	0.00 [b]	<0.01 [ab]
Perch	Floor	%	LB	99.77 [b]	100.00 [a]	100.00 [a]	100.00 [a]
			LW	99.98 [a]	99.99 [a]	100.00 [a]	99.97 [a]
	Across						
Perch	Perch	S	LB	6.39 [cd]	8.98 [ab]	7.79 [bc]	8.04 [abc]
			LW	8.02 [abcd]	9.97 [a]	5.92 [d]	5.77 [d]
Total		S	LB	14.40 [c]	24.29 [b]	23.46 [b]	23.86 [b]
			LW	43.94 [a]	43.66 [a]	41.03 [a]	42.41 [a]
		F	LB	0.12 [bc]	0.05 [cd]	0.02 [d]	0.02 [d]
			LW	0.33 [a]	0.10 [b]	0.04 [cd]	0.04 [cd]
				Light Intensity (lux)			
	Downward			10	30	50	
Perch	Floor	F	LB	0.00 [b]	0.002 [ab]	0.003 [ab]	
			LW	0.01 [a]	0.00 [b]	0.00 [b]	

S—Success. F—Failure. %—Percent success. [a–e] Means within a successful or failed landing with different letters indicate a significant difference ($p < 0.05$).

3.3. Novel Object Test

There was no effect of L on the time taken to peck at the novel object (Table 6). LB pullets had a higher latency to peck (676 s) than LW pullets (212 s, $p < 0.001$, Table 6). There was no interaction between L and S.

Table 6. Latency to peck at novel object (seconds) by Lohmann Brown-Lite (LB) or Lohmann Selected Leghorn Lite (LW) pullets reared in floor pens under light intensities of 10, 30, or 50 lux at 15 wk of age (8 pen replicates per L × S).

Light Intensity (L)				Strain (S)			L × S	
10	**30**	**50**	*p*	**LB**	**LW**	*p*	*p*	**SEM** [1]
397	497	437	0.436	676	212	<0.001	0.415	41.9

[1] SEM—Standard error of mean.

3.4. H/L Ratio

There was no effect of L on H/L ratio (Table 7). LB pullets had a higher H/L ratio than LW pullets (0.26 vs. 0.13, respectively, $p < 0.001$, Table 7). There was no interaction between L and S.

Table 7. Heterophil/lymphocyte ratios of Lohmann Brown-Lite (LB) and Lohmann Selected Leghorn Lite (LW) pullets reared in floor pens under light intensities of 10, 30, or 50 lux at 15 wk of age (12 pen replicates per L × S).

Light Intensity (L)				Strain (S)			L × S	
10	30	50	*p*	LB	LW	*p*	*p*	SEM [1]
0.20	0.18	0.20	0.507	0.26	0.13	<0.001	0.922	0.011

[1] SEM—Standard error of mean.

4. Discussion

Behavioral observations are an important tool in assessing an animal's response to its environment. To understand how S reacts differently to L at different ages, LB and LW pullets were reared to 16 wk. L did not influence the ability to identify behaviors; however S did have an impact. LB pullets' behaviors were consistently more difficult to identify, regardless of L. This may have been due to the dark feather color of LB pullets that made it difficult to distinguish from the bedding when viewed in the infrared videos. Even though it was possible to identify the presence of an LB pullet, the challenge was identifying specifically what behavior the pullet was performing. Conversely, the white feather color of LW pullets provided a contrast against the litter and made for easier identification. Another challenge for identifying pullet behavior was their small size at 4 wk. A decrease in unidentified behavior with age was observed as the pullets' body size increased.

It was hypothesized that pullet activity would increase with L; however, this was not observed. Rather, pullets reared at 10 or 30 lux spent more time jumping or flying than pullets reared at 50 lux at 4 wk, which was in contrast with previous literature that reported increased bird activity with L [6,15]. Interestingly, pullets reared at 50 lux were not occupied with pecking at objects in the environment, whereas previous studies reported increased visual stimulation at high L [24].

Across all recording periods, pullets reared at 10 lux spent more time pecking at the walls than pullets reared at 30 or 50 lux, while pullets reared at 30 and 50 lux were observed spending more time preening. Wall pecking is not a common behavior found or reported in other studies. However, the purpose of this behavior may be an extension of other exploratory behaviors. Kjaer and Vestergaard [25] suggested that low L may lead to a reduced ability to identify environmental cues and thus cause birds to increase the time spent exploratory pecking as compensation. Preening can be visually motivated [26], and higher L of 30 or 50 lux may encourage the pullets to maintain good plumage condition [24].

It is important to mention that there was no effect of L on injurious behavior, similar to a study by Hartini et al. [27] looking at 5 lux versus 60–80 lux. However, the results of this study is in contrast with Kjaer and Vestergaard [25], who reported two to three times more injurious pecking (reported as severe feather pecking in their paper) in pullets reared at 30 lux vs. 3 lux. This may be because of the type of light source used. The study by Kjaer and Vestergaard used incandescent light bulbs [25], while the present study used LED lights. Incandescent lights emit high amounts of red light, which were reported to increase injurious pecking activity [28,29]. The LED lighting used in this experiment was not red-saturated, and LED lights were reported to be preferred by chickens over incandescent lighting [30]. In the present study, feather condition was not measured. However, there was no obvious change in feather condition throughout the trial. Additionally, injurious pecking is multifactorial and is affected by strain, diet, and other environmental and management conditions [31].

The success of pullet jumps was high through all observation periods; however, it is important to note that jumping frequency increased with age and so did jumping accuracy. This supports the importance of preparing pullets for navigating a complex environment by exposing them to the same environment during the rearing period [1,32]. The jumps from the floor to the ramp were highest at 4 wk and decreased with age, highlighting the importance of providing ramps to facilitate movement between landing platforms and tiers [33]. L may also play a role in improving pullet vision for navigational jumps within the environment. Pullets reared at 30 lux performed more jumps than those reared at 10 lux. However, despite this, jumping accuracy was not affected by L, which was in agreeance with Moinard et al. [34], who studied jumping accuracy in hens reared at 5, 10, and 20 lux. Results from the present study suggest that 10 lux is bright enough for pullets to navigate their environment successfully.

Several studies reported increased fear and/or stress levels with increasing L due to increased injurious pecking [3,25]. Results from the present study reported no effect of L on fear or stress responses. This was in agreeance with behavior observations from this study, which reported minimal levels of injurious behavior. Possible explanations for disagreement between studies may be due to type of light source used, evenness of light distribution, and age of birds. However, based on the result of this study, L of 10 to 50 lux did not affect the fear or stress levels of pullets.

Several S differences were reported for behavioral observations, jumping frequency, and fear and stress responses. This may be explained by the characteristic differences between brown- and white-feathered birds. White-feathered pullets are more reactive and flightier than brown-feathered strains [10], which may explain why jumping frequency was higher in LW than LB pullets. LW pullets are also comparatively lighter than LB pullets and can easily generate enough energy to perform aerial ascents within their environment [34,35]. In comparison, LB pullets exhibit more proactive and exploratory characteristics [11,36], as evidenced by the increased time spent on the floor performing exploratory behaviors compared to LW pullets. These S differences could explain the fear and stress responses. LB pullets had a higher latency to peck at the novel object and had higher H/L ratios. Typically, longer latencies to peck at a novel object, or the higher the H/L ratio, are interpreted as indicators of more fear and stress [15,37]. However, LB pullets' higher latency to peck at the object and higher H/L ratio may not be due to a higher fear and stress level, but rather due to different hormonal and behavioral responses to a stressor compared to LW pullets [11].

5. Conclusions

In conclusion, the results of the study suggest that light intensities of 10, 30, or 50 lux result in minor changes in behavior, with a small increase in preening at higher lux and a small increase in wall pecking at lower lux. Light intensity did not impact injurious behavior, fear, or stress levels of pullets up to 16 wk. All pullets increased their time spent preening as they aged. Total number of jumps and jumping accuracy increased with age, supporting the importance of rearing pullets in complex environments, especially if they will be housed in a similar environment during the laying phase. Light intensities above 30 lux may slightly increase jumping frequency; however, 10 lux is sufficient for pullets to jump within their environment successfully.

Author Contributions: Conceptualization, T.W. and K.S.-L.; formal analysis, J.A.C.; funding acquisition, T.W. and K.S.-L.; investigation, J.A.C.; methodology, E.H. and T.S.; supervision, K.S.-L.; validation, E.H. and T.S.; writing—original draft, J.A.C.; writing—review and editing, T.W., E.H., T.S. and K.S.-L. All authors have read and agreed to the published version of the manuscript.

Funding: This research was funded by the Canadian Poultry Research Council and Agriculture and Agri-Food Canada (AAFC) (Grant number 350638). This research is a part of the 2018–2023 Poultry Science Cluster, which is supported by AAFC as part of the Canadian Agricultural Partnership, a federal–provincial–territorial initiative.

Institutional Review Board Statement: The procedures for this experiment were approved by the University of Saskatchewan Animal Care Committee (AUP #19940248), and all birds were cared for as specified in the Guide to the Care and the Use of Experimental Animals by the Canadian Council on Animal Care (2009).

Data Availability Statement: The datasets used and/or analyzed during the current study are available from the corresponding author upon reasonable request.

Acknowledgments: Thank you to Clark's Poultry Inc. in Brandon, MB, for supplying the chicks, and the University of Saskatchewan Poultry Centre staff and students for their technical assistance.

Conflicts of Interest: The authors declare that the research was conducted in the absence of any commercial or financial relationships that could be construed as a potential conflict of interest. The funders had no role in the design of the study; in the collection, analyses, or interpretation of data; in the writing of the manuscript, or in the decision to publish the results.

References

1. Wilkins, L.J.; McKinstry, J.L.; Avery, N.C.; Knowles, T.G.; Brown, S.N.; Tarlton, J.; Nicol, C.J. Influence of housing system and design on bone strength and keel bone fractures in laying hens. *Vet. Rec.* **2011**, *169*, 414. [CrossRef]
2. Taylor, P.E.; Scott, G.B.; Rose, P. The ability of domestic hens to jump between horizontal perches: Effects of light intensity and perch colour. *Appl. Anim. Behav. Sci.* **2003**, *83*, 99–108. [CrossRef]
3. Kristensen, H.H. *The Effects of Light Intensity, Gradual Changes between Light and Dark and Definition of Darkness for the Behaviour and Welfare of Broiler Chickens, Laying Hens, Pullets and Turkeys*; Norwegian Scientific Committee for Food Safety: Frederiksberg C, Denmark, 2008.
4. Widowski, T.M.; Classen, H.L.; Newberry, R.C.; Petrik, M.; Schwean-Lardner, K.; Cottee, S.Y. *Code of Practice for Handling and Care of Pullets, Layers and Spent Fowl: Poultry (Layers): Review of Scientific Research on Priority Issues*; National Farm Animal Care Council: Lacombe, AB, Canada, 2013.
5. Taylor, P.E.; Scott, G.B. The ability of laying hens to negotiate between horizontal perches. *Br. Poult. Sci.* **2002**, *43*, S14–S15. [CrossRef]
6. O'Connor, E.A.; Parker, M.O.; Davey, E.L.; Grist, H.; Owen, R.C.; Szladovits, B.; Demmers, T.G.M.; Wathes, C.M.; Abeyesinghe, S.M. Effect of low light and high noise on behavioural activity, physiological indicators of stress and production in laying hens. *Br. Poult. Sci.* **2011**, *52*, 666–674. [CrossRef]
7. Kozak, M.; Tobalske, B.W.; Springthorpe, D.; Szkotnicki, B.; Harlander-Matauschek, A. Development of physical activity levels in laying hens in three-dimensional aviaries. *Appl. Anim. Behav. Sci.* **2016**, *185*, 66–72. [CrossRef]
8. Dial, K.P.; Jackson, B.E. When hatchlings outperform adults: Locomotor development in Australian brush turkeys (Alectura lathami, Galliformes). *Proc. R. Soc. B Biol. Sci.* **2011**, *278*, 1610–1616. [CrossRef] [PubMed]
9. Tauson, R.; Abrahamsson, P. Foot and keel bone disorders in laying hens: Effects of artificial perch material and hybrid. *Acta Agric. Scand. A Anim. Sci.* **1996**, *46*, 239–246. [CrossRef]
10. Pusch, E.A.; Bentz, A.B.; Becker, D.J.; Navara, K.J. Behavioral phenotype predicts physiological responses to chronic stress in proactive and reactive birds. *Gen. Comp. Endocrinol.* **2018**, *255*, 71–77. [CrossRef]
11. Peixoto, M.R.L.V.; Karrow, N.A.; Newman, A.; Head, J.; Widowski, T.M. Effects of acute stressors experienced by five strains of layer breeders on measures of stress and fear in their offspring. *Physiol. Behav.* **2020**, *228*, 113185. [CrossRef]
12. El-Lethey, H.; Aerni, V.; Jungi, T.W.; Wechsler, B. Stress and feather pecking in laying hens in relation to housing conditions. *Br. Poult. Sci.* **2000**, *41*, 22–28. [CrossRef]
13. Drake, K.A.; Donnelly, C.A.; Dawkins, M.S. Influence of rearing and lay risk factors on propensity for feather damage in laying hens. *Br. Poult. Sci.* **2010**, *51*, 725–733. [CrossRef]
14. Maxwell, M.H.; Robertson, G.W. The avian heterophil leucocyte: A review. *Worlds. Poult. Sci. J.* **1998**, *54*, 155–178. [CrossRef]
15. Hughes, B.O.; Black, A.J. The effect of environmental factors on activity, selected behaviour patterns and "fear" of fowls in cages and pens. *Br. Poult. Sci.* **1974**, *15*, 375–380. [CrossRef]
16. Lohmann Tierzucht. *Layers Management Guide: Cage Housing*; Lohmann Tierzucht: Cuxhaven, Germany, 2018.
17. NFACC. *Code of Practice for the Care and Handling of Pullets and Laying Hens*; Egg Farmers of Canada: Ottawa, ON, Canada, 2017; ISBN 9780993618925.
18. Webster, A.B.; Hurnik, J.F. An ethogram of white leghorn-type hens in battery cages. *Can. J. Anim. Sci.* **1990**, *70*, 751–760. [CrossRef]
19. Estevez, I.; Newberry, R.C.; Keeling, L.J. Dynamics of aggression in the domestic fowl. *Appl. Anim. Behav. Sci.* **2002**, *76*, 307–325. [CrossRef]
20. Nicol, C.J.; Caplen, G.; Edgar, J.; Browne, W.J. Associations between welfare indicators and environmental choice in laying hens. *Anim. Behav.* **2009**, *78*, 413–424. [CrossRef]
21. de Haas, E.N.; Nielsen, B.L.; Buitenhuis, A.J.; Rodenburg, T.B. Selection on feather pecking affects response to novelty and foraging behaviour in laying hens. *Appl. Anim. Behav. Sci.* **2010**, *124*, 90–96. [CrossRef]

22. Ericsson, M.; Fallahsharoudi, A.; Bergquist, J.; Kushnir, M.M.; Jensen, P. Domestication effects on behavioural and hormonal responses to acute stress in chickens. *Physiol. Behav.* **2014**, *133*, 161–169. [CrossRef]
23. Giersberg, M.F.; Spindler, B.; Rodenburg, B.; Kemper, N. The dual-purpose hen as a chance: Avoiding injurious pecking in modern laying hen husbandry. *Animals* **2020**, *10*, 16. [CrossRef]
24. Alvino, G.M.; Blatchford, R.A.; Archer, G.S.; Mench, J.A. Light intensity during rearing affects the behavioural synchrony and resting patterns of broiler chickens. *Br. Poult. Sci.* **2009**, *50*, 275–283. [CrossRef]
25. Kjaer, J.B.; Vestergaard, K.S. Development of feather pecking in relation to light intensity. *Appl. Anim. Behav. Sci.* **1999**, *62*, 243–254. [CrossRef]
26. Widowski, T.M.; Keeling, L.J.; Duncan, I.J.H. The preferences of hens for compact fluorescent over incandescent lighting. *Can. J. Anim. Sci.* **1992**, *72*, 203–211. [CrossRef]
27. Hartini, S.; Choct, M.; Hinch, G.; Kocher, A.; Nolan, J.V. Effects of light intensity during rearing and beak trimming and dietary fiber sources on mortality, egg production, and performance of ISA Brown laying hens. *J. Appl. Poult. Res.* **2002**, *11*, 104–110. [CrossRef]
28. Prescott, N.B.; Wathes, C.M. Spectral sensitivity of the domestic fowl (Gallus g. domesticus). *Br. Poult. Sci.* **1999**, *40*, 332–339. [CrossRef] [PubMed]
29. Rubene, D. Functional Differences in Avian Colour Vision: A Behavioural Test of Critical Flicker Fusion Frequency (CFF) for Different Wavelengths and Light Intensities. Master's Thesis, Uppsala University, Uppsala, Sweden, 2009.
30. Huth, J.C.; Archer, G.S. Comparison of two LED light bulbs to a dimmable CFL and their effects on broiler chicken growth, stress, and fear. *Poult. Sci.* **2015**, *94*, 2027–2036. [CrossRef]
31. van Staaveren, N.; Harlander-Matauschek, A. Cause and prevention of injurious pecking in chickens. In *Understanding the Behaviour and Improving the Welfare of Chickens*; Nicol, C.J., Ed.; Burleigh Dodds Science Publishing: London, UK, 2020; pp. 509–566.
32. Regmi, P.; Deland, T.S.; Steibel, J.P.; Robison, C.I.; Haut, R.C.; Orth, M.W.; Karcher, D.M. Effect of rearing environment on bone growth of pullets. *Poult. Sci.* **2015**, *94*, 502–511. [CrossRef] [PubMed]
33. Stratmann, A.; Fröhlich, E.K.F.; Gebhardt-Henrich, S.G.; Harlander-Matauschek, A.; Würbel, H.; Toscano, M.J. Modification of aviary design reduces incidence of falls, collisions and keel bone damage in laying hens. *Appl. Anim. Behav. Sci.* **2015**, *165*, 112–123. [CrossRef]
34. Moinard, C.; Statham, P.; Haskell, M.J.; McCorquodale, C.; Jones, R.B.; Green, P.R. Accuracy of laying hens in jumping upwards and downwards between perches in different light environments. *Appl. Anim. Behav. Sci.* **2004**, *85*, 77–92. [CrossRef]
35. Tobalske, B.W.; Dial, K.P. Aerodynamics of wing-assisted incline running in birds. *J. Exp. Biol.* **2007**, *210*, 1742–1751. [CrossRef]
36. Fraisse, F.; Cockrem, J.F. Corticosterone and fear behaviour in white and brown caged laying hens. *Br. Poult. Sci.* **2006**, *47*, 110–119. [CrossRef]
37. Jones, R.B. Fear and adaptability in poultry: Insights, implications and imperatives. *Worlds. Poult. Sci. J.* **1996**, *52*, 131–174. [CrossRef]

Article

Commercial Free-Range Laying Hens' Preferences for Shelters with Different Sunlight Filtering Percentages

Md Sohel Rana [1,2,3], Caroline Lee [2], Jim M. Lea [2] and Dana L. M. Campbell [2,*]

1   Department of Animal Science, School of Environmental and Rural Science, University of New England, Armidale, NSW 2351, Australia; ranasoheldvm06@gmail.com
2   Agriculture and Food, Commonwealth Scientific and Industrial Research Organisation (CSIRO), Armidale, NSW 2350, Australia; caroline.lee@csiro.au (C.L.); jim.lea@csiro.au (J.M.L.)
3   Department of Livestock Services, Ministry of Fisheries and Livestock, Dhaka 1215, Bangladesh
*   Correspondence: dana.campbell@csiro.au

Simple Summary: Australian sunlight is intense, and may impact range use by free-range hens. Range design and management are important for optimising commercial layer farms where artificial shelters may offer protection for ranging hens. This study investigated preferences among 34–40-week-old hens for artificial shade cloth shelters of different densities, using two flocks on a commercial farm during the summer. Three types of sunlight-filtering shade cloth shelters, i.e., blocking 50%, 70%, and 90% of ultraviolet (UV) light, each with three replicates, were placed on the range for each flock. The number of hens under each shelter was counted at 30-min intervals using image snapshots from video recordings for 14 to 17 days. An on-site weather station recorded sunlight intensity across different spectra, ambient temperature, and relative humidity. During the day, hens generally preferred the 90%, followed by 70% and 50% sunlight-filtering shelters. However, fewer hens were observed underneath shelters during times of peak sun intensity. Shelter preferences were mostly impacted by ambient temperature in both flocks, with all sunlight spectra having different degrees of effect depending on the shelter type and flock. Overall, shelters comprising higher densities of sunlight-filtering artificial cloth were preferred by hens on the range, but these may not be sufficient to attract more hens outside during intense sunlight and hot climatic conditions.

Citation: Rana, M.S.; Lee, C.; Lea, J.M.; Campbell, D.L.M. Commercial Free-Range Laying Hens' Preferences for Shelters with Different Sunlight Filtering Percentages. *Animals* **2022**, *12*, 344. https://doi.org/10.3390/ani12030344

Academic Editors: Victoria Sandilands and Tina Widowski

Received: 4 December 2021
Accepted: 24 January 2022
Published: 31 January 2022

Publisher's Note: MDPI stays neutral with regard to jurisdictional claims in published maps and institutional affiliations.

Abstract: Extreme sunlight might be aversive to free-range laying hens, discouraging them from going outside. Range enrichment with artificial shelters may protect hens from sunlight and increase range use. The preferences of 34–40-week-old Hy-Line Brown laying hens for artificial shelters were assessed by counting the number of hens under three densities of individual shelters (three replicates/density) from video recordings for 14 to 17 days for two flocks. The artificial shelters used shade cloth marketed as blocking 50%, 70%, and 90% of ultraviolet light, although other sunlight wavelengths were also reduced. Different sunlight spectral irradiances (ultraviolet radiation (UV_{AB}) (288–432 nm), photosynthetically active radiation (PAR) (400–700 nm), and total solar radiation (TSR) (285 nm–3000 nm), ambient temperature, and relative humidity were recorded with an on-site weather station. There was a significant interaction between sunlight-filtering shelter and time of day (both Flocks, $p < 0.0001$), i.e., hens preferred shelters with the highest amount of sunlight-filtering at most time points. Regression models showed that the most variance in shelter use throughout the day resulted from the ambient temperature in both flocks, while sunlight parameters had different degrees of effect depending on the shelter type and flock. However, fewer hens under the shelters during the midday period suggest that during periods of intense sunlight, hens prefer to remain indoors, and artificial structures might not be sufficient to attract more hens outside.

Keywords: Australia; chicken; free-range; preference; radiation; range enrichment; shelter; ultraviolet

1. Introduction

The important positive attributes of free-range (including organic) layer farming are the birds' access to an outdoor range, exposure to natural daylight (sunlight), ability to move freely, increased space to better regulate social interactions, and opportunities for expression of natural behaviours [1,2]. Free-range systems may also have some potential risks such as parasitic infections [3], increased disease exposure [4], heat stress [5], and predation [6,7]. However, welfare benefits such as reduced plumage damage and reduced footpad dermatitis can be seen in individuals that range more [8,9]. The number of hens using the range in the first few weeks following the opening of the pop holes is typically low which gradually increases upon adaptation to the outdoor environment [10,11]. However, a range of external factors impact the daily outdoor range use of hens, even after acclimatisation, including weather conditions [12–14], season [15], time of day [16,17], and range enrichments [18,19]. Sometimes, hens may hesitate to venture outside or only use certain areas of the range closer to the pop holes [16,20–22]. The distribution of hens on the range can depend on range features such as vegetation and other enrichments [17,22]. A lower use of the outdoor range at the flock level may lead to increased feather and injurious pecking [23,24] or crowding of hens closer to the shed may cause smothering problems leading to bird mortality [25].

The outdoor range needs to be attractive to increase use by hens, i.e., offering different kinds of natural or artificial shelters and/or shades within the range [10,18,26]. These shelters may increase hens' ranging by serving as protection from predators [27,28] or diffusing intense sunlight [17,29]. Laying hens might exhibit preferences for specific types of natural shelter options [30] as their ancestors were accustomed to dense vegetation. However, for range areas that may not have established vegetation, artificial shelters can provide protection to increase range usage and/or improve range use distribution. The exact features of these artificial shelters are likely to impact the extent to which hens use them [17,29] and the most preferred features need to be better understood.

Artificial shelters that provide protection from sunlight may be particularly important for free-range hens in climates with more extreme sunlight conditions such as those experienced in Australia during the summer months. The sunlight spectrum contains all forms of ultraviolet (UV) radiation: UVA (315–400 nm), UVB (280–315 nm), and UVC (100–280 nm), of which only UVA and UVB reach the earth's surface. Hens can visually perceive UVA and UVB has a physiological effect on the synthesis of vitamin D_3 in featherless skin [31]. However, high intensities and/or overexposure of UV radiation may have damaging effects [32,33], and, thus, hens might avoid direct sunlight at its most intense. Intense sunlight can also be visibly bright where the photosynthetically active radiation (PAR) wavelengths (400–700 nm) may be visually aversive and infrared wavelengths (>700 nm) are associated with heat. Different sunlight wavelengths as well as associated ambient climatic temperature and humidity variables may all impact the motivation of hens to use artificial shelters.

Behaviourally, studies in both free-range layers and meat chickens show that birds range more and are more active during the early morning and late afternoon periods compared to around midday [16,22,34]. Previous studies assessing shelter preferences on commercial farms within Australia have shown hens preferred higher density (% of UV filtering) shade cloth structures that filtered the most UV radiation [17] although preferences varied with time of day [29]. Hens also preferred artificial horizontal structures including those with one vertical side rather than vertical shelters alone [29]. However, shelter height, orientation, and cover density were all factors that affected hen preferences [29]. These studies to date highlight the complexity around optimal artificial shelter design. Further confirmation of hen preferences for artificial shelter cloth densities in relation to different sunlight wavelengths and ambient climatic variables is needed for optimising free-range systems in hot climates.

This study was conducted to assess the use by hens of different sunlight filtering shade cloth shelters in relation to different sunlight wavelengths on the range of a commercial free-range laying hen farm in Australia. The study hypothesised that hens would prefer the shelters that blocked a greater amount of sunlight, particularly when there was high sunlight intensity.

2. Materials and Methods

2.1. Animals and Husbandry

The study was conducted using two individual flocks (Flock-A, and Flock-B) at a single commercial free-range laying hen farm (comprised of multiple sheds and associated range areas) during the summer months (December 2020 to March 2021) in Queensland, Australia. Both flocks, comprising approximately 20,000 Hy-Line Brown laying hens each were studied from 34 to 40 weeks of age. The birds were from the same hatchery and reared indoors for 16 weeks (10–15 Lux) with the same resources, feed, and housing management as per the national laying hen guidelines [35] before shifting into the free-range facility. From 16 to 20 weeks, the hens were housed inside the indoor aviary with standard farm management protocols and resource access as per the national laying hen guidelines [35] and artificial lighting of approximately 70 Lux. At 20 weeks of age, hens were provided range access via pop holes (09:00–20:00 h). Hens were given 14 weeks of range acclimation before the study commenced.

2.2. Study Sites

The two study sites within the larger farm property each had distinct land layout and vegetation within the range but identical resources inside the sheds and the same management practices. Both sites had an indoor shed, which was longitudinal in the east-west position with an outdoor range at both the north and south face. Hens within the shed could only access the range on either the north or south face due to an internal shed division, thus each shed actually contained 40,000 hens total. The south side of each shed was used for this study. The indoor sheds included an aviary system, furnished with feeders, drinkers, nest boxes and perches. Feed and water were provided *ad libitum* inside the shed only. The base of the shed sidewalls (0.62 m) was made from solid materials (poly panel) and the upper parts were covered by curtains up to the ceiling. The indoor shed temperature and relative humidity were maintained both mechanically by lowering and raising the curtains and automatically with fans throughout the study periods. When the curtains were raised (between 23 and 29 °C), sunlight could enter the barn, although the shed was positioned so this was minimised during the summer months. Each of the indoor sheds measured 120 L × 20 W × 8 m H with an indoor stocking density of 9 hens/m^2. The outdoor stocking density was 1500 hens/ha (equivalent to 0.15 hens/m^2). Pop-holes for range access were 0.55 m in height and located in the sidewalls. There was a total of 14 pop holes (6 m L × 0.62 m W) on each side, but typically, only half were opened for the full shed length. The range area adjacent to the shed wall (2.5 m length) was covered with compact gravel, then the immediate range area (12 m length) was covered with heavy weed fabric, followed by approximately 25 m length of uncovered (dirt) area, and the rest of the range was covered with grass. The total range area was approximately 13 hectares in size and thus the grassed area was expansive but typically few hens were observed in the farthest range areas (producer communication to DLMC, 2020). A number of trees were establishing within the range area, planted at varying distances from the shed past the gravel and fabric-covered areas (Figure 1). The boundaries of the range area were wire fences. During the observation periods, the daylight hours in the study sites were 04:57–18:51 h (at the beginning) and 05:14–18:54 h (at the end) for Flock-A, and 05:29–18:47 h (at the beginning) and 05:46–18:28 h (at the end) for Flock-B. The average minimum and maximum outdoor ambient temperatures in Flock-A were recorded as 24.1 ± 0.10 °C and 26.6 ± 0.10 °C respectively, and average relative humidity was 51.4 ± 0.27%; in Flock-B, the average minimum and maximum outdoor ambient temperatures were recorded as 24.0 ± 0.11 °C and 27.3 ± 0.11 °C respectively, and average relative humidity was 49.3 ± 0.17%.

(a)   (b)

Figure 1. Experimental set-up in two different sheds of a commercial free-range farm: (**a**) Flock-A, and (**b**) Flock-B.

2.3. Experimental Set-Up

To test the preferences of hens for shade cloth shelters of different densities, three types of shade cloth shelters with three replicates each were used: (i) 50% UV block (Coolaroo, 484866, Shade cloth, Rainforest), (ii) 70% UV block (Garden Shield, SC303610CG, HDPE, Cottage Green where supplier labelling indicated 30% UV block but controlled testing showed it was actually 70% UV block), and (iii) 90% UV block (Coolaroo, 486921, Shade cloth, Rainforest) (Figure 1).

The UV filtering percentages of the treatment shelter cloths were confirmed using an Ocean Insight Flame-S-XR1 Spectroradiometer (200–1025 nm; Quark Photonics, Melbourne, VIC, Australia) set with an integration time of 180,000 μs and integration range from 280–1000 nm (Figure 2). Measurements were taken at a distance of 20 cm with each type of shade cloth placed over a set of three Exo Terra® (Rolf C. Hagen, Montreal, QC, Canada) pet reptile bulbs (Reptile UVB200, 25W, PT2341) used as a standard, controlled source of UV radiation. Although the shade cloths are marketed as blocking UV radiation, they also filtered out solar radiation in the visible spectrum (Figure 2). Each shelter (4 m L × 3 m W × 1 m H) was positioned in a straight line parallel with the shed 10.5 m away from the pop holes. This distance was selected to avoid the shadow of the shed and to entice the hens farther out onto the range. Shelters were placed 3 m apart following the repeating pattern of 90%, 70%, and 50% UV block shade cloth in Flock-A and 70%, 90%, and 50% UV block shade cloth in Flock-B (Figure 1). The structure of the shelter was made of galvanised steel and shade cloth was stretched tight over the frame to minimise its movement in the wind with a small apex along the centre. Temperature and humidity loggers (Tinytag Plus 2, TGP-4500; Gemini Data Loggers Ltd., West Sussex, UK) were placed under each shelter on the rear left post at 300 mm height with automated logging at 15 min intervals. The position of these loggers resulted in them sometimes being shaded and sometimes being under direct sunlight, depending on the position of the sun. A high-resolution security camera system (Hikvision DS-7608NI-I2-8P CCTV NVR Recorder) was installed with a camera (Hikvision DS-2CD2355FWD-I2 CCTV 6MP Turret cameras) on a stand 1.6 m in front of each shelter to capture the entire shelter and the shadows that were cast during the day (Figure S1). Each IP camera was individually cabled back to a small enclosure mounted within the range that contained a Hikvision Ethernet POE Switch (Model DS-3e0109P-E(C)) that powered the cameras as well as a set of NanoBeam®–ac's (model NBE-5AC-Gen2; Ubiquiti Inc., New York, NY, USA) that wirelessly routed the cameras back to the NVR system set up in the site office. An MEA weather station (Green

Brain, 41 Vine Street, Magill, SA, Australia) was set-up on the respective farm site for recording sunlight and climatic variables and recorded weather data every 15 min over the study periods. The weather station was mounted on a post (user supplied) at a height of 1 m (height as instructed by the manufacturer) and included different sensors (UV3pAB UV sensor (288–432 nm), QS5 PAR pyranometer (400–700 nm), and SR-05 pyranometer (285–3000 nm)) for recording sunlight variables including ultraviolet radiation (UV_{AB}) (W/m^2), photosynthetically active radiation (PAR) ($\mu mol/m^2/s$), and total solar radiation (TSR) (W/m^2), respectively. The TSR included UV_{AB}, PAR and infrared (IR) wavelengths and was used to extract IR (700 nm–3000 nm) (W/m^2). Additionally, an air temperature and relative humidity sensor recorded the ambient temperature (°C), relative humidity (%), barometric pressure (mBar), dew point (°C), voltage (V), and vapor pressure deficit (kPa). As the study was primarily focused on the hen preferences for different shelters relative to sunlight variables, only solar radiation spectra, air temperature, and relative humidity data were considered in the final analyses.

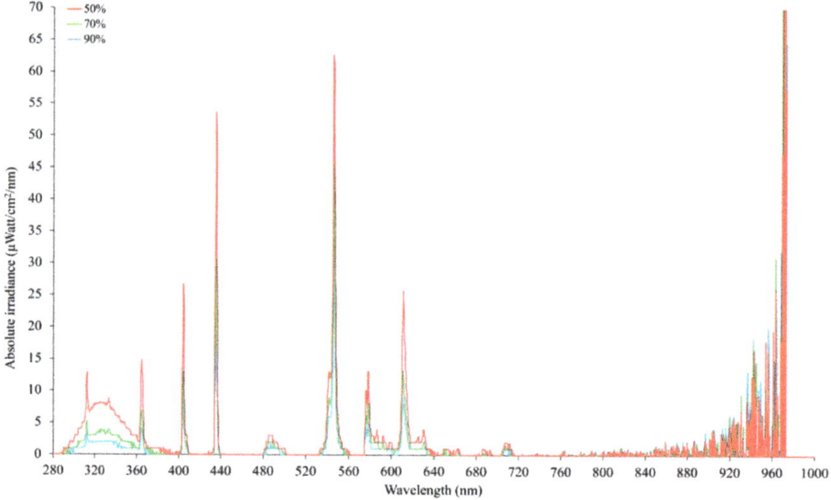

Figure 2. Spectral irradiance under different UV-filtering shade cloths (50%, 70%, and 90% UV block) as measured by an Ocean Insight Flame-S-XR1 Spectroradiometer at a distance of 20 cm with each type of shade cloth placed over a set of three Exo Terra® (Rolf C. Hagen, Montreal, QC, Canada) pet reptile bulbs (Reptile UVB200, 25W, PT2341).

2.4. Observations and Data Collection

The shelters were installed when the hens were 34 weeks of age with 2 weeks allowed for habituation to the range shelters before the study observations began. Recording was continuous during daylight hours for approximately 5 weeks for Flock-A and 4 weeks for Flock-B. Due to temporary failures in video recording, a total of 14 days videos for Flock-A and 17 days for Flock-B were analysed and these days were not consecutive within the recording period. For assessing shelter preferences, image snapshots from video records were taken at 30 min intervals from 30 min after pop hole opening (i.e., 09:30) until just before sunset (i.e., 18:30). The images were imported into Image-J 1.53a software (Wayne Rasband, National Institute of Health, MA, USA) and an observer counted the number of hens both under the individual shelters and on top of the shelters. When there were two observers, both researchers discussed the counts on common snapshots to ensure agreement. Observers were not blinded to the shade cloth densities given the differing darkness of the shadows cast by the shelters, but each observer conducted counts for all densities to minimise observer bias for a specific treatment density. On sunny days, the area for counting the hens under the shelter was defined by the shadow that the shelter cast (the exact position of the shadow varied throughout the day) (Figure S1). On cloudy days without a prominent shadow, the counting area was considered as the area directly underneath the shelter frame. If the individual hens could not be clearly identified due to crowding under the shelter, the number of hens was estimated in the group by counting the birds within a certain area and then multiplying that number by the counted area (this occurred on 41 occasions out of 5301 observations for both flocks).

2.5. Data and Statistical Analyses

All observations for each flock were analysed separately. A combined total of 5301 observations were made over the 14-day period in Flock-A (2394 observations), and 17-day period in Flock-B (2907 observations) to count both the hens underneath and on top of the shelters. The number of hens counted in each observation was matched with the corresponding weather parameters during the 15-min period directly prior to the observation time point. Weather parameters included the UV_{AB}, PAR, TSR, ambient temperature and relative ambient humidity, and temperature and relative humidity readings from the loggers underneath the shelters. The hen count data contained a considerable number of '0' values (when no hens were under or on top of the shelters) and were not normally distributed, thus these data were log (x + 1) transformed to include the '0' values in the analyses as well as to approach data normality. To test the preferences of the hens to be underneath the shelters during the study period, data were analysed using JMP® 14.0 (SAS Institute, Cary, NC, USA) with α level set at 0.05. General linear mixed models (GLMM) were applied with the different UV-filtering percentages, time of day, and their interaction included as fixed effects and shelter replicate nested within UV-filtering percentage as a random effect. A separate model with the same parameters was fitted to assess the preferences of the hens to be on top of the shelters. While the different sunlight filtering percentages were not predicted to affect hen preferences on top of the shelters, there may have been social influences if shelters with more hens underneath them, also had more hens located on top. The studentised model residuals were visually inspected for confirming homoscedasticity. Where significant differences were present, post hoc Student's t-tests were applied to the least squares means with Bonferroni corrections to the α level to account for multiple post-hoc comparisons. The means of the temperature and relative humidity underneath the shelter were plotted, along with the mean ambient temperature and humidity during the day. However, these data were not statistically analysed as their positioning on the rear leg of the shelters resulted in the loggers sometimes being under direct sunlight which meant that they were not always an accurate measure of the temperature experienced by a hen when under the shaded part of the shelter.

To investigate the effects of sunlight variables on shelter use by hens during the day (presence under the shelter regardless of shelter type), an overall linear regression model was constructed for each flock using a summarised dataset where values within each UV-filtering percentage were averaged across all three replicates for each time point for each day (n = 798 per UV-filtering percentage in Flock-A, and n = 969 per UV-filtering percentage in Flock-B). Before setting the model, IR spectrum values were extracted from the TSR readings by subtracting UV_{AB} and PAR. A conversion value ($\mu mol/m^2/s$ to W/m^2) as described by Thimijan and Heins [36], was applied to the PAR readings so all measures were in the same units for calculating the IR values. The number of hens underneath the UV-filtering shelters were included as the dependent variable, whereas sunlight variables (UV_{AB}, PAR and IR), ambient temperature, and relative ambient humidity were included as independent variables in the model. Prior to running the model in R statistical software [37], the collinearity among the independent variables were checked through determination of variance inflation factors (VIF). Due to collinearity (VIF $\geq$ 10) among the sunlight variables, the ridge regression [38] was chosen to best fit the predictors into the model using the 'lmridge' package in R [39]. The relative contributions of the predictors in the regression model were estimated by the R package 'relaimpo' [40]. All independent variables were initially included in the model with nonsignificant variables ($p \geq 0.10$)

removed through backward elimination until the model of best fit was produced based on the adjusted R^2 values. To specifically determine how sunlight and weather variables may affect the use of the different shelter types, individual linear ridge regression models were performed separately for each UV-filtering percentage with the number of hens underneath included as the dependent variable, and the sunlight variables (UV_{AB}, PAR, and IR), ambient temperature, and relative ambient humidity included as independent variables. Nonsignificant variables ($p \geq 0.10$) were removed through backward elimination. The raw values are plotted in the figures.

3. Results

3.1. Shelter Preferences

There was a significant interaction between UV-filtering shelter and time of day for hen preferences in both Flock-A ($F_{36,\,2331} = 3.49$, $p < 0.0001$), and Flock-B ($F_{36,\,2844} = 2.63$, $p < 0.0001$) (Figure 3). In general, at most observation points throughout the day, more hens were seen under the 90% UV-filtering shelters in both flocks, but at some time points their preferences were similar for all filtering percentages ($p > 0.001$) (Figure 3).

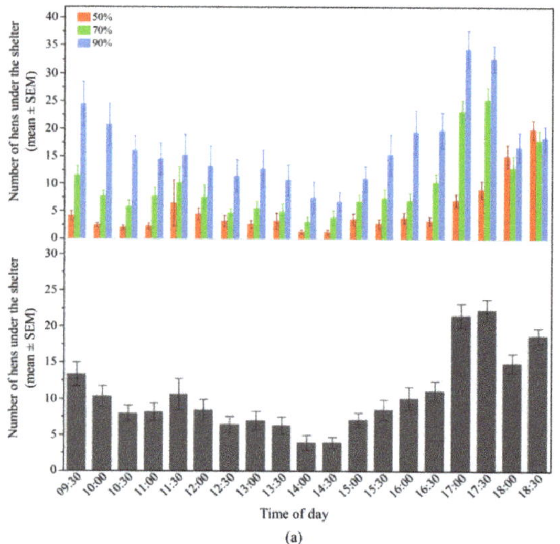
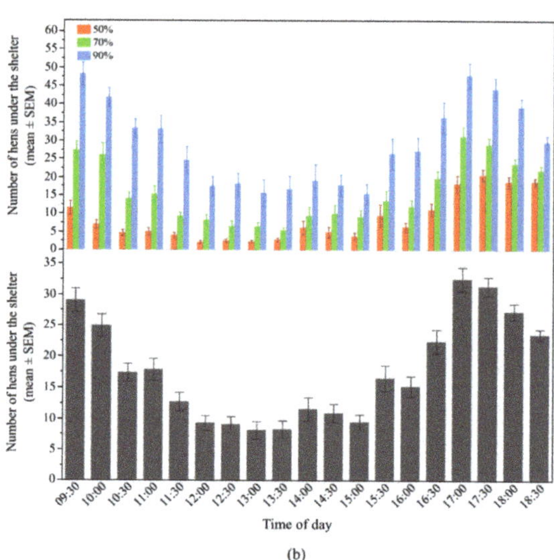

Figure 3. The mean (±SEM) number of hens underneath the shelters during the day: (**a**) Flock-A: with different UV-filtering percentages (50%, 70%, and 90%) (**top**), under all UV-filtering shelters (**bottom**); (**b**) Flock-B: with different UV-filtering percentages (50%, 70%, and 90%) (**top**), under all UV-filtering shelters (**bottom**). Note the different Y-axis scales between Flock-A and Flock-B. Raw values are presented with analyses conducted on transformed data.

Overall, more hens were found underneath the 90% UV-filtering shelters (LSM mean ± SEM, Flock-A: 16.9 ± 2.67 hens; Flock-B: 29.1 ± 1.52 hens), followed by the 70% (LSM mean ± SEM, Flock-A: 9.7 ± 2.67 hens; Flock-B: 15.7 ± 1.52 hens) then 50% UV-filtering shelters (LSM mean ± SEM, Flock-A: 5.2 ± 2.67 hens; Flock-B: 8.4 ± 1.52 hens) in both study flocks (Flock-A: $F_{2,\,6} = 16.25$, $p = 0.004$, and Flock-B: $F_{2,\,6} = 134.09$, $p < 0.0001$). The use of the shelter shade by hens varied throughout the day in both Flock-A ($F_{18,\,2331} = 44.64$, $p < 0.0001$) and Flock-B ($F_{18,\,2844} = 75.11$, $p < 0.0001$), with peaks in the morning and in the late afternoon, compared to the midday ($p < 0.003$) (Figure 3).

In contrast, there was no significant interaction between UV-filtering shelter and time of day for the number of hens on top of the shelters in Flock-A ($F_{36,\,2331} = 0.89$, $p = 0.65$); whereas a significant interaction was found in Flock-B ($F_{36,\,2844} = 2.68$, $p < 0.0001$) (Figure 4).

In Flock-B, throughout the day, there was a general pattern of more hens on top of the 90% UV-filtering shelters in the morning and late afternoon relative to both the 50% and 70% shelters ($p > 0.001$) (Figure 4).

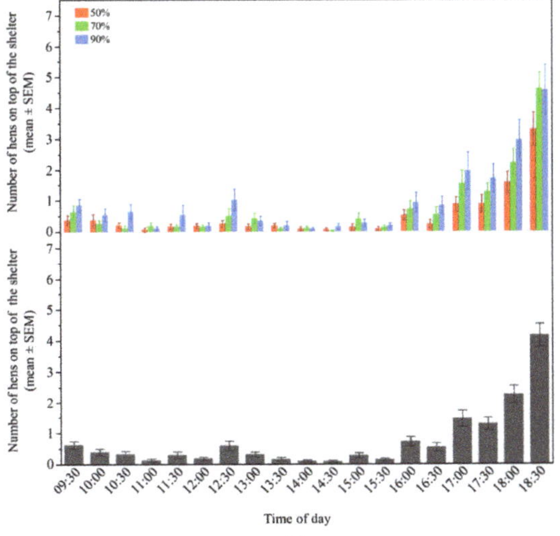 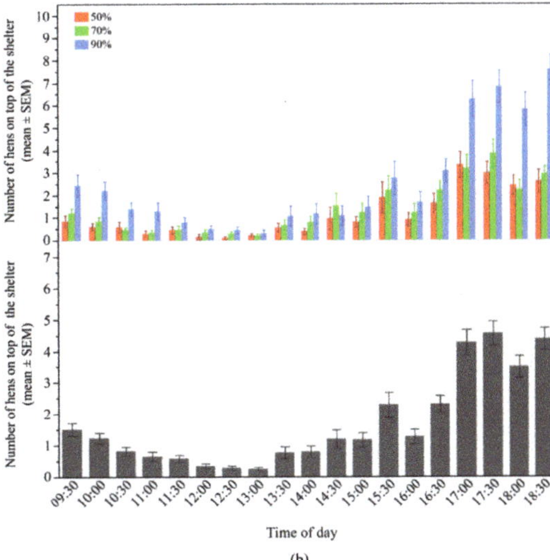

Figure 4. The mean ($\pm$SEM) number of hens on top of the shelters throughout the day: (**a**) Flock-A: with different UV-filtering percentages (50%, 70%, and 90%) (**top**), on all UV-filtering shelters (**bottom**); (**b**) Flock-B: with different UV-filtering percentages (50%, 70%, and 90%) (**top**), on all UV-filtering shelters (**bottom**). Note the different Y-axis scales between Flock-A and Flock-B for the top graphs. Raw values are presented with analyses conducted on transformed data.

Overall, there was no difference for the number of hens on top of the shelters between the different UV-filtering percentages in Flock-A (LSM mean $\pm$ SEM, 50%: 0.52 $\pm$ 0.18, 70%: 0.75 $\pm$ 0.18, 90%: 0.96 $\pm$ 0.18, $F_{2,6} = 1.37$, $p = 0.32$), but the time of day had an effect on the number of hens throughout the day ($F_{18,2331} = 41.78$, $p < 0.0001$), with a gradually increasing trend after 17:00 compared to the rest of the day ($p < 0.003$) (Figure 4). In Flock-B, more hens were found on top of the 90% UV-filtering shelter with no differences between the 50% and 70% shelters (LSM mean $\pm$ SEM, 50%: 1.14 $\pm$ 0.24, 70%: 1.38 $\pm$ 0.24, 90%: 2.54 $\pm$ 0.24, $F_{2,6} = 9.14$, $p = 0.02$). Time of day had an effect on the number of hens on top of the shelters ($F_{18,2844} = 36.56$, $p < 0.0001$) with more hens observed in the late afternoon ($p < 0.003$) (Figure 4).

The temperature and humidity loggers underneath the shelters were intended to provide measurements on ambient conditions the hens may have been experiencing. However, the placement of loggers at hen eye height on one of the rear posts of the shelters resulted in the loggers sometimes being under direct sunlight and sometimes being under the shelter shade. Figure 5 displays the temperature and relative humidity readings under each shelter type relative to the ambient temperature and relative ambient humidity readings obtained from the weather station which was placed 1 m above ground. The temperature under the shelters was higher than the ambient temperature, whereas relative humidity was lower than the relative ambient humidity during the daytime (Figure 5). The temperatures and relative humidity under the different shelter types were visually similar, but these data were not statistically analysed, as the loggers did not capture data as originally intended.

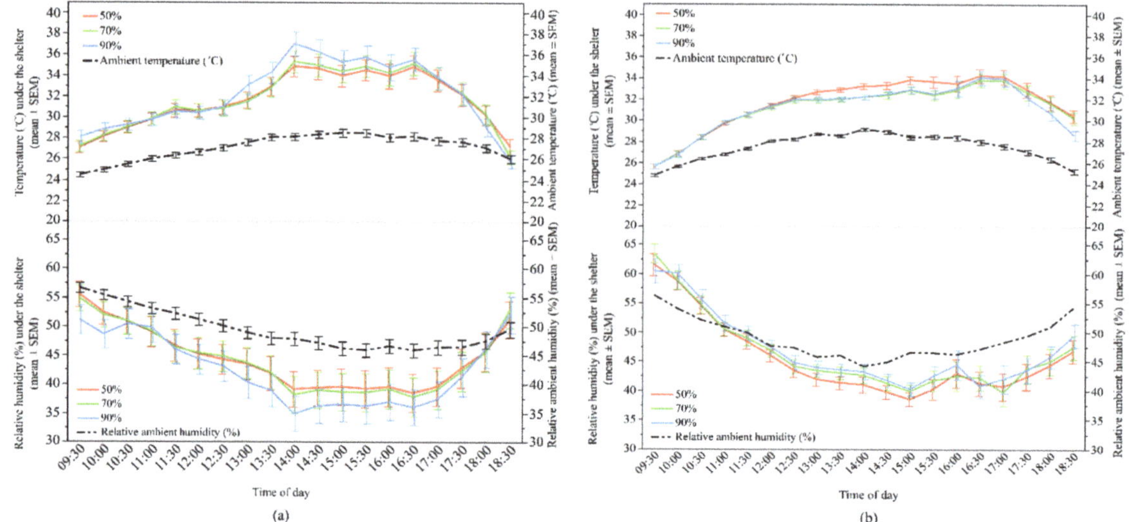

Figure 5. The mean (±SEM) temperature and relative humidity throughout the day underneath the shelters of different UV-filtering percentages (50%, 70%, and 90%) and ambient temperature and relative humidity: (**a**) Flock-A: temperature under the shelters and ambient temperature (**top**), relative humidity under the shelters and relative ambient humidity (**bottom**); (**b**) Flock-B: temperature under the shelters and ambient temperature (**top**), relative humidity under the shelters and relative ambient humidity (**bottom**). Raw values are presented with analyses conducted on transformed data.

3.2. Sunlight Effects

A ridge regression model for each flock was performed to investigate the relationship between the number of hens underneath the shelters across all the UV-filtering percentages and the sunlight variables, ambient temperature and relative ambient humidity. The best-fit model results are presented in Table 1. In Flock-A, the model accounted for 34.21% of the variance in the use of all the UV-filtering shelters throughout the day. The ambient temperature, UV_{AB}, IR, and relative ambient humidity contributed significantly to the model ($F_{3.35, 794.24} = 120.50$, $p < 0.0001$). However, all these predictors had a negative correlation with the number of hens under the shelters throughout the day (Table 1).

In Flock-B, the model accounted for 35.77% of the variance in the number of hens under the shelters with respect to sunlight and weather variables considered within the model. The majority of the variance was explained by the ambient temperature (49.01%), however IR, UV_{AB} and PAR also significantly contributed to the model ($F_{2.68, 965.98} = 146.64$, $p < 0.0001$, Table 1). The ambient temperature, UV_{AB}, and IR were negatively correlated, and PAR was positively correlated with the number of hens underneath the shelters (Table 1).

The separate ridge regression models for each UV-filtering percentage showed differences in the relative impacts of the sunlight and weather variables on the number of hens underneath the shelters. For the 50%, 70%, and 90% UV-filtering shelter preferences, both sunlight and weather variables accounted for 51.71% (Flock-A: $F_{2.79, 263.03} = 108.58$, $p < 0.0001$) and 57.94% (Flock-B: $F_{2.53, 320.19} = 156.77$, $p < 0.0001$) of the variance for the 50% shelters, 40.35% (Flock-A: $F_{2.79, 263.03} = 71.33$, $p < 0.0001$) and 44.29% (Flock-B: $F_{2.68, 319.98} = 71.26$, $p < 0.0001$) of the variance for the 70% shelters, and 35.16% (Flock-A: $F_{3.08, 262.54} = 51.13$, $p < 0.0001$) and 37.77% of the variance (Flock-B: $F_{2.68, 319.98} = 56.45$, $p < 0.0001$) for the 90% shelters (Table 2).

Table 1. Two ridge regression analyses (ridge parameter, k = 0.02) on the number of hens under the shelter throughout the day. Only variables that significantly contributed to the most parsimonious model are presented.

Flock	Predictor [1]	β- Coefficient (Standardised) ‡	t-Value	p-Value	Adjusted R^2 and Model's F-Statistics	Relative Weight of the Predictors in the Model
Flock-A	Ambient temperature	−0.70	−12.61	<0.0001	R^2-adjusted = 0.34	33.54%
	Relative ambient humidity	−0.46	−8.29	<0.0001	$F_{3.35,\ 794.24}$ = 120.50, $p < 0.0001$	13.86%
	UV$_{AB}$	−0.15	−2.17	0.03		25.36%
	IR	−0.24	−3.37	0.001		27.24%
Flock-B	Ambient temperature	−0.41	−15.90	<0.0001	R^2-adjusted = 0.36	49.01%
	UV$_{AB}$	−0.27	−3.78	<0.001	$F_{2.68,\ 965.98}$ = 146.64, $p < 0.0001$	16.88%
	PAR	0.10	2.07	0.04		16.84%
	IR	−0.19	−3.70	<0.001		17.28%

‡ β-coefficients (standardised) of the predictor variables were estimated separately using the ridge regression coefficient in 'R' as the original ridge package did not include the 'β-coefficient' value in the regression outputs.
[1] UV$_{AB}$ (ultraviolet radiation A and B wavelengths), PAR (photosynthetically active radiation), IR (infrared radiation).

Table 2. Multiple ridge regression analyses (ridge parameter, k = 0.02) on the number of hens under different UV-filtering shelters throughout the day. Only variables that significantly contributed to the most parsimonious model are presented.

UV-Filtering Shelter	Flock	Predictor [1]	B-Coefficient (Standardised) ‡	t-Value	p-Value	Adjusted R^2 and Model's F-Statistics	Relative Weight of the Predictors in the Model
50%	A	Ambient temperature	−0.59	−7.16	<0.0001	R^2-adjusted = 0.52	24.62%
		Relative ambient humidity	−0.36	−4.42	<0.0001	$F_{2.79,\ 263.03}$ = 108.58, $p < 0.0001$	10.52%
		PAR	−0.56	−13.25	<0.0001		64.86%
	B	Ambient temperature	−0.40	−11.24	<0.0001	R^2-adjusted = 0.58	34.26%
		UV$_{AB}$	−0.29	−3.18	<0.01	$F_{2.53,\ 320.19}$ = 156.77, $p < 0.0001$	32.43%
		IR	−0.28	−3.08	<0.01		33.31%
70%	A	Ambient temperature	−0.77	−8.48	<0.0001	R^2-adjusted = 0.40	38.52%
		Relative ambient humidity	−0.46	−5.11	<0.0001	$F_{2.79,\ 263.03}$ = 71.33, $p < 0.0001$	15.64%
		UV$_{AB}$	−0.41	−8.87	<0.0001		45.85%
	B	Ambient temperature	−0.44	−10.63	<0.0001	R^2-adjusted = 0.44	45.57%
		UV$_{AB}$	−0.33	−2.85	<0.01	$F_{2.68,\ 319.98}$ = 71.26, $p < 0.0001$	18.10%
		PAR	0.13	1.66	0.10		17.91%
		IR	−0.22	−2.70	0.01		18.42%
90%	A	Ambient temperature	−0.93	−9.99	<0.0001	R^2-adjusted = 0.35	40.75%

Table 2. *Cont.*

UV-Filtering Shelter	Flock	Predictor [1]	B-Coefficient (Standardised) [‡]	*t*-Value	*p*-Value	Adjusted R^2 and Model's F-Statistics	Relative Weight of the Predictors in the Model
		Relative ambient humidity	−0.68	−7.44	<0.0001	$F_{3.08,\ 262.54} = 51.13$, $p < 0.0001$	18.85%
		PAR	0.20	1.84	0.07		19.14%
		IR	−0.53	−4.92	<0.0001		21.26%
	B	Ambient temperature	−0.54	−12.28	<0.0001	R^2-adjusted = 0.38	71.58%
		UV_{AB}	−0.31	−2.51	0.01	$F_{2.68,\ 319.98} = 56.45$, $p < 0.0001$	9.20%
		PAR	0.24	2.85	<0.01		9.58%
		IR	−0.15	−1.78	0.08		9.63%

[‡] β-coefficients (standardised) of the predictor variables were estimated separately using the ridge regression coefficient in 'R' as the original ridge package did not include the 'β-coefficient' value in the regression outputs.
[1] UV_{AB} (ultraviolet radiation A and B wavelengths), PAR (photosynthetically active radiation), IR (infrared radiation).

The ambient temperature significantly affected the preferences of the hens for each shelter type in both flocks (Flock-A: all $p < 0.0001$; Flock-B: all $p < 0.0001$) (Figure 6); this parameter was the greatest contributing factor for use of the 70% and 90% UV-filtering shelters. Temperature accounted for 38.52% and 40.75% of the variation in Flock-A for the 70% and 90% shelters respectively, and 45.57% and 71.58% of the variation in Flock-B for the 70% and 90% shelters, respectively. The results indicated that increased ambient temperature resulted in fewer hens under the shelters (Table 2).

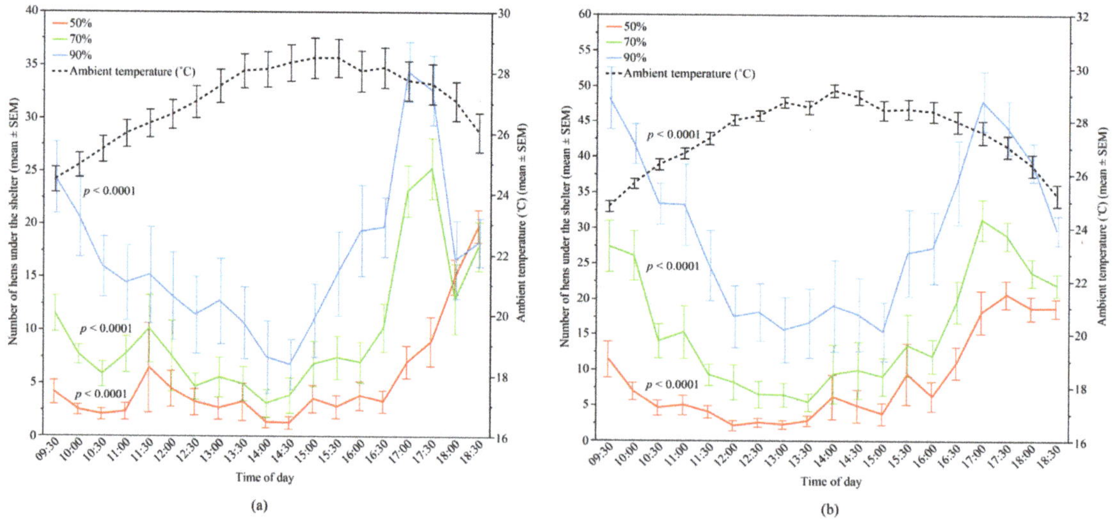

Figure 6. The mean (±SEM) number of hens under the different UV-filtering shelters (50%, 70%, and 90%) and the mean (±SEM) ambient temperature throughout the day: (**a**) Flock-A; (**b**) Flock-B ($p > 0.10$ indicates the variable had no significant effect and was removed from final model). Note the different Y-axis scales between Flock-A and Flock-B. Raw values are presented with analyses conducted on transformed data.

The relative ambient humidity also significantly contributed to the preferences of the hens for each shelter type in Flock-A (all $p < 0.0001$), but did not show an effect in Flock-B (Figure 7). However, in Flock-A, the relative contribution of the relative ambient humidity was less than 20% in the models of 50%, 70% and 90% UV-filtering shelters (accounting for 10.52%, 15.64%, and 18.85% of the variation, respectively), and had a negative correlation with the number of hens under the shelters (Table 2).

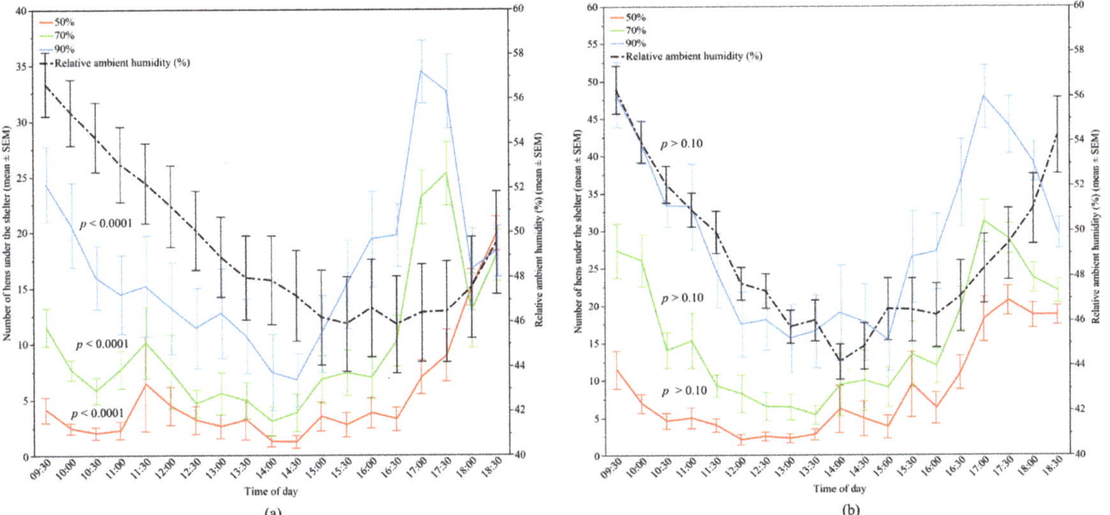

Figure 7. The mean (±SEM) number of hens under the different UV-filtering shelters (50%, 70%, and 90%) and the mean (±SEM) relative ambient humidity throughout the day: (**a**) Flock-A; (**b**) Flock-B ($p > 0.10$ indicates the variable had no significant effect and was removed from final model). Note the different Y-axis scales between Flock-A and Flock-B. Raw values are presented with analyses conducted on transformed data.

UV_{AB} radiation only had a significant effect for the 70% UV-filtering shelter preferences ($p < 0.0001$) in Flock-A where it was the most contributory effect (45.57% variation) in that specific model (Figure 8). In contrast, UV_{AB} radiation showed a significant relationship with the use of all shelter types in Flock-B (all $p \leq 0.01$) (Figure 8). The relative contribution of UV_{AB} among the predictors in Flock-B for 50%, 70%, and 90% UV-filtering shelter was 32.43%, 18.10%, and 9.20%, respectively, with the number of hens under the shelter decreasing with increasing UV_{AB} radiation (Table 2).

In Flock-A, PAR had a significant negative correlation with the use by the hens of the 50% UV-filtering shelter ($p < 0.0001$) showing the greatest contributory effect (64.86% variation, Table 2) in the model, and a positive trend for the 90% shelters ($p = 0.07$) but no association with use of the 70% UV-filtering shelters (Figure 9). Whereas, in Flock-B, PAR was a significant contributing variable for use by the hens of the 90% UV-filtering shelters ($p < 0.01$), and it had a trend effect for the 70% shelters ($p = 0.10$), but no significant contribution for the 50% UV-filtering shelters (Figure 9). While the relative weight of PAR in the models of 90% and 70% UV-filtering shelter preferences was 9.58% and 17.91%, respectively, this had a positive relationship with the number of hens under the respective shelters, indicating increases in PAR also increased shelter use by the hens (Table 2).

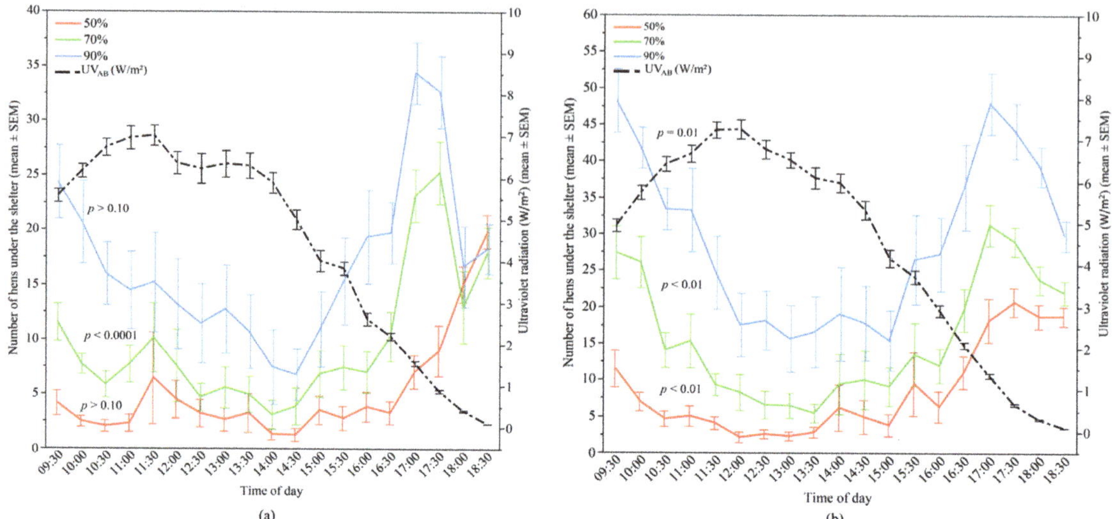

Figure 8. The mean (±SEM) number of hens under the different UV-filtering shelters (50%, 70%, and 90%) and the mean (±SEM) ultraviolet (UV_{AB}) radiation throughout the day: (**a**) Flock-A; (**b**) Flock-B ($p > 0.10$ indicates the variable had no significant effect and was removed from final model). Note the different Y-axis scales between Flock-A and Flock-B. Raw values are presented with analyses conducted on transformed data.

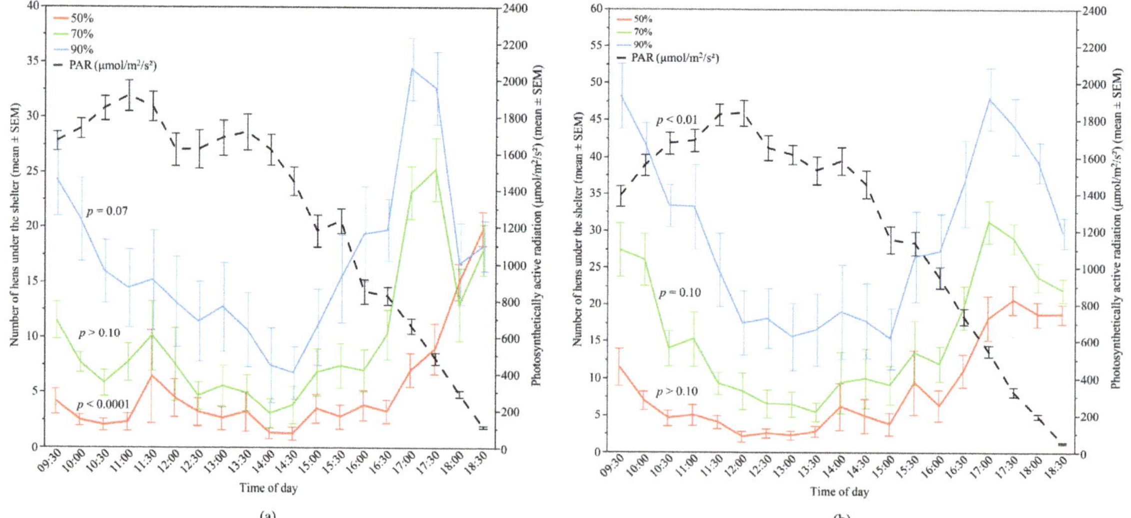

Figure 9. The mean (±SEM) number of hens under the different UV-filtering shelters (50%, 70%, and 90%) and the mean (±SEM) photosynthetically active radiation (PAR) throughout the day: (**a**) Flock-A; (**b**) Flock-B ($p > 0.10$ indicates the variable had no significant effect and was removed from final model). Note the different Y-axis scales between Flock-A and Flock-B. Raw values are presented with analyses conducted on transformed data.

IR significantly affected shelter use of only the 90% UV-filtering shelters ($p < 0.0001$) in Flock-A. However, in Flock-B, IR significantly influenced shelter use of both the 70%

and 50% UV-filtering shelters (both $p \leq 0.01$) and had a trend of an effect for the 90% UV-filtering shelters ($p = 0.08$) (Figure 10). However, a negative correlation between IR and use of shelters indicated that the number of hens under the shelters decreased when IR increased (Table 2).

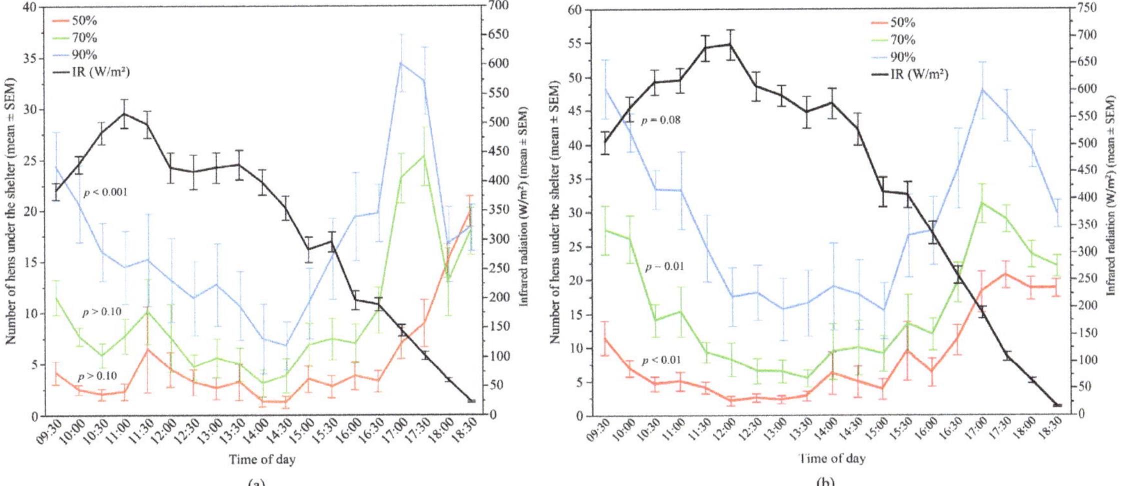

Figure 10. The mean (±SEM) number of hens under the different UV-filtering shelters (50%, 70%, and 90%) and the mean (±SEM) infrared radiation (IR) throughout the day: (**a**) Flock-A; (**b**) Flock-B ($p > 0.10$ indicates the variable had no significant effect and was removed from final model). Note the different Y-axis scales between Flock-A and Flock-B. Raw values are presented with analyses conducted on transformed data.

4. Discussion

This study assessed the use by hens of different sunlight filtering shade cloth shelters on the range of a commercial farm in Australia. Different sunlight wavelengths were also measured directly on-farm to determine if shelter preferences were dependent on ambient conditions. The results showed that hens had clear preferences for shelters with the highest density, i.e., those that blocked the greatest amount of sunlight. However, relationships with temperature, humidity, and sunlight wavelengths were generally negative, with fewer hens under the shelters as the values of the weather parameters increased. This may have been a result of reduced ranging at times of peak sun intensity, which is consistent with the findings of previous studies.

Previous studies have shown that outdoor range enrichments will increase range use by hens and improve their distribution outside [10,18,19,27]. However, the effects of these enrichments can vary depending on the structural design, i.e., type (artificial/natural), location, height, orientation, and density [16,17,29,30]. Similar to other studies that have been conducted on commercial farms within Australia [17,29], the hens in our study showed clear preferences for the higher densities of the shade cloth with a linear relationship between use of the shelter and percentage of sunlight it filtered. These results confirm that hens can differentiate between shaded environments and will preferentially select the environment that provides the greatest amount of shaded protection. Anecdotal observations in the current study indicated that hens were sometimes crowded under the shade provided by trees on the range just beyond the artificial shelters, while shelters were comparatively empty. Formal counts were not made on this, as the camera position did not enable clear observations of hen numbers under the trees. This result aligns with the jungle fowl origins of domestic chickens, as well as previous observations that the greatest numbers of hens on commercial farms are attracted to natural shelter provided by trees [10], or will preferably gather under dense vegetation and established trees [30]. While hens may not always seek shelter on the range and may use the sunlight for sunbathing [41] and warmth [12], conditions of extreme heat and intense sunlight are likely more aversive than enticing. Temperatures were not taken under the trees on the range in this study, but they may have provided a cooler environment than under the shelters as a result of blocking more sunlight, as well as evaporative cooling from transpiration. While the temperatures were recorded as hotter under the shelters than the ambient temperature, the temperature loggers were sometimes in direct sunlight, and thus, we are limited in the conclusions we can draw from these results.

Environmental factors that explained the number of hens under the shelters were ambient temperature, UV_{AB}, and IR in both flocks, indicating that the overall use of shelters decreased with increasing intensity of these factors. The predominant influencing factor was ambient outdoor temperature. These results are in contrast to what was predicted, given that previous studies have shown that hens increase their use of shaded areas as temperatures increase, with few birds in nonshaded areas during the summer [18]. Richards et al. [12] reported that the percentage of hens ranging gradually decreased as temperatures increased above 17 °C, although the study was conducted in the UK, where much lower temperatures overall are experienced than those in the current study. Slow-growing broilers will increase their use of shelters as solar radiation increases [42,43], which is expected, given the damaging impact of UV light [32,33] even though UV radiation was not found to be a predictor of range use in fast-growing broilers [34]. The hens in this study showed variation in shelter use throughout the day, regardless of shelter type, corresponding with typical patterns of range use reported in other studies [12,15,19], including observations in different flocks of the same farm as the current study [44]. Thus, the negative relationships between the environmental predictor variables were likely a reflection of fewer hens on the range during peak sun periods. General range use was not measured in this study, but the observed patterns of shelter use suggest that in regions of intense sunlight such as many regions of Australia, hens prefer to remain inside during the midday period, regardless of the presence or absence of artificial shelters on the range. Further studies could assess if more trees [10], different designs of artificial shelters [29] or additional range enrichments [18,26] could entice hens outside. If temperature is a key variable affecting the shelter preferences of hens, then shelter size may be another variable to consider, as well as the extent to which temperature varies under the edges or centre of the shelters. Alternatively, remaining inside the shed could prevent heat stress; in some regions of Australia, range use will be prevented on days of high heat to prevent bird mortality [44]. Further study assessing temperatures under different shelter types, established trees, shrubs, and inside sheds will confirm the different microclimates which exist in a free-range system and how they affect hen locational preferences.

The positive relationships seen between shelter use and the PAR wavelengths demonstrates that hens were using shelters to avoid bright light. This may have been comfort-related, i.e., the same way humans will wear sunglasses or could be motivated to seek cover rather than being exposed to bright light. The 90% filtering shelter would have reduced the visibility of hens from above to a greater degree, and hens may have used it as protection from aerial predators [26,28]. This could also explain the higher use of shelters in the late afternoon, when sunlight wavelengths greatly decreased in intensity, but hens may still have been seeking protection from aerial predators. In contrast to this, hens also increased in numbers on top of the shelters in the late afternoon/evening, but this may have been related to a nighttime desire to roost [45]. While hens were kept inside the shed overnight, the setting sun may have stimulated motivation to seek elevation for those hens still ranging as sunset approached. These observations, in conjunction with environmental parameters only accounting for part of the variation in shelter use, indicate the interplay of many factors regarding a hen's decision to reside under a shelter versus in the open range area.

5. Conclusions

This study found that higher densities of sunlight-filtering artificial cloth shelters are preferred by hens, and that temperature is a key variable affecting shelter use. All wavelengths of sunlight had some effect on shelter use, but the effect varied among the shelter densities and flocks in this study. The low use of all shelters during the midday period and negative relationships with temperature, humidity, UV_{AB} and IR suggest that the shelters may not be sufficient for attracting more hens to the range in periods of intense sunlight and hot temperatures, during which hens are typically observed to range less. Range enrichments of both artificial and natural shelters may encourage more hens outside. In the absence of established trees on the range providing a larger canopy cover and reduced temperatures underneath, cooler conditions inside the shed may be preferable, but further research is needed to confirm this.

Supplementary Materials: The following supporting information can be downloaded at: https://www.mdpi.com/article/10.3390/ani12030344/s1, Figure S1: Snapshots of one of the 90% UV-filtering shelters in Flock-B showing use of the shelter and the immediate surrounding range area across one day.

Author Contributions: Conceptualisation, C.L. and D.L.M.C.; Data curation, M.S.R.; Formal analysis, M.S.R.; Funding acquisition, C.L. and D.L.M.C.; Investigation, M.S.R. and J.M.L.; Methodology, M.S.R., C.L., J.M.L. and D.L.M.C.; Resources, C.L., J.M.L. and D.L.M.C.; Supervision, C.L. and D.L.M.C.; Writing—original draft, M.S.R.; Writing—review & editing, C.L. and D.L.M.C. All authors have read and agreed to the published version of the manuscript.

Funding: This research was funded by Australian Eggs, grant number: 31HS902CO and the APC was funded by Australian Eggs. M.S. Rana was supported by an IPRA postgraduate scholarship through the University of New England and a McIlrath Trust Scholarship through the Commonwealth Scientific and Industrial Research Organisation (CSIRO).

Institutional Review Board Statement: The study was conducted according to the guidelines of the Declaration of Helsinki, and approved by the Commonwealth Scientific and Industrial Research Organisation (CSIRO) Wildlife, Livestock and Laboratory Animal, Animal Ethics Committee (AEC approval number: 2020-27) although husbandry and management of the birds fell under the responsibility of the commercial farm.

Informed Consent Statement: Not applicable.

Data Availability Statement: Data that support this study will be made available upon any reasonable request to the corresponding author.

Acknowledgments: Thank you to Don Foster (CSIRO) for construction of the shelters, Tim Dyall (CSIRO) and Troy Kalinowski (CSIRO) for assistance with set-up and data collection, Andrew Cohen-Barnhouse for the spectroradiometer measurements, and Rowyn Hellyar (RDH Integration Services, Toowoomba, QLD) for camera system installation. The authors express great appreciation to the commercial farm staff for their willingness to host and support this research project.

Conflicts of Interest: The authors declare no conflict of interest. The funders had no role in the design of the study; in the collection, analyses, or interpretation of data; in the writing of the manuscript, or in the decision to publish the results.

References

1. Knierim, U. Animal welfare aspects of outdoor runs for laying hens: A review. *NJAS Wagening. J. Life Sci.* **2006**, *54*, 133–145. [CrossRef]
2. Miao, Z.; Glatz, P.; Ru, Y. Free-range poultry production—A review. *Asian Australas. J. Anim. Sci.* **2005**, *18*, 113–132. [CrossRef]
3. Shifaw, A.; Feyera, T.; Walkden-Brown, S.W.; Sharpe, B.; Elliott, T.; Ruhnke, I. Global and regional prevalence of helminth infection in chickens over time: A systematic review and meta-analysis. *Poult. Sci.* **2021**, *100*, 101082. [CrossRef] [PubMed]
4. Courtice, J.M.; Mahdi, L.K.; Groves, P.J.; Kotiw, M. Spotty Liver Disease: A review of an ongoing challenge in commercial free-range egg production. *Vet. Microbiol.* **2018**, *227*, 112–118. [CrossRef] [PubMed]

5. Singh, M.; Ruhnke, I.; de Koning, C.; Drake, K.; Skerman, A.G.; Hinch, G.N.; Glatz, P.C. Demographics and practices of semi-intensive free-range farming systems in Australia with an outdoor stocking density of ≤1500 hens/hectare. *PLoS ONE* **2017**, *12*, e0187057. [CrossRef] [PubMed]

6. Moberly, R.; White, P.; Harris, S. Mortality due to fox predation in free-range poultry flocks in Britain. *Vet. Rec.* **2004**, *155*, 48–52. [CrossRef] [PubMed]

7. Bestman, M.; Bikker-Ouwejan, J. Predation in organic and free-range egg production. *Animals* **2020**, *10*, 177. [CrossRef]

8. Bari, M.S.; Downing, J.A.; Dyall, T.R.; Lee, C.; Campbell, D.L.M. Relationships between rearing enrichments, range use, and an environmental stressor for free-range laying hen welfare. *Front. Vet. Sci.* **2020**, *7*, 480. [CrossRef]

9. Rodriguez-Aurrekoetxea, A.; Estevez, I. Use of space and its impact on the welfare of laying hens in a commercial free-range system. *Poult. Sci.* **2016**, *95*, 2503–2513. [CrossRef]

10. De Koning, C.; Kitessa, S.M.; Barekatain, R.; Drake, K. Determination of range enrichment for improved hen welfare on commercial fixed-range free-range layer farms. *Anim. Prod. Sci.* **2019**, *59*, 1336–1348. [CrossRef]

11. Campbell, D.L.M.; Dyall, T.R.; Downing, J.A.; Cohen-Barnhouse, A.M.; Lee, C. Rearing enrichments affected ranging behavior in free-range laying hens. *Front. Vet. Sci.* **2020**, *7*, 446. [CrossRef] [PubMed]

12. Richards, G.; Wilkins, L.; Knowles, T.; Booth, F.; Toscano, M.; Nicol, C.; Brown, S. Continuous monitoring of pop hole usage by commercially housed free-range hens throughout the production cycle. *Vet. Rec.* **2011**, *169*, 338. [CrossRef] [PubMed]

13. Pettersson, I.C.; Freire, R.; Nicol, C.J. Factors affecting ranging behaviour in commercial free-range hens. *World's Poult. Sci. J.* **2016**, *72*, 137–150. [CrossRef]

14. Nicol, C.; Pötzsch, C.; Lewis, K.; Green, L. Matched concurrent case-control study of risk factors for feather pecking in hens on free-range commercial farms in the UK. *Br. Poult. Sci.* **2003**, *44*, 515–523. [CrossRef] [PubMed]

15. Gilani, A.M.; Knowles, T.G.; Nicol, C.J. Factors affecting ranging behaviour in young and adult laying hens. *Br. Poult. Sci.* **2014**, *55*, 127–135. [CrossRef]

16. Chielo, L.; Pike, T.; Cooper, J. Ranging behaviour of commercial free-range laying hens. *Animals* **2016**, *6*, 28. [CrossRef]

17. Rault, J.L.; Van De Wouw, A.; Hemsworth, P. Fly the coop! Vertical structures influence the distribution and behaviour of laying hens in an outdoor range. *Aust. Vet. J.* **2013**, *91*, 423–426. [CrossRef]

18. Nagle, T.A.; Glatz, P.C. Free range hens use the range more when the outdoor environment is enriched. *Asian Australas. J. Anim. Sci.* **2012**, *25*, 584–591. [CrossRef]

19. Hegelund, L.; Sørensen, J.T.; Kjaer, J.; Kristensen, I.S. Use of the range area in organic egg production systems: Effect of climatic factors, flock size, age and artificial cover. *Br. Poult. Sci.* **2005**, *46*, 1–8. [CrossRef]

20. Hartcher, K.M.; Hickey, K.A.; Hemsworth, P.H.; Cronin, G.M.; Wilkinson, S.J.; Singh, M. Relationships between range access as monitored by radio frequency identification technology, fearfulness, and plumage damage in free-range laying hens. *Animal* **2016**, *10*, 847–853. [CrossRef]

21. Campbell, D.L.M.; Hinch, G.N.; Downing, J.A.; Lee, C. Fear and coping styles of outdoor-preferring, moderate-outdoor and indoor-preferring free-range laying hens. *Appl. Anim. Behav. Sci.* **2016**, *185*, 73–77. [CrossRef]

22. Larsen, H.; Cronin, G.M.; Gebhardt-Henrich, S.G.; Smith, C.L.; Hemsworth, P.H.; Rault, J.L. Individual ranging behaviour patterns in commercial free-range layers as observed through RFID tracking. *Animals* **2017**, *7*, 21. [CrossRef] [PubMed]

23. Bestman, M.; Verwer, C.; Brenninkmeyer, C.; Willett, A.; Hinrichsen, L.; Smajlhodzic, F.; Heerkens, J.; Gunnarsson, S.; Ferrante, V. Feather-pecking and injurious pecking in organic laying hens in 107 flocks from eight European countries. *Anim. Welf.* **2017**, *26*, 355–363. [CrossRef]

24. Jung, L.; Brenninkmeyer, C.; Niebuhr, K.; Bestman, M.; Tuyttens, F.A.; Gunnarsson, S.; Sørensen, J.T.; Ferrari, P.; Knierim, U. Husbandry conditions and welfare outcomes in organic egg production in eight European countries. *Animals* **2020**, *10*, 2102. [CrossRef]

25. Barrett, J.; Rayner, A.; Gill, R.; Willings, T.; Bright, A. Smothering in UK free-range flocks. Part 1: Incidence, location, timing and management. *Vet. Rec.* **2014**, *175*, 19. [CrossRef]

26. Zeltner, E.; Hirt, H. Factors involved in the improvement of the use of hen runs. *Appl. Anim. Behav. Sci.* **2008**, *114*, 395–408. [CrossRef]

27. Zeltner, E.; Hirt, H. Effect of artificial structuring on the use of laying hen runs in a free-range system. *Br. Poult. Sci.* **2003**, *44*, 533–537. [CrossRef]

28. Hegelund, L.; Sørensen, J.; Hermansen, J. Welfare and productivity of laying hens in commercial organic egg production systems in Denmark. *NJAS Wagening. J. Life Sci.* **2006**, *54*, 147–155. [CrossRef]

29. Larsen, H.; Rault, J.-L. Preference for artificial range enrichment design features in free-range commercial laying hens. *Br. Poult. Sci.* **2021**, *62*, 311–319. [CrossRef]

30. Larsen, H.; Cronin, G.; Smith, C.L.; Hemsworth, P.; Rault, J.-L. Behaviour of free-range laying hens in distinct outdoor environments. *Anim. Welf.* **2017**, *26*, 255–264. [CrossRef]

31. Rana, M.S.; Campbell, D.L.M. Application of ultraviolet light for poultry production: A review of impacts on behavior, physiology, and production. *Front. Anim. Sci.* **2021**, *2*, 699262. [CrossRef]

32. Lewis, P.D.; Gous, R.M. Responses of poultry to ultraviolet radiation. *World's Poult. Sci. J.* **2009**, *65*, 499–510. [CrossRef]

33. Weihs, P.; Schmalwieser, A.W.; Schauberger, G. UV effects on living organisms. In *Environmental Toxicology*; Springer: New York, NY, USA, 2013; pp. 609–688.

34. Taylor, P.S.; Hemsworth, P.H.; Groves, P.J.; Gebhardt-Henrich, S.G.; Rault, J.-L. Ranging behaviour of commercial free-range broiler chickens 1: Factors related to flock variability. *Animals* **2017**, *7*, 54. [CrossRef] [PubMed]
35. Primary Industries Standing Committee. *Model Code of Practice for the Welfare of Animals—Domestic Poultry*, 4th ed.; CSIRO Publishing: Collingwood, VIC, Australia, 2002.
36. Thimijan, R.W.; Heins, R.D. Photometric, radiometric, and quantum light units of measure: A review of procedures for interconversion. *Hort. Sci.* **1983**, *18*, 818–822.
37. R-Core-Team. *R: A Language and Environment for Statistical Computing*; R Foundation for Statistical Computing: Vienna, Austria, 2020. Available online: https://www.r-project.org/index.html (accessed on 11 August 2020).
38. Schreiber-Gregory, D.N. Ridge Regression and multicollinearity: An in-depth review. *Model Assist. Stat. Appl.* **2018**, *13*, 359–365. [CrossRef]
39. Ullah, M.I.; Aslam, M.; Altaf, S. lmridge: A comprehensive R package for Ridge regression. *R J.* **2018**, *10*, 326–346. [CrossRef]
40. Grömping, U. Relative importance for linear regression in R: The package relaimpo. *J. Stat. Softw.* **2006**, *17*, 1–27. [CrossRef]
41. Duncan, I.J.; Widowski, T.M.; Malleau, A.E.; Lindberg, A.C.; Petherick, J.C. External factors and causation of dustbathing in domestic hens. *Behav. Process.* **1998**, *43*, 219–228. [CrossRef]
42. Stadig, L.M.; Rodenburg, T.B.; Ampe, B.; Reubens, B.; Tuyttens, F.A.M. Effect of free-range access, shelter type and weather conditions on free-range use and welfare of slow-growing broiler chickens. *Appl. Anim. Behav. Sci.* **2017**, *192*, 15–23. [CrossRef]
43. Stadig, L.; Rodenburg, T.; Ampe, B.; Reubens, B.; Tuyttens, F. Effects of shelter type, early environmental enrichment and weather conditions on free-range behaviour of slow-growing broiler chickens. *Animal* **2017**, *11*, 1046–1053. [CrossRef]
44. Rana, M.S.; Lee, C.; Lea, J.M.; Campbell, D.L.M. Relationship between sunlight and range use of commercial free-range hens in Australia. *PLoS ONE* **2021**. submitted.
45. Olsson, I.A.S.; Keeling, L.J. Night-time roosting in laying hens and the effect of thwarting access to perches. *Appl. Anim. Behav. Sci.* **2000**, *68*, 243–256. [CrossRef]

Article

Do Hens Use Enrichments Provided in Free-Range Systems?

Victoria Sandilands [1,*], Laurence Baker [1], Jo Donbavand [1] and Sarah Brocklehurst [2]

[1] Scotland's Rural College (SRUC), Easter Bush Campus, Midlothian EH25 9RG, UK; laurence.baker@sruc.ac.uk (L.B.); jo.donbavand@sruc.ac.uk (J.D.)
[2] BioSS, JCMB, The Kings Buildings, Peter Guthrie Tait Road, Edinburgh EH9 3FD, UK; sarah.brocklehurst@bioss.ac.uk
[*] Correspondence: vicky.sandilands@sruc.ac.uk

Citation: Sandilands, V.; Baker, L.; Donbavand, J.; Brocklehurst, S. Do Hens Use Enrichments Provided in Free-Range Systems?. *Animals* **2022**, *12*, 995. https://doi.org/10.3390/ani12080995

Academic Editor: Sabine Gebhardt-Henrich

Received: 21 February 2022
Accepted: 29 March 2022
Published: 12 April 2022

Publisher's Note: MDPI stays neutral with regard to jurisdictional claims in published maps and institutional affiliations.

Simple Summary: Free-range hens are typically given enrichments to encourage foraging and reduce injurious pecking. Of four enrichments (Lucerne hay, pecking blocks, pelleted feed scattered in litter, and jute ropes) provided to eight commercial flocks of free-range hens, pecking blocks and bales provided consistent interest to hens, based both on observations of hens in the vicinity of the enrichments (doing anything), interacting with the enrichments, and least walking/running or standing near the enrichments. Hens were most interested in pelleted feed at the time of scatter, but pelleted feed was consistently of greater interest than ropes, which hens seemed to find least attractive. Ropes were no more attractive to hens than no enrichment at all. Feather scores (a proxy measure for feather pecking) worsened with age, but differences between treatments were small and variable between ages, possibly due to lack of data and/or hens mixing between treatments. While ropes were by far the cheapest enrichment to provide, behaviour at ropes was indistinguishable from behaviour away from any enrichments, and thus did not sufficiently encourage foraging and other desirable behaviours. A balance between encouraging positive hen behaviour and cost to the producer needs to be taken into account in the practical use of any enrichment.

Abstract: Hens in free-range systems are given enrichments to increase foraging and limit injurious pecking, but the efficacy of enrichment types requires investigation. We studied hen behaviour and feather cover in eight commercial free-range flocks each given access to four enrichments within the shed. Sheds were split into quarters, in which two enrichments (jute ropes (R) + other) were installed. Other enrichments were: lucerne hay bales (B), pecking blocks (PB), pelleted feed (PF), or further R (control). Hens were observed at three ages, at three times per age (-1, 0, ≥ 1 h relative to PF application), in 1 m diameter circle locations around ropes (ControlR), Enrich (B, PB, PF, R), and Away from each enrichment. Feather scores were recorded at all ages/times, at the Away location only. Significantly more birds were in Enrich locations where PB, B, and PF were available, and least near R, ControlR, and Away locations ($p < 0.001$). Proportions of birds interacting with enrichments were significantly higher for PB, B, and PF than R ($p < 0.001$), but enrichments did not generally affect proportions of birds foraging in the litter, apart from a significant decrease ($p < 0.001$) in PF birds foraging in the Enrich location because they were directing behaviour at PF instead. Feather scores worsened with age ($p < 0.001$) but were not consistently affected by enrichment. Enrichment replacement rates varied between farms. Enrichments costs were highest for PB and cheapest for R. Enrichments except R were used by hens, but with no obvious effect on feather cover. A balance has to be struck between enrichment benefits to hens and economics, but evidence suggested that hens did not benefit from R.

Keywords: hay; pecking blocks; scattered feed; ropes; injurious pecking; feather scores; cost; behaviour; welfare

1. Introduction

The use of enrichments in captive animal housing is commonplace, with the main aim to improve animal welfare. Newberry [1] defined environmental enrichment as modifications to the environment that improve the biological functioning of captive animals, such as by improving health, which is one of the key components of good welfare. Injurious pecking, which includes gentle and severe feather pecking, cannibalism, and vent pecking [2] is one such laying-hen behaviour that damages the health of its victims. Feather pecking, the most common form of injurious pecking [3], is thought to be redirected foraging (i.e., food seeking) behaviour [4–6]. In alternative (e.g., barn and free-range egg production) systems in the UK and EU, one-third of the floor area must be litter (which is defined as any friable material that enables hens to show natural behaviour) [7]; commonly, this is provided as sawdust, wood shavings, straw, or a mixture [3]. While litter provides a foraging substrate, and is undoubtedly better than no substrate at all (for a review, see [8]), there is likely to be little positive feedback from litter that consists of only non-nutritive material and bird faeces. In addition, pressure to ban beak trimming as a means of controlling damage when feather pecking develops means that there is even greater interest in preventing the behaviour from starting. Therefore, increasing appropriate pecking behaviour by providing pullets and laying hens with suitable enrichments to peck at inside of loose-housed sheds is becoming more commonplace in these systems. While providing enrichments in these systems does not guarantee reduction of feather pecking and plumage damage, in many instances they have a positive effect (for a review, see [9]).

It is generally accepted that indestructible items such as balls, cones, and hanging CDs are unsuitable for encourage foraging behaviour in laying hens. They lack fundamental characteristics that elicit foraging; namely, that they do not fall apart on pecking, and there is no nutritional benefit to them [1,6]. That said, commercial free-range systems do use hanging ropes as enrichment [10], which are manipulable and fray, although they are non-nutritive, and have been demonstrated to reduce feather pecking in cages and floor pens in some studies [11] and to have no effect in others [12]. Previous work has shown that substrates that are destructible and provide nutrients are more likely to improve foraging and/or reduce injurious pecking than those that are/do not [6,9,13,14]; however, this is not consistently true (e.g., [15,16]). Destructible enrichments include long-cut straw, pecking blocks/stones, alfalfa hay bales, silage, and carrots [14,16–18]. Nutritive enrichments can potentially have an impact on feed consumption, egg production, and/or egg quality. While Schreiter et al. [14] found no effect of alfalfa hay or pecking stones on daily feed consumption, body mass, or egg-shell-breaking strength, they did find that these affected albumen quality. Steenfeldt et al. [17] found egg production increased with two out of three foraging materials provided, but intake of these was extremely high (33–48% of total feed intake). In contrast, Cronin et al. [19] found no effect of straw enrichment on laying performance.

The aim of this study was to investigate hen use of four commonly provided destructible enrichments in commercial free-range flocks, and their effects on feather scores as a proxy measure of feather pecking. We hypothesised that hens provided with rope-only enrichment would have the worst feather cover and show the least foraging behaviour around the enrichment. We also estimated the cost of enrichments, based on replacement rates seen.

2. Materials and Methods

This work was approved by SRUC's animal ethics committee (study number AU AE 36-2018, approved on 30 May 2019).

2.1. Housing and Birds

Eight flocks (A–H) of free-range hens at four different farms (2 flocks/farm) were recruited for the study via British Free Range Egg Producers Association (BFREPA) membership. All farms were based in Scotland, and pullets arrived at the farms at 15–16 weeks of age. The free-range sheds contained multitier structures (all Natura Step, Big Dutchman International GmbH, Vechta, Germany) of three levels (the litter floor and two tiers), and were split into four 4000-bird colonies ('quarters') by high fencing installed within the shed. Quarters were identified as Q1 (nearest the annex where staff first entered) to Q4 (Figure 1). Hens could access the range area through popholes in the outer wall of the shed, but the range area was not split into quarters, thus hens could potentially exit one quarter and enter another. There were 16,000 hens in total per flock. Normal commercial practice was undertaken at each farm, apart from the provision of enrichments. The bird strains used were all brown egg layers (Lohmann Brown, 3 flocks; Bovan Brown, 3 flocks; H&N Brown Nick, 2 flocks).

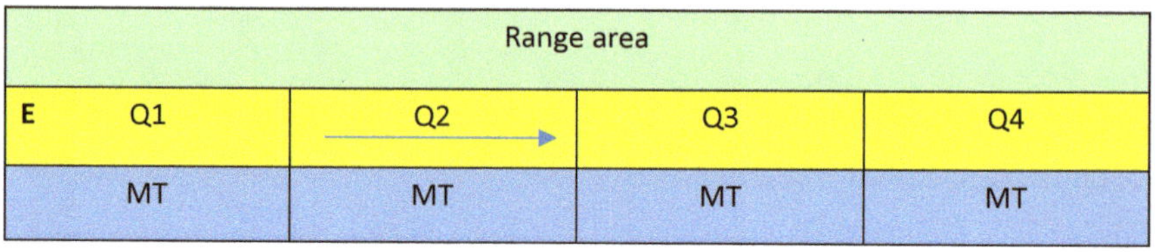

Range area			
E Q1	Q2 →	Q3	Q4
MT	MT	MT	MT

Figure 1. Overhead schematic of free-range hen sheds used in the study showing the four quarters, Q1–Q4. Q1 was always the quarter nearest to the annex, where staff would enter (**E**). The blue area is where the multitier (MT) structure was positioned, the yellow area was the litter, and birds could reach the range via popholes from the litter area. The arrow shows the direction that staff would walk, scattering litter from side-to-side in the relevant pelleted feed (PF) treatment/quarter.

2.2. Enrichment Treatments

Some schemes for egg production assurance require that at least two enrichments are provided for every 1000 hens, some of which must be destructible (e.g., RSPCA, 2017). Therefore, we used this level and these types of enrichments in our study, all provided inside the sheds. All flocks were provided with 2 different enrichment items per 1000 hens, and thus 8 enrichments per quarter. Enrichments were installed shortly after pullets arrived at the laying farm, and farm staff were advised of a replacement schedule based on the estimated time each enrichment should last; however, staff were advised to replace as often as necessary to ensure that hens were never without the enrichments (apart from pelleted feed, which was always given twice a day—see below). There were four different enrichments used (i–iv below, and Figure 2), all of which were destructible:

i.   Lucerne (alfalfa) hay bales. Analysed content: 16.7% crude protein, 90.4% dry matter. Four bales provided per quarter (1 per 1000 hens), which were placed into hay nets and suspended over the litter (some farms placed them on the floor initially, and hung them up after approximately 3 days). Bales weighed approximately 15 kg and measured $65 \times 45 \times 35$ cm or 102,375 cm^3 per bale. Cost: from GBP 6.50 per bale. Estimated to last 3 weeks.

ii.   Pecking blocks (Pickblock^{TM} medium, Crystalyx$^®$ Products GmbH, Münster, Germany), compact hard edible blocks made of grains (rye, maize, wheat), calcium carbonate, oyster shells, dextrose, molasses, wheat gluten feed, and lucerne meal; crude protein 5.8%; weight 5 kg; dimensions $23 \times 16.5 \times 13$ cm, or 4934 cm^3 per block. Provided at 1 block per 500 hens, thus 8 blocks per quarter, which were placed in pairs onto slats or plastic bucket lids (to stop them from getting damp) on top of the litter. Cost: approximately GBP 7 per block. Estimated to deteriorate at 1 g/hen/day, and thus expected to last approximately 10 days.

iii. Pelleted feed formulated for layers (Farmgate Layers Pellets, ForFarmers UK Ltd., Dumfriesshire, UK). Analysed content: 16.0% protein, 86.2% dry matter. Provided 2 kg twice a day, scattered from side-to-side covering a roughly 0.5 m width, down the centre of the litter area (Figure 1), thus providing 1 g pellets/hen/day. Staff were provided with plastic jugs marked with a 'fill' line to the correct weight, and feed was stored in plastic bins within the shed quarter for ease of use and rodent control.

The timing of scattering was arranged to coincide with staff inspections/collections of floor eggs, and ranged from farm-to-farm between 09:00–11:30 (scatter 1), and 13:00–16:30 (scatter 2). Cost: GBP 8.38 per 20 kg bag, or GBP 419/tonne. Estimated to last up to a few hours.

iv. Jute ropes (Ropes Direct, Norfolk, UK). Four ropes (8 mm diameter, cut into 30 cm lengths and looped in half; approximately 15.1 cm^3 in volume per rope) were attached initially by polypropylene string (flocks A, B), and then cable ties (all flocks) to the first platform or alighting rails of the multitier structure, evenly spread along the structure. Cost: just over GBP 0.08 per 30 cm, or GBP 0.33 for 4 rope pieces. Estimated to last 6 months.

(a)   (b)

(c)   (d)

Figure 2. Enrichments: (**a**) alfalfa hay bales prior to hanging in the hay nets (yellow); (**b**) pecking blocks paired and on slats; (**c**) pelleted feed scattered from a plastic jug; (**d**) jute ropes.

The rope was considered the standard (control) enrichment, so combinations of enrichment treatments (known as classification factor '**treatment**') for each quarter were:

1. 4 bales and 4 ropes (B);
2. 4 pairs of pecking blocks and 4 ropes (PB);
3. 4 kg pelleted feed and 4 ropes (PF);
4. 8 ropes (R).

Due to their predicted destruction/intake rate, edible enrichments (e.g., B, PB, and PF) were not expected to have an influence on feed intake or egg production and egg quality (none of which was measured here). Enrichments were offered in a balanced design over all quarters with all treatments provided to each flock, by placing enrichments in each shed based on two Latin squares (Table 1).

Table 1. Enrichment treatments (B, PB, PF, R), and their layout per flock, according to which quarter of the shed the items were provided in (quarters 1–4; quarter 1 was the section nearest the annex door).

Flock	Bales and Ropes (B)	Pecking Blocks and Ropes (PB)	Pelleted Feed and Ropes (PF)	Ropes Only (R)
A	Q1	Q2	Q3	Q4
B	Q3	Q1	Q4	Q2
C	Q2	Q4	Q1	Q3
D	Q4	Q3	Q2	Q1
E	Q3	Q4	Q1	Q2
F	Q2	Q1	Q4	Q3
G	Q1	Q3	Q2	Q4
H	Q4	Q2	Q3	Q1

2.3. Behaviour Observations and Feather Scores

Observations were due to take place during three visits at 34, 52, and 70 weeks of age (i.e., every 18 weeks) (known as classification factor '**age**'). Actual flock visits took place when birds were 33 weeks 5 days–34 weeks 6 days old, 51 weeks 6 days–53 weeks old, and 70 weeks 1 day–71 weeks 4 days old, but for simplicity, they are still referred to as visits at 34, 52, and 70 weeks of age throughout. The two flocks on a single farm were observed on two consecutive days, by one of two people. Popholes were open during observations. Bird behaviour was recorded using scan-sampling methods, at times relative to the first scatter of pelleted feed: -1, 0, and 1 h (known as classification factor '**time**'). The observer always began in the quarter with the pelleted feed treatment and then moved up the quarters (e.g., if PF was in Q2, then they observed in the order Q2, Q3, Q4, Q1). The observer entered the shed quarter and positioned themselves between the treatment enrichments (B, PB, PF, R) and the control enrichment (R), and remained quiet for 3 min to allow the hens to settle. The observer then scan-sampled a 1 m diameter area around three locations: the treatment enrichment ('Enrich'), the control (ropes) enrichment ('ControlR'), plus a 1 m diameter area away from either enrichment ('Away') (known as factor '**location**'). For R, both the 'Enrich' and the 'ControlR' observations were at ropes. A count of birds within each of the three circles and their behaviours (Table 2) were recorded.

Table 2. List of mutually exclusive behaviours. The first two behaviours could not be assessed for location 'away' (because there were no enrichments there).

Behaviour
* Interacting with (e.g., peck, pull, scratch at) enrichment (or in litter where feed was scattered, PF treatment),
* At, but not interacting with, enrichment: birds were located within 1 m diameter of the enrichment, but were not in contact with it
Stand/sit: birds were holding still and performing no other behaviour
Forage: peck/scratch at litter (but not at location where feed is scattered, PF treatment)
Walk/run: birds were in locomotion
Dustbathe: birds were in a prone position, while raking litter with their beaks, or tossing/rubbing litter onto the plumage
Feather peck: gentle or vigorous pecks at the plumage of other birds, often repetitive until the target bird withdrew
Aggressive peck: forceful, downward pecks directed towards the head or neck
Perch: birds standing or sitting on perch rails
Other: any other behaviour

* Only collected at locations ControlR and Enrich.

The counts were repeated three times in 15 min (e.g., at 3, 8, and 13 min). Thus, a total of 324 observations per flock were made (i.e., 3 locations × 3 observations per time relative to scatter × 3 times relative to scatter × 4 quarters × 3 ages). The observer moved to the next quarter after 15 min, so that all four quarters per flock were observed within each 1 h period.

Feather scores (i.e., the recording of feather damage on a scale of 0–5, where 0 = no damage, 1 = slight damage/loss with no bare skin, up to 5 = 1–2 cm^2 haemorrhage or >5 × 5 cm^2 bare skin with <1 cm^2 haemorrhage [20]) of five body locations (neck, back, tail, breast, and both wings) were carried out remotely (i.e., without handling, [21]) on 10 birds in the Away location once per scan sampling time (−1, 0, 1) per treatment (i.e., quarter) at each age (thus 5 feather scores/bird × 10 birds/time × 3 sampling times × 4 treatments × 3 ages = 1800 scores/flock).

Due to a combination of heightened biosecurity related to avian influenza and COVID-19, some visits to flocks were prevented. As a result, no feather scores or behaviour data were collected at age 52 weeks for flocks G and H, and no behaviour data were collected at age 70 weeks for flocks C, D, E, and F. Feather scores for C, D, E, and F at age 70 weeks were recorded from photographs taken by the farm staff of the birds in the Away location, from 10 birds. However, data from photos were judged to be unreliable, as they did not follow patterns seen in other flocks, with higher scores than expected at some body locations and lower than expected at other body locations. Therefore, the data from photos were omitted from all means shown and analyses.

2.4. Statistical Analyses

Behaviour data were analysed with the following fixed effects: age (34, 52, 70 weeks), time (−1, 0, 1 h), location (ControlR, Enrich, Away), and treatment (R, B, PB, PF) (and their interactions). For R, both the 'Enrich' and the 'ControlR' observations were at ropes, so one was randomly assigned to ControlR and one to Enrich to give the full complement of three locations to allow a full factorial statistical analyses of behaviour data. Random effects were flock, shed quarter, location within shed quarter, and interactions of these spatial effects with age, time within age, and scans within time within age, but many of these effects were negligible, and so were dropped from some models in order to achieve convergence.

With hen behaviour, three elements were analysed:

1. Total counts of birds in each location (ControlR, Enrich, Away) at a scan engaged in all behaviours (because total birds in a particular location might indicate a desire to be there);
2. Counts of birds engaged in each particular behaviour in each location at a scan;
3. Proportions of birds engaged in each behaviour (i.e., counts of birds performing a behaviour/total birds in that location per scan).

Results for counts of birds engaged in particular behaviours are not shown because results were similar for counts and proportions.

Feather scores were summed over all body sites per bird, and total feather score was analysed. Fixed effects were age (34, 52, 70 weeks), time relative to scatter (−1, 0, 1), treatment (R, B, PB, PF) (and their interactions). Random effects were flock, shed quarter, and interactions of these spatial effects with age and time within age. Analyses focused on total feather scores, but some analysis is also reported from analysing feather scores from individual sites using LMMs fitted to feather scores (not transformed) or GLMMs applied to a binary data feather score >0, adding site and interactions with site to the fixed effects.

To analyse proportions, generalised linear mixed models (GLMMs) were fitted to binomial counts with appropriate binomial totals, logit link function (i.e., for proportion p, $\log_e (p/1 - p)$), binomially distributed errors, and dispersion fixed at 1. To analyse counts, GLMMs were fitted to the counts with log link function, Poisson distributed errors, and dispersion fixed at 1. Where data were sparse, GLMMs with all effects included would not converge, so random and fixed effects in these models were simplified. Linear mixed models (LMMs) with all effects included were fitted to the total feather score after log transformation (i.e., $\log_e$ (total feather score + 1)) and were used as approximations in addition to simplified GLMMs for binomial data and counts. With LMMs, proportion data were first angular-transformed to a degrees scale (see Equation (1) below) to normalise the distribution of residuals; i.e., for proportion p:

$$\frac{180}{\pi} sin^{-1}(\sqrt{p}) \tag{1}$$

While counts and total feather scores were natural log transformed. Where high-level interactions were substantial, lower-level effects are not reported.

Due to the large number of tests being carried out, the results focus on highly significant effects and make clear when results were marginal. In some instances where interactions were marginal, lower-level associated effects are also shown. The p-values were based on approximate F tests when available, but otherwise were based on Wald tests; statistics for F tests are given in the results as $F_{ndf,ddf}$, where ndf is the numerator degrees of freedom (the number of effects to be estimated, which is the number of levels for a categorical factor less 1) and ddf is the denominator degrees of freedom; or for Wald tests as $Wald_{ndf}/ndf$ to make this comparable with the F statistic. Tables and figures either show raw means along with standard deviations (SDs) or model estimates $\pm$ standard errors (SEs) obtained from the LMMs and GLMMs as well as these estimates back transformed onto the original scale (e.g., proportions or counts) to aid interpretation.

For replacement of enrichments, the mean and SD over flocks ($n = 8$) of the mean days between replacement of each enrichment per flock was calculated. We also briefly investigated the above-reported statistical models of behaviour data, including covariates on days since last replacement and the cumulative amounts of enrichments replaced at each visit, generating p-values for the covariates tested last after all other fixed effects and examined estimated coefficients. All data were compiled and linked in Excel. Genstat 18 was used for data processing and all statistical analyses.

3. Results

Mean mortality across flocks was 4.8% (range: 2.60–7.98%). Observation times relative to scatter feed application were in reality 1.5–0.47 h before scatter (still called −1 h for simplicity), 0.0–0.17 h (0 h) at scatter, and 1.0–3.0 h postscatter (hereafter referred to as $\geq$1 h).

Overall mean proportions of birds observed in behaviours, according to location and treatment, are shown in Table 3. In the area where only ropes were available (ControlR) and in the Away location, most hens were observed standing/sitting, followed by foraging and walking/running. Hens observed in ControlR showed low proportions of birds interacting with the enrichments (ropes). In the Enrich location, the mean proportions of hens in R treatments were mostly standing/sitting, whereas with other treatments, much higher proportions of birds were interacting with the enrichments. All proportions of hens observed in dustbathing, feather pecking, perching, and other were low.

Table 3. The mean over scans of proportions of hens observed by location and treatment in various behaviours. All behaviours were mutually exclusive, and rows within location by treatment add up to 1.0. At ControlR, the only enrichment to interact with was rope; at Away, there were no enrichments. Figures in red are values $\geq$0.500; figures in blue are values between 0.100 and 0.499.

Location	Treatment				Behaviour						
		Interacting *	At But Not Interacting *	Stand/Sit	Forage	Walk/Run	Dustbathe	Feather Peck	Aggressive Peck	Perch	Other
ControlR	R	0.052	0.000	0.509	0.180	0.170	0.009	0.006	0.000	0.004	0.070
	B	0.033	0.000	0.525	0.203	0.141	0.010	0.004	0.000	0.015	0.069
	PB	0.060	0.000	0.552	0.153	0.147	0.006	0.005	0.000	0.006	0.071
	PF	0.032	0.005	0.564	0.143	0.184	0.004	0.006	0.001	0.006	0.055

Table 3. *Cont.*

Location	Treatment	Interacting *	At But Not Interacting *	Stand/Sit	Forage	Walk/Run	Dustbathe	Feather Peck	Aggressive Peck	Perch	Other
Enrich	R	0.048	0.000	0.517	0.166	0.182	0.009	0.003	0.000	0.009	0.066
	B	0.370	0.094	0.218	0.235	0.047	0.002	0.004	0.000	0.000	0.030
	PB	0.599	0.111	0.083	0.169	0.020	0.000	0.001	0.001	0.000	0.016
	PF	0.378	0.437	0.063	0.027	0.038	0.001	0.002	0.001	0.000	0.053
Away	R	NA	NA	0.452	0.238	0.197	0.017	0.012	0.000	0.000	0.085
	B	NA	NA	0.445	0.266	0.170	0.023	0.011	0.001	0.003	0.082
	PB	NA	NA	0.489	0.217	0.191	0.008	0.005	0.002	0.000	0.088
	PF	NA	NA	0.512	0.218	0.190	0.019	0.001	0.002	0.000	0.058

* With enrichment; NA = not applicable, because there were no enrichments to interact with.

3.1. Counts of Birds (Over All Behaviours)

On average over scans, there were 6.8–17.2 hens observed per location by treatment (Table 4).

Table 4. The mean $\pm$ SD over scans of total counts of birds observed over all behaviours, according to location (1 m diameter around the control ropes (ControlR), the enrichment (Enrich), or away from either (Away)) and enrichment treatment (i.e., ropes (R), bales (B), peck blocks (PB), or pelleted feed (PF)) provided in shed quarters. The estimated means (back-transformed from GLMM) are shown in brackets (which are adjusted for missing data).

	Location		
Treatment	ControlR	Enrich	Away
R	7.4 $\pm$ 3.6 (6.6)	6.8 $\pm$ 3.1 (6.0)	8.2 $\pm$ 3.1 (7.2)
B	7.1 $\pm$ 3.4 (6.4)	13.1 $\pm$ 4.4 (12.0)	8.7 $\pm$ 3.3 (7.8)
PB	7.2 $\pm$ 3.6 (6.4)	17.2 $\pm$ 4.0 (17.2)	8.3 $\pm$ 3.1 (7.6)
PF	6.8 $\pm$ 3.4 (6.1)	10.8 $\pm$ 3.4 (10.1)	7.3 $\pm$ 3.0 (6.5)

There was a highly significant interaction between time, location, and treatment in the total numbers of birds observed over all behaviours (Wald$_{12}$/ndf = 4.62 by GLMM, $p < 0.001$) (Figure 3a). There were more birds in the Enrich locations when the enrichments were not R, with the most birds observed with PB, then B, then PF. When feed was scattered (time 0), the number of birds went up only for PF in the Enrich location (and correspondingly went down for PF at the ControlR and Away locations, as hens moved away from these areas to the enrichment area), and then returned to -1 levels by time ≥ 1. In contrast, the numbers of birds in all locations with PB, B, and R remained constant across the three observation times.

The total numbers of birds observed, regardless of location, were similar between the different treatments at age 34 weeks, but treatment differences increased with age; at age 70 weeks, the greatest number of birds were observed for PB and the least for R (Figure 3b) (marginally significant interaction age.time.treatment, Wald$_{12}$/ndf = 1.94 by GLMM; $p = 0.026$). Other effects of bird age were also marginal, but on average, the total birds observed declined with age at all locations (interaction of age.location, Wald$_4$/ndf = 2.69 by GLMM; $p = 0.030$) (Table 5).

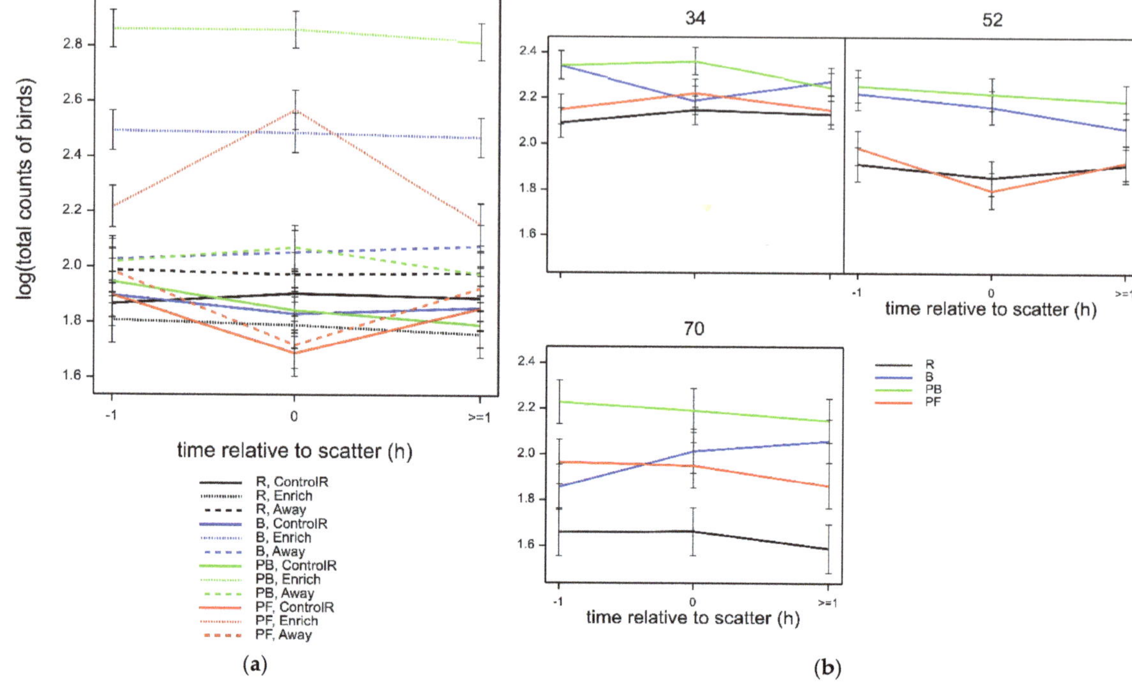

Figure 3. (**a**) Mean ± SE log (total counts of birds) observed over all behaviours in various locations (ControlR, Enrich, Away), according to enrichment treatments (R, B, PB, PF) and the time of observation relative to scatter of pelleted feed (−1, 0, ≥1), estimated from GLMM, with standard error (SE) bars shown. (**b**) Mean log (total counts of birds) observed over all behaviours with different enrichment treatments (R, B, PB, PF), according to bird age (34, 52, 70 weeks) and the time of observation relative to scatter of pelleted feed (−1, 0, ≥1), estimated from GLMM, with SE bars shown.

Table 5. Mean ± SE log (total counts of birds) (back-transformed shown in parentheses) observed over all behaviours by age (34, 52, and 70 weeks) and location (ControlR, Enrich, Away), estimated from GLMM.

	34 Weeks	52 Weeks	70 Weeks
ControlR	1.94 ± 0.10 (7.0)	1.90 ± 0.11 (6.7)	1.73 ± 0.12 (5.6)
Enrich	2.47 ± 0.10 (11.9)	2.33 ± 0.10 (10.2)	2.27 ± 0.11 (9.7)
Away	2.25 ± 0.10 (9.5)	1.91 ± 0.11 (6.7)	1.79 ± 0.12 (6.0)

3.2. Behaviour

3.2.1. Interacting with Enrichments (ControlR and Enrich Locations Only)

Of the total birds observed, the mean proportion of birds interacting with the enrichments in the Enrich locations was higher for PF at scatter-feeding time (0), then PB, then B, and was lowest for R (highly significant interaction time.location.treatment; $F_{6,621}$ = 8.44 by GLMM; $p < 0.001$); however, proportions were consistent across all three times for PB and B, whereas interaction with PF dropped at times −1 and ≥1 (Figure 4). Observations of birds in all treatments in the ControlR locations, plus R birds in the Enrich location, showed similarly low proportions of birds interacting with R, compared to B, PB, and PF birds in the Enrich location. All interactions with bird age, and the main bird age effect, were not statistically significant for the mean proportion of birds (all $p > 0.05$).

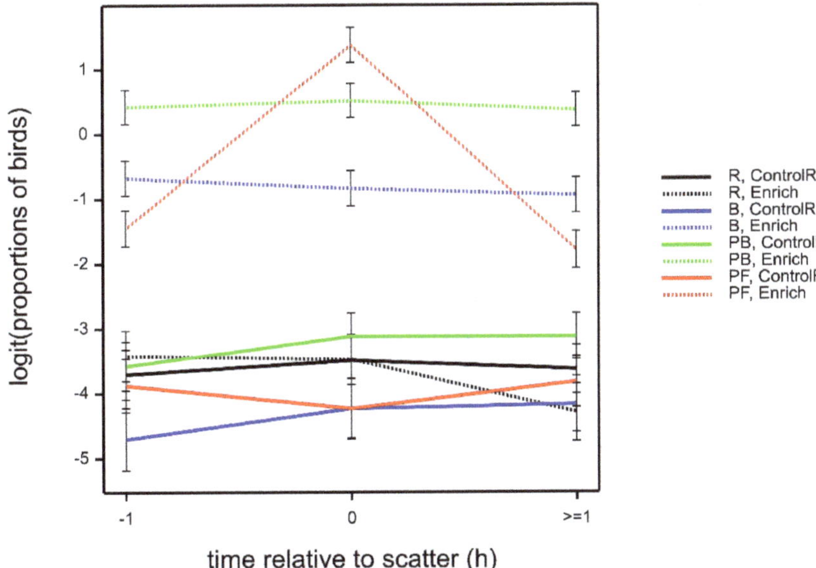

Figure 4. Mean ± SE logit (proportions of birds) observed interacting with enrichments, by time relative to scatter (−1, 0, ≥1) and location (ControlR, Enrich), estimated from GLMM. In all ControlR locations, the only enrichments present were ropes (R); in Enrich locations, there were ropes (R), bales + ropes (B), peck blocks + ropes (PB), or pelleted feed + ropes (PF).

3.2.2. At (but Not Interacting with) Enrichments (ControlR and Enrich Locations Only)

The proportion of birds at, but not interacting with, the enrichments was highest with PF outside of scatter feeding time (i.e., at time −1 and ≥1), then PB, then B in the Enrich locations, with a much lower proportion for R (highly significant interaction time.location.treatment; $F_{6,122}$ = 13.41 by LMM; $p < 0.001$) (Figure 5). The proportion of birds at, but not interacting with, the PB and B enrichments was consistent across all three observation times. There was a weak and inconsistent effect of age (marginally significant interaction location.age.treatment; $F_{6,63}$ = 2.49 by LMM; $p = 0.032$) (data not shown). The other three-way interactions were not statistically significant.

3.2.3. Stand/Sit

There were some marginally significant three-way interactions in stand/sit behaviour that were largely due to hens in PF treatment at location Enrich: the proportion of PF Enrich birds observed in stand/sit was both greatest at time ≥ 1 (time.location.treatment, $Wald_{12}/ndf$ = 1.92 by GLMM; $p = 0.027$) and lowest at age 34 weeks (age.location.treatment, $Wald_{12}/ndf$ = 2.15 by GLMM; $p = 0.012$) (data not shown). Averaged over other fixed effects, the proportion of birds observed in stand/sit behaviour increased with age (predicted means ± SE logit (back-transformed proportions) 34 weeks −0.80 ± 0.09 (0.31), 52 weeks −0.53 ± 0.10 (0.37), 70 weeks −0.28 ± 0.12 (0.43); $Wald_2/ndf$ = 8.69 by GLMM; $p < 0.001$). There was a highly significant interaction of location.treatment in the proportion of birds observed in stand/sit behaviour ($Wald_6/ndf$ = 34.94 by GLMM; $p < 0.001$): the greatest proportions of hens standing/sitting were seen in those locations where there were no enrichments (i.e., Away) or only rope enrichments (i.e., location ControlR, and treatment R in Enrich; whilst for PF, PB, and B, significantly fewer hens were standing/sitting at location Enrich (Figure 6).

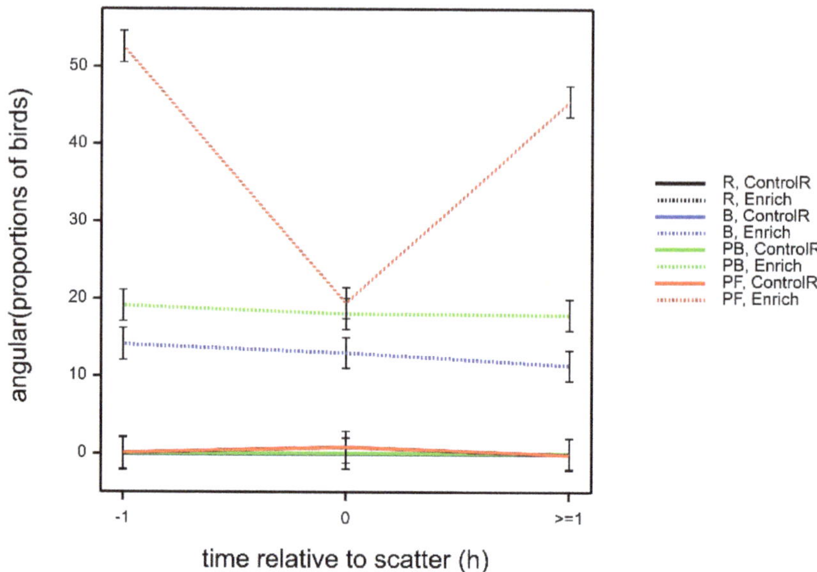

Figure 5. Mean ± SE angular (proportions of birds) observed at, but not interacting with, enrichments in Enrich and Control R locations, according to treatment (R, B, PB, PF) and time relative to scatter (−1, 0, ≥1), estimated from LMM. (Note that estimates are all 0 for R, Enrich and for R, B and PB at ControlR).

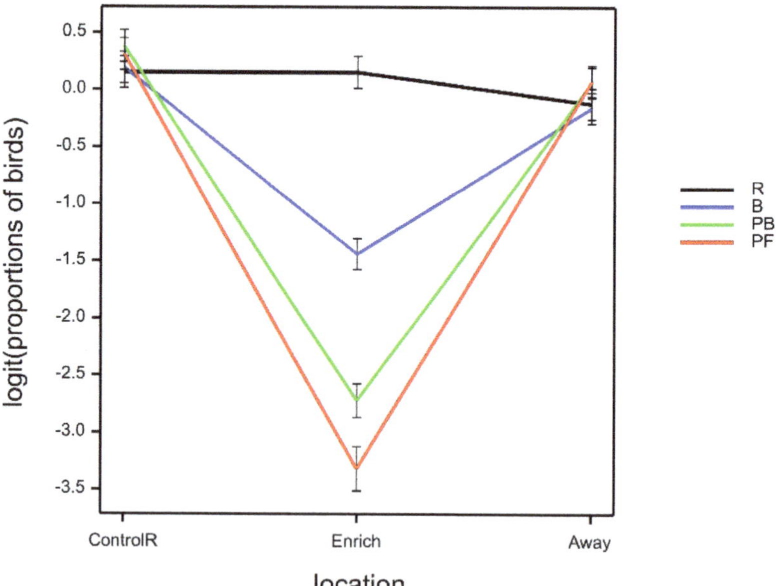

Figure 6. Mean ± SE logit (proportions of birds) observed in stand/sit behaviour, according to location (ControlR, Enrich, Away) and treatment (R, B, PB, PF), estimated from GLMM.

3.2.4. Forage

There was a weak three-way interaction of time.location.treatment on the proportion of birds observed foraging (excluding the PF scatter area) ($Wald_{12}/ndf$ = 1.81 by GLMM;

$p = 0.041$) that was solely due to a decrease in PF birds foraging at litter (other than where feed was scattered) at time 0 in the Enrich location, but this was merely due to no birds foraging on anything other than PF scattered at this time (data not shown). There was a highly significant location.treatment interaction on the proportion of birds foraging (Wald$_6$/ndf = 9.49 by GLMM; $p < 0.001$), again due to a decrease in PF birds foraging at litter (other than where feed was scattered) at time 0 (Figure 7). Foraging decreased with bird age (predicted means $\pm$ SE logit (back-transformed proportions) 34 weeks -1.64 ± 0.10 (0.16), 52 weeks -1.72 ± 0.10 (0.15), 70 weeks -2.09 ± 0.13 (0.11); Wald$_2$/ndf = 8.42 by GLMM; $p < 0.001$).

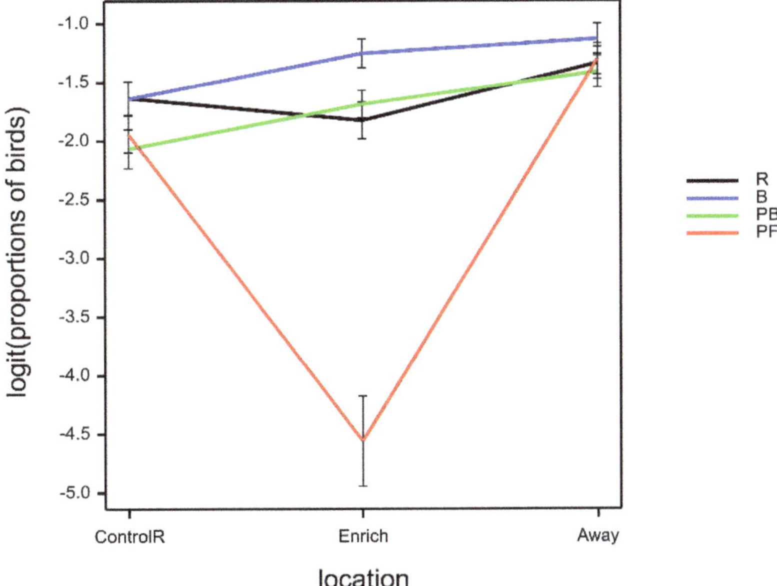

Figure 7. Mean $\pm$ SE logit (proportions of birds) observed in foraging behaviour, according to location (ControlR, Enrich, Away) and treatment (R, B, PB, PF), estimated from GLMM.

3.2.5. Walk/Run

The proportion of birds observed in walk/run behaviours was marginally affected by the interaction of time.location.treatment (Wald$_{12}$/ndf = 2.10 by GLMM; $p = 0.014$) largely due to the influence of PF and time relative to scatter, for which walking/running declined then increased at the enrichment and commensurately increased then declined at ControlR; whilst for the other treatments, behaviour remained broadly steady with the times observed relative to scatter (Figure 8a). There was a highly significant location.treatment interaction on the proportion of birds observed in walk/run behaviour, where birds were observed walking/running least in the Enrich area with all treatments except R, while hens seen in treatment R, and at all treatments in locations ControlR and Away, were all similar (Wald$_6$/ndf = 13.20 by GLMM; $p < 0.001$) (Figure 8b). Walking/running decreased marginally with bird age (predicted means $\pm$ SE logit (back-transformed proportions) 34 weeks -2.10 ± 0.09 (0.11), 52 weeks -2.38 ± 0.11 (0.08), 70 weeks -2.59 ± 0.13 (0.07); Wald$_2$/ndf = 4.47 by GLMM; $p = 0.011$).

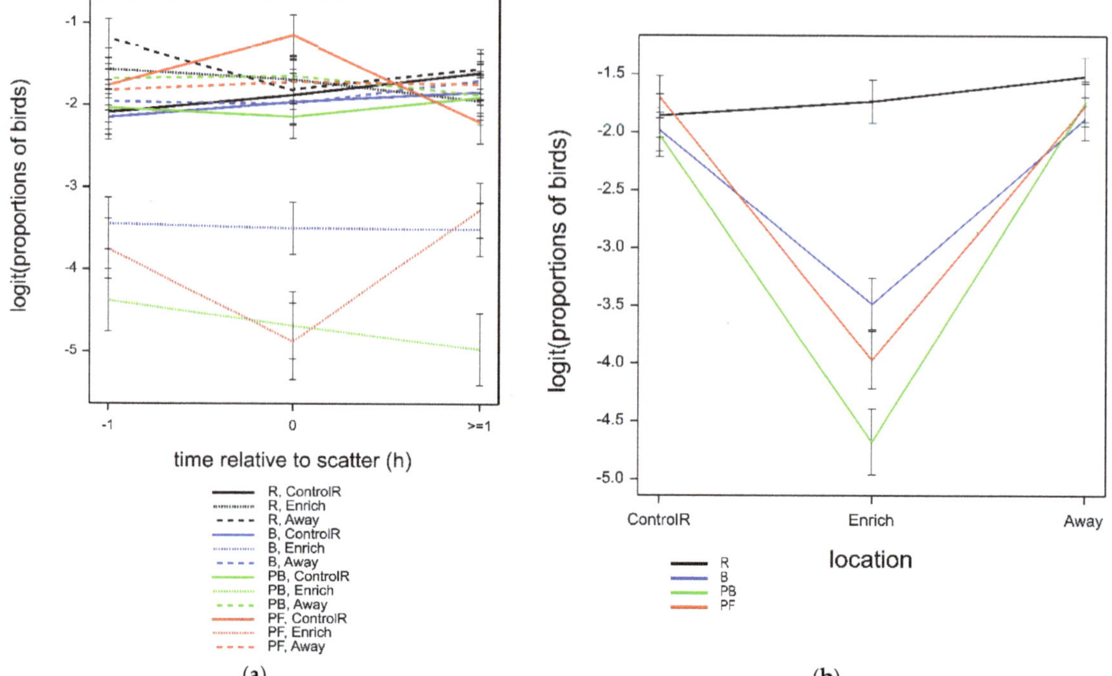

Figure 8. (**a**) Mean ± SE logit (proportions of birds) observed in walk/run behaviour, according to time (−1, 0, ≥1), location (ControlR, Enrich, Away) and treatment (R, B, PB, PF), estimated from GLMM. (**b**) Mean ± SE logit (proportions of birds) observed in walk/run behaviour, according location (ControlR, Enrich, Away) and treatment (R, B, PB, PF), estimated from GLMM.

There were very few counts of birds seen dustbathing, feather pecking, aggressive pecking, perching, or in 'other' behaviours, so these are not reported further.

3.3. Feather Scores

Feather scores were low (i.e., little damage) at bird ages 34 and 52 weeks, with only tails having some damage (Table 6). Feather scores were highest at age 70 weeks, with a mean total feather score of 2, but mean feather scores at each body site were each less than 1. The prevailing effects on feather score were due to age and (when examining the individual scores) from the tails (where scores were highest; scores were lowest at breast, and in between for neck, back, and wings) (site.bird.age interaction, $Wald_8/ndf = 15.42$ by LMM; $p < 0.001$). Many interactions would not converge due to sparse data or were not significant ($p > 0.05$) in the GLMMs applied to individual sites data, so this is not reported further.

Total mean feather scores were significantly affected by the interaction of treatment and age, whereby feather scores were lowest for B hens at 52 weeks of age, but were higher than PF at 70 weeks of age ($F_{6,43} = 3.8$ by LMM; $p = 0.004$) (Figure 9a), but in reality, these differences were small (back-transformed means: age 52 weeks, B 0.14 versus other treatments (range) 0.19–0.24; age 70 weeks, B 1.80 versus PF 1.33), and furthermore, the difference between 52 and 70 weeks may have been influenced by the lack of data from four out of eight flocks at age 70 weeks. There was a further interaction between age and time (Figure 9b), with no differences between times at ages 34 or 52 weeks, but with more hens seen with poorer feather scores at time ≥1 compared to time −1 at 70 weeks ($F_{4,2058} = 4.7$ by LMM; $p = 0.001$), but again, in reality, differences were small (back-transformed means age 70 weeks: 1.35–1.69) and may have been influenced by missing data from half of the flocks at age 70 weeks.

Table 6. Mean ± SD feather score by bird age and body location (overall treatments and flocks) and mean ± SD total feather score (FS).

	34 Weeks	52 Weeks	70 Weeks
Neck	0.000 ± 0.000	0.003 ± 0.053	0.398 ± 0.536
Back	0.000 ± 0.000	0.001 ± 0.037	0.362 ± 0.520
Tail	0.024 ± 0.153	0.257 ± 0.444	0.664 ± 0.495
Breast	0.000 ± 0.000	0.000 ± 0.000	0.145 ± 0.358
Wings	0.000 ± 0.000	0.000 ± 0.000	0.436 ± 0.580
Total FS	0.024 ± 0.153	0.261 ± 0.452	2.004 ± 1.790

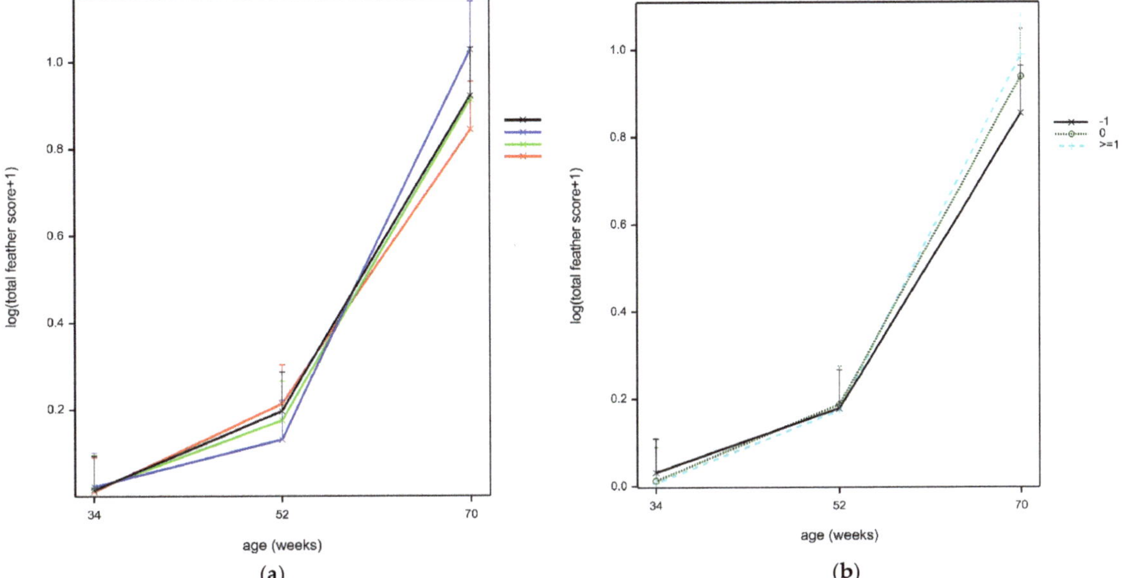

Figure 9. (**a**) Mean ± SE log (total feather score + 1) by age (34, 52, 70) and treatment (R, B, PB, PF), estimated from LMM. (**b**) Mean ± SE log (total feather score) by age (34, 52, 70) and time relative to scatter (−0, 0, ≥1), estimated from LMM.

3.4. Replacement Frequency and Cost

Enrichments were replaced regularly by the farms based on their judgement of depletion. As a result, rates of replacement varied widely from flock to flock (Figure 10) apart from with PF, which was scattered twice a day in every flock (not shown). For example, replacement of PB pairs was highest in flocks A and B (which were on the same farm). Replacement of ropes was understandably higher in the treatment R, where there were twice as many ropes as in B, PB, or PF, but was lowest in flocks A, B, E, and F in all quarters. When covariates on days since last replacement and the cumulative amounts of enrichments replaced were tested last in the above-reported statistical models of behaviour data, as would be expected, the more recently items had been replaced, the more interest was shown by the birds. These covariates were often statistically significant with estimated coefficients in the expected direction, but no further details of this modelling are reported, as these covariates were observational, and the full range of their scales was only sparsely represented in the data.

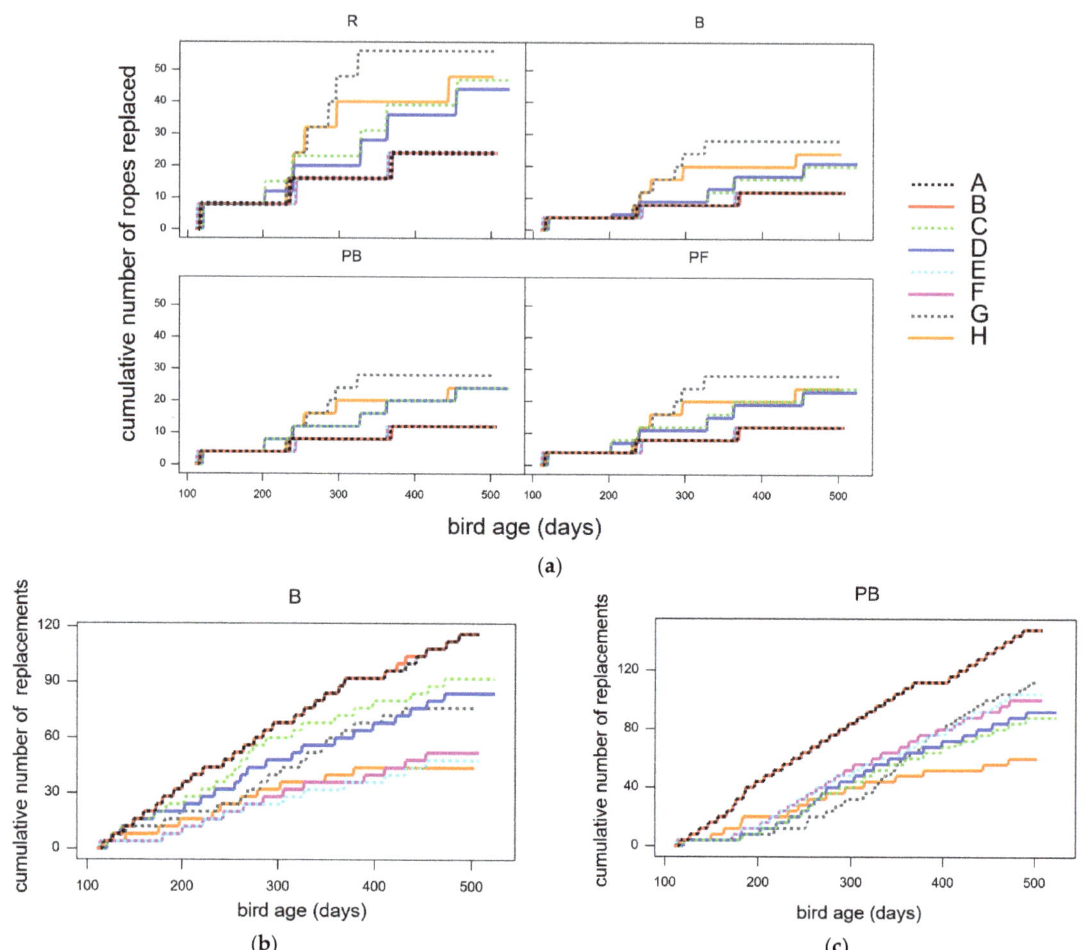

Figure 10. The cumulative number of enrichments replaced in each flock (A–H) over bird age (days). (**a**) Replacement of all ropes (in all treatments, R, B, PB, PF); (**b**) replacement of all hay bales (B); and (**c**) replacement of all pecking block pairs (PB).

The estimated and actual rates of enrichment replacements, and the total costs for use, are shown in Table 7. Flocks studied here were followed to 70+ weeks of age; however, flocks are likely to be housed for longer than this, depending on production. Therefore, the following cost estimates were based on the actual mean rate of replacement shown, in 16,000-hen flocks housed from 16 to 80 weeks (ignoring varying rates in mortality), thus needing enrichments for 64 weeks. Note that flocks are often expected to be given a variety of enrichments. Here, we estimated the costs based on providing each enrichment per 16,000 hens. However, where required (e.g., by accreditation schemes), flock managers would have to choose combinations of the enrichments shown to determine the total cost per flock. For example, RSPCA Assured require two items of permanent, destructible enrichment for every 1000 hens [22], so two items below would have to be added together (and pelleted feed might not be permitted, if not considered permanent, despite it being of greater interest than ropes).

Table 7. The estimated and actual mean rate of replacement for the four enrichments in eight flocks, with standard deviation (SD) given, and the total cost of using each enrichment in a flock of 16,000 hens, housed for 64 weeks (16–80 weeks of age), based on the actual mean rate of replacement seen here. Costs do not include local taxes or shipping.

	Bales	Pecking Blocks	Pelleted Feed	Rope
Estimated replacement	21 days	10 days	Twice a day	180 days
Mean replacement (*n* = 8)	21.9 days	14.9 days	Twice a day	96.6 days
SD (*n* = 8)	8.8	4.9	0	36.3
Cost of 1 item (GBP)	GBP 6.50/bale	GBP 7.00/block	GBP 8.38/20 kg bag (GBP 419/tonne)	8.295 *p*/30 cm (GBP 27.65/100 m reel)
No. required for 16,000 hens	16	32	16 kg	16
Cost as 1 enrichment for 16,000 hens	GBP 104.00	GBP 224	GBP 6.70	GBP 1.33
Number of times item would need replacing in 64 weeks	20	30	448	5
Total cost	GBP 2080	GBP 6720	GBP 3008 *	GBP 6.64

* Only 7168 kg needed, but feed can only be bought in bags of 20 kg, so 7080 kg = GBP 3008.

With all enrichments used, the mean replacement rate varied widely from flock to flock: standard deviation values were 33–40% of the mean values. However, it was still clear that while Lucerne bales, pecking blocks, and pelleted feed generated the most interest in hens, ropes were by far the cheapest enrichment to provide. The most expensive was pecking blocks, followed by pelleted feed, then bales.

4. Discussion

The expected benefits of providing destructible enrichments are to encourage birds to direct pecking behaviours away from other hens, fulfil natural behaviour, and improve feather cover. In this study, we considered both interacting (i.e., pecking, scratching, pulling) with the enrichments plus foraging behaviour in the litter (which excluded the PF scatter area in that treatment). While foraging behaviour alone showed little differences between treatments, apart from a drop in foraging with PF as hens were drawn to the scattered feed area, all the nonrope enrichments achieved the desired goal of encouraging interaction at the enrichments, which would hopefully benefit feather cover. However, feather-cover responses were unclear and probably exacerbated by the loss of data, plus hens were able to move out of popholes in one quarter, and re-enter the shed at another quarter, thus potentially mixing some birds between the treatments. Previous research suggests that bird mixing was unlikely to have a large effect on our feather score data, since only small proportions of flocks are typically seen on range [23], particularly with large ($\geq$16,000) hen flocks [24]. Feather cover did worsen with age, as expected, but feather cover was generally good (overall total feather scores on average of 2 or less), which is highly desirable. It may be that since evidence of feather pecking (via feather scores) was low in these flocks, there were only small differences gained from different enrichments, and a better comparator would be to have a treatment with no enrichments at all. However, that was not possible in these commercial flocks, which were required to provide enrichments by the accreditation schemes. Another theory is that enrichments may benefit hens with access to an outdoor area less than barn-system hens, which have the same indoor design as free-range hens, but no range access. For example, Heerkens [25] found plumage damage was worse in commercial barn flocks than in free-range flocks. However, given previous evidence of small proportions of free-range hens using the range, and evidence of feather pecking in free-range hens [3,10,25] particularly where range use was low [26], then appropriate enrichments are still likely to benefit birds in this system.

In this study, ropes were least useful for hens, based on the lack of hens observed in the vicinity of, and interacting with, ropes. A high proportion of hens were seen standing/sitting in both the ControlR and Away locations with all treatments, but this was significantly lower with B, PB, and PF compared to R in the Enrich locations, probably related to the commensurate increase in birds interacting with enrichments (other than R) in Enrich, which occupied 0.370–0.599 of the mean proportion of hens observed. This suggests that, of the four enrichments studied, ropes were no more attractive to hens than no enrichment at all. In contrast, interacting directly with the enrichments was significantly greater with PB and B at all observation times, while PF interaction peaked at feed scatter (with a concurrent decline in hens in the vicinity of, but not engaging with, the PF enrichment), but declined within an hour, probably because most pellets had been consumed by then, but hens were still showing an interest in PF at other times $(-1, \geq 1)$ compared to R or areas where there were no enrichments.

Previous work showed that using string (white polypropylene bailing twine) reduced both gentle and severe feather pecks, and elicited pecking at the string, in layer chicks housed on litter floors from 1–63 days of age, but also that the later that strings were introduced, the more negligible the effects [11]. It is unknown if hens used in our study had experienced string during rearing, but if their first introduction was upon entering the laying house, then it may be less surprising that they did not interact with ropes. String (or rope) may be more effective in wire-floor systems (e.g., enriched cages) where hens will encounter them more easily (due to the smaller overall space) than in litter-based systems where the much larger litter-covered floor, and other enrichments presented there, encourage foraging behaviour more effectively due to their size and/or position. For example, McAdie et al. [11] found that with hens reared in cages, presenting string at point of lay was effective at reducing feather damage. Of the four enrichments used here, all were destructible, but ropes had no nutritional value (unlike the other three). This, combined with their comparatively low volume (15.1 cm^3 each versus hay bales of 102,375 cm^3, and pairs of pecking blocks of 9868 cm^3), may have combined to make them not only unattractive, but also comparatively difficult to locate. Given the lack of interest around ropes shown here, there was little supporting evidence to suggest that increasing the number of rope bundles would bring any benefit to hen behaviour, at least in free-range systems.

The mean counts of birds seen in any location, engaged in all behaviours, ranged from 6.8–17.2 birds. Given that the observation locations of 1 m diameter each provided an area of 7854 cm^2, then on the basis of stocking density for hens in free-range systems of 9 hens/m^2 (equivalent to 1111 cm^2 per hen), this would have comfortably allowed space for 7 hens. In locations where there were ropes (all treatments in ControlR, and treatment R in Enrich) or no enrichments (Away) there were on average about the number of hens expected based on this stocking density, with 6.8–8.7 hens seen. In contrast, where there were B, PB, or PF enrichments (in location Enrich), we observed on average 10.8–17.2 hens, suggesting that birds were attracted to these enrichments. Bird attraction to the area was highest (and consistent) with PB, then B, whereas PF showed a decline in attraction outside of scattering, presumably because scattered feed was depleted. However, PF interest was still higher than that around ropes, suggesting that scattering of feed had long-lasting effects.

Adding feed or grain to the litter has been used previously to encourage foraging and reduce feather damage. Blokhuis and van der Haar [27] applied grain to litter pens of rearing pullets (40 g per pen of 12 pullets, three times per week), and found a significant increase in ground scratching compared to pullets supplemented with straw or nothing, plus the effects of grain carried over into the laying phase, in which hens had less feather damage compared to control hens. We saw a distinct rise in hens interacting with PF at the time of scatter, but also that this interacting behaviour was maintained long after the feed was presumably depleted, compared to the area around the ropes (ControlR), and also based on the high proportions of hens in the PF Enrich location. One criticism of using feed or grain is that it is not permanently available to hens, due to rapid depletion rates, but evidence here suggested that it elicited interacting behaviours more effectively than ropes (which were permanently available). However, in another study that examined relationships between management practices and feather pecking in over 111 free-range flocks in the UK, spreading feed on the floor was a significant risk factor for severe development of feather pecking [3]. In our study, effects of enrichments on feather cover were weak. Although hens interacted least with ropes, hens from rope treatment quarters showed intermediate levels of feather damage by age 70 weeks, similar to hens with pecking blocks (which was one of the most attractive enrichments to interact with), but damage overall was low across all treatments.

Providing enrichments comes at a cost to the producer, and must be balanced against benefits to the birds. If enrichments are impractical or costly, producers are unlikely to implement them [28]; however, if enrichments affect feather loss (which in turn increases feed costs, and can lead to mortality and reduced egg production) and/or fulfil an accrediting body's requirements [22], they are more likely to be adopted. We could not conduct a full economic analysis in this study (e.g., examining the effects of enrichments on mortality, egg production, egg quality, and feed consumption), because mortality and egg production were not collected by shed quarter. However, based on enrichment costs alone, while rope was the cheapest enrichment by far over the lifetime of flocks, it was also the least effective in terms of effects on behaviour, and indistinguishable from behaviours observed in locations away from all enrichments in this study. In all shed quarters, we tested rope and another enrichment (or rope and rope, for control), but we did not test all combinations of the four treatments (e.g., B and PF, PB and PF, etc.) It may be that such combinations would have further benefits on behaviour and feather scores, but it is likely that the costs of these would be prohibitive to many producers. Therefore, given the requirements of some accreditation schemes for two different enrichments, rope + another is potentially a good compromise between interest for hens and reasonable costs, but it should be acknowledged that ropes are least likely to be of use.

5. Conclusions

Ropes are unsuitable enrichment for hens, in terms of encouraging interaction with the enrichment, but are inexpensive. In contrast, pecking blocks and alfalfa hay bales promoted interaction, but are comparatively expensive. Enrichments should be selected based on a balance between their efficacy and cost, in which case alfalfa bales are potentially the best choice from those studied here, but future studies that measure mortality, egg production, and egg quality according to enrichment type would be beneficial, to determine if enrichment costs are offset by other benefits.

Author Contributions: Conceptualisation, V.S.; methodology, V.S., L.B. and S.B.; data collection, L.B. and J.D.; formal analysis, S.B.; data curation, L.B. and J.D.; writing—original draft preparation, V.S. and S.B.; writing—review and editing, J.D., S.B. and V.S.; funding acquisition, V.S. All authors have read and agreed to the published version of the manuscript.

Funding: This research was funded by the Scottish government (RESAS Programme 2016–2021) and by the British Free Range Egg Producers Association (BFREPA).

Institutional Review Board Statement: This work was approved by SRUC's animal ethics committee (study number AU AE 36-2018, approved on 30 May 2019).

Informed Consent Statement: Not applicable.

Data Availability Statement: The data presented in this study are openly available in Zenodo at https://doi.org/10.5281/zenodo.5996957 (accessed on 31 March 2022).

Acknowledgments: We are grateful for the participation of the four farms that each provided two flocks for the study, and to those providers that donated enrichments (J Wanstall & Sons for lucerne hay bales, ScotMin and Crystalyx for pecking blocks, and ForFarmers UK Ltd. for layer pellets) to support the research.

Conflicts of Interest: The authors declare no conflict of interest.

References

1. Newberry, R.C. Environmental enrichment: Increasing the biological relevance of captive environments. *Appl. Anim. Behav. Sci.* **1995**, *44*, 229–243. [CrossRef]
2. Lambton, S.L.; Nicol, C.J.; Friel, M.; Main, D.C.J.; McKinstry, J.L.; Sherwin, C.M.; Walton, J.; Weeks, C.A. A bespoke management package can reduce levels of injurious pecking in loose-housed laying hen flocks. *Vet. Rec.* **2013**, *172*, 423. [CrossRef] [PubMed]
3. Lambton, S.L.; Knowles, T.G.; Yorke, C.; Nicol, C.J. The risk factors affecting the development of gentle and severe feather pecking in loose housed laying hens. *Appl. Anim. Behav. Sci.* **2010**, *123*, 32–42. [CrossRef]
4. Blokhuis, H.J.; Arkes, J.G. Some observations on the development of feather-pecking in poultry. *Appl. Anim. Behav. Sci.* **1984**, *12*, 145–157. [CrossRef]
5. Rodenburg, T.B.; Koene, P. Feather pecking and feather loss. In *Welfare of the Laying Hen*; Perry, G.C., Ed.; CABI Publishing: Cambridge, UK, 2004; pp. 227–238.
6. Dixon, L.M.; Duncan, I.J.H.; Mason, G. What's in a peck? Using fixed action pattern morphology to identify the motivational basis of abnormal feather-pecking behaviour. *Anim. Behav.* **2008**, *76*, 1035–1042. [CrossRef]
7. European Union. *Council Directive 1999/74/EC of 19 July 1999 Laying Down Minimum Standards for the Protection of Laying Hens*; European Union: Brussels, Belgium, 1999; pp. 1–5.
8. Rodenburg, T.B.; Van Krimpen, M.M.; De Jong, I.C.; De Haas, E.N.; Kops, M.S.; Riedstra, B.J.; Nordquist, R.E.; Wagenaar, J.P.; Bestman, M.; Nicol, C.J. The prevention and control of feather pecking in laying hens: Identifying the underlying principles. *Worlds Poult. Sci. J.* **2013**, *69*, 361–374. [CrossRef]
9. Schreiter, R.; Damme, K.; Borell, E.; Vogt, I.; Klunker, M.; Freick, M. Effects of litter and additional enrichment elements on the occurrence of feather pecking in pullets and laying hens—A focused review. *Vet. Med. Sci.* **2019**, *5*, 500–507. [CrossRef]
10. Coton, J.; Guinebretière, M.; Guesdon, V.; Chiron, G.; Mindus, C.; Laravoire, A.; Pauthier, G.; Balaine, L.; Descamps, M.; Bignon, L.; et al. Feather pecking in laying hens housed in free-range or furnished-cage systems on French farms. *Br. Poult. Sci.* **2019**, *60*, 617–627. [CrossRef]
11. McAdie, T.M.; Keeling, L.J.; Blokhuis, H.J.; Jones, R.B. Reduction in feather pecking and improvement of feather condition with the presentation of a string device to chickens. *Appl. Anim. Behav. Sci.* **2005**, *93*, 67–80. [CrossRef]
12. Hartcher, K.M.; Tran, K.T.N.; Wilkinson, S.J.; Hemsworth, P.H.; Thomson, P.C.; Cronin, G.M. The effects of environmental enrichment and beak-trimming during the rearing period on subsequent feather damage due to feather-pecking in laying hens. *Poult. Sci.* **2015**, *94*, 852–859. [CrossRef]
13. D'Eath, R.B.; Arnott, G.; Turner, S.P.; Jensen, T.; Lahrmann, H.P.; Busch, M.E.; Niemi, J.K.; Lawrence, A.B.; Sandøe, P. Injurious tail biting in pigs: How can it be controlled in existing systems without tail docking? *Animal* **2014**, *8*, 1479–1497. [CrossRef]
14. Schreiter, R.; Damme, K.; Freick, M. Edible environmental enrichments in littered housing systems: Do their effects on integument condition differ between commercial laying hen strains? *Animals* **2020**, *10*, 2434. [CrossRef] [PubMed]
15. Rudkin, C. Feather pecking and foraging uncorrelated–the redirection hypothesis revisited. *Br. Poult. Sci.* **2021**, 1–9. [CrossRef] [PubMed]
16. Iqbal, Z.; Drake, K.; Swick, R.A.; Taylor, P.S.; Perez-Maldonado, R.A.; Ruhnke, I. Effect of pecking stones and age on feather cover, hen mortality, and performance in free-range laying hens. *Poult. Sci.* **2020**, *99*, 2307–2314. [CrossRef] [PubMed]
17. Steenfeldt, S.; Kjaer, J.B.; Engberg, R.M. Effect of feeding silages or carrots as supplements to laying hens on production performance, nutrient digestibility, gut structure, gut microflora and feather pecking behaviour. *Br. Poult. Sci.* **2007**, *48*, 454–468. [CrossRef]
18. Aerni, V.; El-Lethey, H.; Wechsler, B. Effect of foraging material and food form on feather pecking in laying hens. *Br. Poult. Sci.* **2000**, *41*, 16–21. [CrossRef]
19. Cronin, G.M.; Hopcroft, R.L.; Groves, P.J.; Hall, E.J.S.; Phalen, D.N.; Hemsworth, P.H. Why did severe feather pecking and cannibalism outbreaks occur? An unintended case study while investigating the effects of forage and stress on pullets during rearing. *Poult. Sci.* **2018**, *97*, 1484–1502. [CrossRef]
20. Savory, C.J.; Mann, J.S. Feather pecking in groups of growing bantams in relation to floor litter substrate and plumage colour. *Br. Poult. Sci.* **1999**, *40*, 565–572. [CrossRef]
21. Bright, A.; Jones, T.A.; Dawkins, M.S. A non-intrusive method of assessing plumage condition in commercial flocks of laying hens. *Anim. Welf.* **2006**, *15*, 113–118.
22. RSPCA. *RSPCA Welfare Standards for Laying Hens*; RSPCA: Horsham, UK, 2017.
23. Bubier, N.E.; Bradshaw, R.H. Movement of flocks of laying hens in and out of the hen house in four free range systems. *Br. Poult. Sci. Suppl.* **1998**, *39*, S5–S6. [CrossRef]
24. Gebhardt-Henrich, S.G.; Toscano, M.J.; Fröhlich, E.K.F. Use of outdoor ranges by laying hens in different sized flocks. *Appl. Anim. Behav. Sci.* **2014**, *155*, 74–81. [CrossRef]
25. Heerkens, J.L.T.; Delezie, E.; Kempen, I.; Zoons, J.; Ampe, B.; Rodenburg, T.B.; Tuyttens, F.A.M. Specific characteristics of the aviary housing system affect plumage condition, mortality and production in laying hens. *Poult. Sci.* **2015**, *94*, 2008–2017. [CrossRef] [PubMed]
26. Nicol, C.J.; Potzsch, C.; Lewis, K.; Green, L.E. Matched concurrent case-control study of risk factors for feather pecking in hens on free-range commercial farms in the UK. *Br. Poult. Sci.* **2003**, *44*, 515–523. [CrossRef] [PubMed]

27. Blokhuis, H.J.; van der Haar, J.W. Effects of pecking incentives during rearing on feather pecking of laying hens. *Br. Poult. Sci.* **1992**, *33*, 17–24. [CrossRef] [PubMed]
28. Jones, R.B. Environmental enrichment: The need for practical strategies to improve poultry welfare. In *Welfare of the Laying Hen. Poultry Science Symposium Series*; No. 27; Perry, G.C., Ed.; CABI: Wallingford, UK, 2004; pp. 215–226.

Article

Effect of Environmental Complexity and Stocking Density on Fear and Anxiety in Broiler Chickens

Mallory G. Anderson [1], Andrew M. Campbell [1], Andrew Crump [2], Gareth Arnott [3], Ruth C. Newberry [4] and Leonie Jacobs [1,*]

[1] Department of Animal and Poultry Sciences, Virginia Tech, Blacksburg, VA 24061, USA; mallory8@vt.edu (M.G.A.); drewc1@vt.edu (A.M.C.)
[2] Centre for Philosophy of Natural and Social Science, London School of Economics and Political Science, London WC2A 2AE, UK; andrewcrump94@gmail.com
[3] School of Biological Sciences, Queen's University Belfast, Belfast BT9 5DL, UK; g.arnott@qub.ac.uk
[4] Department of Animal and Aquacultural Sciences, Faculty of Biosciences, Norwegian University of Life Sciences, 1432 Ås, Norway; ruth.newberry@nmbu.no
* Correspondence: jacobsl@vt.edu

Citation: Anderson, M.G.; Campbell, A.M.; Crump, A.; Arnott, G.; Newberry, R.C.; Jacobs, L. Effect of Environmental Complexity and Stocking Density on Fear and Anxiety in Broiler Chickens. *Animals* **2021**, *11*, 2383. https://doi.org/10.3390/ani11082383

Academic Editors: Victoria Sandilands and Tina Widowski

Received: 28 June 2021
Accepted: 10 August 2021
Published: 12 August 2021

Publisher's Note: MDPI stays neutral with regard to jurisdictional claims in published maps and institutional affiliations.

Simple Summary: Broiler chickens are conventionally housed in monotonous environments at high stocking densities, which can negatively affect their welfare. This study evaluated the impact of environmental complexity and stocking density on anxiety and fear in broilers. Through behavioral testing, we found that broilers housed at higher densities responded less fearfully than those housed at the lower density, which is contradicting to expectations and previous research. Broilers housed in complex environments exhibited responses consistent with reduced anxiety compared to broilers housed in monotonous environments, suggesting improved welfare for broilers housed in the complex environment.

Abstract: Barren housing and high stocking densities may contribute to negative affective states in broiler chickens, reducing their welfare. We investigated the effects of environmental complexity and stocking density on broilers' attention bias (measure of anxiety) and tonic immobility (measure of fear). In Experiment 1, individual birds were tested for attention bias ($n = 60$) and in Experiment 2, groups of three birds were tested ($n = 144$). Tonic immobility testing was performed on days 12 and 26 ($n = 36$) in Experiment 1, and on day 19 ($n = 72$) in Experiment 2. In Experiment 1, no differences were observed in the attention bias test. In Experiment 2, birds from high-complexity pens began feeding faster and more birds resumed feeding than from low-complexity pens following playback of an alarm call, suggesting that birds housed in the complex environment were less anxious. Furthermore, birds housed in high-density or high-complexity pens had shorter tonic immobility durations on day 12 compared to day 26 in Experiment 1. In Experiment 2, birds from high-density pens had shorter tonic immobility durations than birds housed in low-density pens, which is contrary to expectations. Our results suggest that birds at 3 weeks of age were less fearful under high stocking density conditions than low density conditions. In addition, results indicated that the complex environment improved welfare of broilers through reduced anxiety.

Keywords: broiler chicken; affective state; environmental complexity; stocking density; anxiety; fear; animal welfare; attention bias; tonic immobility

1. Introduction

Environmental enrichment can be defined as "a modification of the environment of captive animals, thereby increasing the animal's behavioral possibilities and leading to improvements of their biological function" [1]. Although results vary depending on the outcome variables assessed, the addition of different structures to the environment adds complexity and can have enriching effects for livestock, including broiler chickens [2–4]. These provisions are therefore typically referred to as enrichments.

Fear and anxiety raise welfare concerns because they generate negative affect and, if chronically aroused, highlight an animal's inability to cope with its environment [5,6]. Fear is a short-term emotional response motivating flight from, or freezing in response to, a currently present, immediate threat to survival, while anxiety is a longer-term emotional response motivating vigilance (i.e., alertness) in response to perceived potential threat and is amplified by adverse pre-and postnatal life experiences [5,7–10]. These systems have evolved as adaptive mechanisms promoting survival in dangerous situations through temporary activation of sympathetic and hypothalamic-pituitary-adrenal axis activity and suspension of growth-promoting parasympathetic activity [5]. However, excessive fear in broilers can be maladaptive, provoking panicked escape behaviors that cause injury, pain, and suffocation [11]. In addition, high levels of fear and anxiety impair the birds' ability to cope with environmental change, such as handling, transport, and loud noises, and have been linked with a worsened feed conversion ratio [12,13]. In many studies, fear in birds is measured using a tonic immobility (TI) test. TI is an anti-predator freezing response (feigning death) which prey species exhibit as a last resort when captured [14]. Longer TI durations have revealed higher levels of fear in broilers handled roughly compared to gently [15], manually caught compared to mechanically caught [16], or heat-stressed [12] or shocked [17] prior to testing compared to control. A TI test could provide valuable insight into broiler fear levels when handled after rearing in environments varying in complexity and stocking density.

Level of anxiety can be evaluated through an attention bias (AB) test. AB describes the differential, affect-mediated allocation of attention towards one stimulus compared to others [18]. In particular, anxious (vigilance) affective states can increase AB towards a stimulus [18]. Humans with clinical anxiety show a greater AB towards threatening stimuli than those without anxiety [19–21], and studies involving macaques [22], sheep [6,23], cattle [24], and laying hens [25] have validated AB testing as a measure of anxiety level, where animals receiving an anxiogenic drug spent more time looking towards a threatening stimulus and showed increased vigilance behavior compared to control animals. For example, after receiving an anxiogenic drug, laying hens exposed to a conspecific alarm call were slower to feed, faster to vocalize, and exhibited increased locomotion, compared to hens that received a saline injection [25]. These findings suggest that relatively anxious hens allocate more attention to a perceived threat, suggesting that this test could possibly serve as a tool to measure anxiety levels in broilers also. Although studies have reported successful differentiation of AB in animals, others have found unexpected or null results [26–28]. To our knowledge, however, AB in broilers has not been previously tested.

Typical broiler chicken housing lacks complexity, such as provision of perches or preferred dustbathing substrate, limiting the expression of diverse natural behaviors, potentially contributing negatively to broiler welfare and performance [1,29–32]. High stocking density is another welfare concern in broilers. For instance, high stocking densities can lead to poor foot health [3,11,33] and may increase fear (response to a detected threat) [5]. Lack of environmental complexity has also been associated with fear in broilers [11]. However, behavioral indices of fear were not affected when birds were housed with or without access to string or barrier perches at various stocking densities [34–36], raising questions about how stocking density affects fearfulness of broilers housed in a complex environment.

A reported benefit of adding perches as an enrichment for broilers is that the birds were less aggressive and experienced fewer disturbances while resting compared to broilers without perches [29,35]. For broilers, low perching platforms are used more than single linear perches, probably because heavy birds find them easier to balance on [37], and they were found to reduce avoidance of people, suggesting they reduced fear [38]. Moreover, while broilers are conventionally provided with a single type of litter over the whole floor, adding additional substrate materials can be enriching given that they vary in their value for different functions. For example, sand has been found to increase dustbathing behavior and activity levels compared to rice hull, paper, or wood shaving substrates [39], and adding maize roughage increased foraging behavior compared to wood shavings alone [32]. In addition, broilers housed with novel objects exhibited shorter durations of tonic immobility following acute stressors (sound, heat, and crating stress) compared to the control (no added objects), indicating decreased fearfulness [40]. Given this evidence, increasing environmental complexity with perches, sand, and novel objects would enhance broiler welfare through reduced anxiety and fearfulness.

Potential combined effects of environmental complexity and stocking density on fear and anxiety in broilers have not previously been examined experimentally. Our objective was to investigate the impact of complex housing conditions and stocking density on fearfulness, as measured through a TI test, and anxiety, using an AB test. We hypothesized that broilers housed in a high-complexity, low-density environment would experience the lowest levels of fear and anxiety, whereas broilers from a low-complexity, high-density environment would experience the highest levels of fear and anxiety, with a low-complexity, low-density environment and a high-complexity, high-density environment showing intermediate results. In particular, we predicted that higher levels of fear and anxiety would be reflected by longer TI durations and stronger AB to perceived threatening stimuli.

2. Materials and Methods

2.1. Birds, Treatments, and Housing

Two experiments were conducted. In each, 1620 male Ross 708 chicks (total $n = 3240$), vaccinated against Marek's disease, were obtained at day 0 from a commercial hatchery (Elizabethtown, PA, USA). Upon arrival to the research facility, chicks were randomly allocated to one of four treatment groups in a 2×2 factorial design with environmental complexity (low-complexity (LC) vs. high-complexity (HC)) and stocking density (low-density (LD) vs. high-density (HD)) as factors at pen level. Each treatment group was replicated three times (12 pens in total), distributed in a randomized complete block design.

All pens (14.5 m^2) contained standard pine shavings as bedding (approximately 10 cm depth), four hanging galvanized tube feeders (~12 kg capacity; no longer in production, but similar to "Flex" chicken feeder unit, SKU# CO30131, Hog Slat, Newton Grove, NC, USA), and three water lines (Valco Industries, Inc., New Holland, PA, USA), each with three nipple drinkers. All birds had ad libitum access to water and commercially-formulated broiler chicken feed (starter day 0–14, grower day 15–28, and finisher day 29–50). The birds were fed a corn/soy-based diet which met their nutritional requirements [41]. Birds had access to three heat lamps/pen and 24 h light in the first 7 days, followed by a light:dark schedule of 18L:6D, with a light intensity of approximately 15 lux during light hours. Due to a technical issue in Experiment 1, birds received 24 h light for 7 additional days during week 2 of age. House temperature was gradually decreased from 35 °C on day 1 to 21 °C on day 50 by assessing bird comfort. Comfort was evaluated based on behaviors indicative of heat or cold stress (panting or huddling respectively), bird activity (birds are active and alert when a person enters the facility), and bird distribution (birds are showing a somewhat homogenous distribution throughout the pen). In Experiment 1, all birds received a therapeutic dose of antibiotics via the water lines from day 33–40 in response to a pathogen exposure.

2.2. Environmental Complexity

HC pens contained four functional spaces (Figure 1a), including space for "feeding" (approximately 3 m^2), "comfort" (approximately 3 m^2), "resting" (approximately 3 m^2), and "exploration" (approximately 4.3 m^2). The feeding, comfort, and resting spaces included a water line. The feeding space contained four feeders and one third of a medium PECKstone™ (Proteka, Inc., Lucknow, ON, Canada) broken into smaller pieces. The comfort space contained a wooden-frame dust bath (180 cm L × 91 cm W × 10 cm H) filled with 68 kg of playground sand (QUIKRETE, Atlanta, GA, USA) that was raked and partially replaced when depleted. The resting space in Experiment 1 included three perches (182.9 cm L × 30.5 cm W × 8.5 cm H) constructed of 1.9 cm diameter PVC pipe, which was sprayed with textured black spray paint (Rust-Oleum, Vernon Hills, IL, USA) to enhance grip while perching (Figure 2a). Birds had access to 7.6 cm of linear perch space/bird in high-density pens, and 15.2 cm/bird in low-density pens. In Experiment 2, the PVC pipes were replaced with three wide wooden perches forming a platform (121.9 cm L × 45.7 cm W × 7.6 cm H; Figure 2b), providing 76 cm^2 of space/bird in the low-density pens, and 39 cm^2 of space/bird in the high-density pens. The exploration space contained a pair of enrichment objects, starting on day 2 of age. Six objects were randomly paired into three groups of two, combining a nutritional and an occupational enrichment object, and these pairs were rotated every three days according to a randomized schedule to maintain variation and novelty (Table 1). The LC pens had a similar set-up to the HC pens with four spaces, but without the peck stones, dust bath, perching platforms, or enrichment objects to differentiate the spaces into different functional areas (Figure 1b).

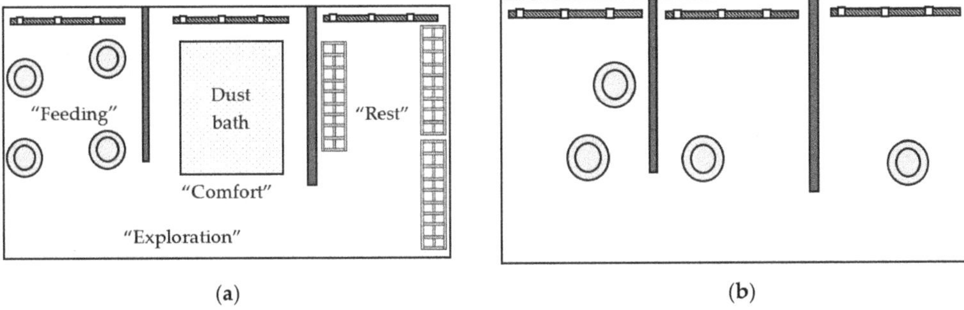

Figure 1. (a) High-complexity pen with four functional spaces for "feeding", "comfort", "resting", and "exploration". The feeding space contained four feeders (O) and pecking stones, the "comfort" space included a sand dust bath ( ), the resting space contained three perches ( ), and the exploration space contained varying pairs of enrichment objects. The feeding, comfort, and resting spaces each contained a water line with three nipple drinkers ( ). (b) Low-complexity (control) pen, containing four feeders and three water lines.

Figure 2. Photograph of the perch design in high-complexity pens in (a) Experiment 1 (*n* = 3/pen) and (b) Experiment 2 (*n* = 3/pen).

Table 1. Pairs of enrichment objects rotated every 3 days in high-complexity pens.

Nutritional Enrichment	Occupational Enrichment
Hanging bundles of white string	Free-moving metal ball (20.3 cm diameter) [1] filled with alfalfa hay
Yellow treat dispenser (7.6 cm diameter) [2] filled with whole-grain oats	Colored ball (5.8 cm diameter) [3]
Laser light (5 min, 2×/day) [4]	Experiment 1: Kong toy (5.6 cm diameter) [5] filled with iceberg lettuce Experiment 2: half a head of cabbage hung at bird height

[1] Darice, Strongsville, OH, USA; [2] Lixit Corp., Napa, CA, USA; [3] Click N' Play, USA; [4] Ethical Products, Inc., Bloomfield, NJ, USA; [5] KONG, Golden, CO, USA.

2.3. Stocking Density

The HD pens were stocked with 180 chicks/pen, resulting in 42.1 kg/m^2 at day 50 in Experiment 1, and 42.6 kg/m^2 in Experiment 2 (Table 2). The LD pens were stocked with 90 chicks/pen and reached a density of 23.8 kg/m^2 at day 50 in Experiment 1, and 23.3 kg/m^2 in Experiment 2 (Table 2).

Table 2. Mean pen stocking density (kg/m^2), and birds/m^2, at day 1, 29, and 50 in Experiments 1 and 2.

	Experiment 1					
Stocking Density	**Day 1**		**Day 29**		**Day 50**	
	Kg/m^2	**Birds/m^2**	**Kg/m^2**	**Birds/m^2**	**Kg/m^2**	**Birds/m^2**
High	0.52	13.85	18.93	13.14	42.08	12.31
Low	0.26	6.92	9.81	6.71	23.83	6.29
	Experiment 2					
Stocking Density	**Day 1**		**Day 29**		**Day 50**	
	Kg/m^2	**Birds/m^2**	**Kg/m^2**	**Birds/m^2**	**Kg/m^2**	**Birds/m^2**
High	0.46	12.41	19.90	12.23	42.64	11.56
Low	0.23	6.21	10.22	5.97	23.31	5.79

2.4. Experiment 1—Attention Bias Test

A square testing arena was constructed with two plastic, perforated folding partitions (approximately 124.5 cm L × 124.5 cm W × 91.4 cm H) with pine shavings on the floor and a feeder containing commercial feed, oats, and mealworms (Figure 3). The arena was located in a separate room adjacent to, but separate from, the broilers' home pens.

AB testing (modified from [18,25,42]) was performed with five randomly selected birds/pen (n = 60 birds across pens) on days 30, 32, and 33 of age. The testing order of pens was randomized. Each bird was tested separately by one observer, another person was present to move birds to and from the testing arena. The test started when the bird was placed in the AB arena. Immediately thereafter, an 8 second (s) conspecific alarm call was played from portable speakers (FUGOO, Van Nuys, Irvine, CA, USA) at full volume (95 dB). The alarm call was recorded from a chicken signaling a ground predator, which previous playback experiments have found to elicit a vigilance response [42]. Following the alarm call, latency to begin feeding was recorded. If the bird began feeding at any point during the test, it was allowed approximately 10 s to feed, then the alarm call was played a second time, and latency to resume feeding was recorded. The test ended when the bird resumed feeding a second time (maximum test duration of 300 s). Birds that never began feeding received a maximum latency to begin feeding score of 300 s and those failing to resume feeding received no score (missing data). Additional live-recorded variables included latency to first vocalization and occurrence (yes/no) of vigilance behaviors in the 30 s following the first alarm call (visibly stretching neck, looking around, freezing, and erect posture) [25]. Each of the four vigilance behavior characteristics (erect posture, neck stretching, looking around, and freezing) were scored as either 0 (not observed) or 1 (observed), giving a vigilance score between 0 (no vigilance behavior observed) and 4 (all vigilance behaviors observed at least once) for each bird tested. Videos were used to record latency to first step from when the alarm call playback ended, as a potential additional indicator of anxiety to determine how long the birds remained in a motionless state after the alarm call playback [25,43].

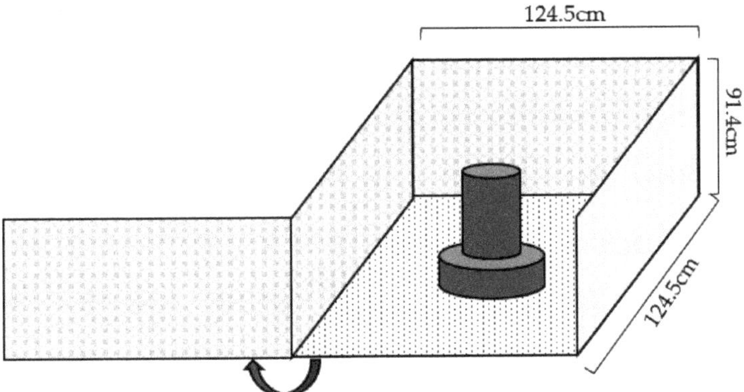

Figure 3. Diagram of the attention bias (AB) arena used in Experiments 1 and 2. A familiar feeder (exact same as provided in pens) was placed in the center of the arena and wood shavings were provided as litter.

2.5. Experiment 2—Attention Bias Test

After Experiment 1, the AB test was modified with an increased sample size, a group testing approach rather than testing individual birds, and allowing more time in the test arena if most (but not all) birds began feeding after the first alarm call was played. The AB test was performed on days 32, 33, and 38 of age with 12 randomly selected birds/pen (n = 144 birds across all pens) by two observers. These observers were trained by the researcher collecting data for Experiment 1. Inter-rater agreement was tested for latency to feed of 12 birds and was excellent among the three observers (Cronbach's alpha of 0.933). The order of pens was randomized for testing. Birds were tested in groups of 3 (4 tests/pen) to avoid isolation stress [44]. The same location, arena, feeder, feed, and alarm call were used as described for Experiment 1 (Figure 3). Prior to placement in the arena, two out of three birds were marked with livestock marker (All-Weather Paintstik, LA-CO Industries, Inc., Elk Grove Village, IL, USA) for individual identification. Immediately after three birds were placed into the arena, the 8 s conspecific alarm call was played. Latency to begin feeding (s) from the feeder was then recorded for each individual bird (observer 1 recorded two birds, observer 2 recorded the third bird). Thereafter, the test procedure had four possible outcomes depending on how many birds began feeding and the time-point that they started feeding within the first 300 s of the test.

If all three birds fed from the feeder at least once during the 300 s testing period, they were allowed 5 s to feed before the second alarm call playback. Thereafter, the second alarm call was played. If all three birds fed from the feeder between 270–300 s, birds were allowed to feed for 5 s starting from when the last bird fed, the second alarm call was played, and the test time was extended to 420 s. Latency to resume feeding was recorded for each individual bird (observer 1 recorded two birds, observer 2 recorded the third bird).

If at the end of the 300 s testing period, two out of three birds fed from the feeder, they were allowed 5 s to feed starting from when the last bird fed, then the second alarm call was played and the testing time was extended to 420 s. Latency to resume feeding was recorded for each individual bird (observer 1 recorded two birds, observer 2 recorded the third bird). The bird that did not feed received a maximum latency score of 300 s for latency to begin feeding and no score for latency to resume feeding.

If one of the tree birds fed from the feeder during the testing period, latency to begin feeding was recorded for the bird that began feeding, and the second alarm call was not played. The other two birds received a maximum latency score of 300 s.

If none of the three birds fed from the feeder during the testing period, all three birds received a maximum latency score of 300 s.

Video recordings were also used to record latency to step (s) and occurrence (yes/no) of vigilant behaviors within 30 s following the first alarm call. Each of the four vigilance behavior characteristics (erect posture, neck stretching, looking around, and freezing) were scored as either 0 (not observed) or 1 (observed), giving a vigilance score between 0 (no vigilance behavior observed) and 4 (all vigilance behaviors observed at least once) for each bird tested. It was not feasible to record latency to first vocalization because birds were tested in groups.

2.6. Tonic Immobility Test

In both experiments, a single observer performed TI testing in the hallway area of the house, directly adjacent to the birds' home pens. In Experiment 1, TI testing was performed on three randomly-marked birds/pen ($n = 36$) on day 12 of age. Birds were marked on their back with livestock marker (All-Weather Paintstik, LA-CO Industries, Inc., Elk Grove Village, IL, USA). The same marked birds were tested again on day 26 of age. In Experiment 2, TI testing was performed on six randomly selected birds/pen ($n = 72$) on day 19 of age. TI was induced by the handler carefully placing the bird on his back in a V-shaped cradle, placing one hand over the sternum and applying gentle pressure while cupping the other hand over the head (modified from [45]). After 15 s, the handler lifted her hands from the bird, moved out of the bird's line of sight, and recorded latency until righting response (TI duration [s]). If the bird attempted to right himself within 10 s after the hands were lifted, TI was considered not induced and the handler repeated the restraint procedure (maximum of three induction attempts). If TI could not be induced, the bird received the minimum score of 0 s. If birds remained in TI for the full 300 s testing period, a maximum latency score of 300 s was given.

2.7. Statistical Analysis

Data were analyzed in JMP Pro 15 (SAS Institute Inc., Cary, NC, USA). Data residuals were assessed for their distribution by visual inspection of normal quantile plots. An overview of the distribution of data residuals and subsequent statistical approaches is shown in Table 3. The sample for resumption of feeding in the Experiment 1 AB test was too low for statistical analysis, so raw means are presented. For normally distributed data (see Table 3), with the exception of AB data in Experiment 2, general linear mixed-effects models were used, with complexity (HC/LC), stocking density (HD/LD), and their interaction as fixed effects, and pen as a random factor. For AB test data, age was not considered a factor, as treatment groups were randomized across testing days. Normally distributed AB data in Experiment 2 were analyzed using general linear mixed-effects models, with complexity (HC/LC), stocking density (HD/LD), and their interaction as fixed effects, and testing group nested within pen as a random factor. No significant interaction effect between complexity and density was found for any response variables, so the interaction term was removed from the models. Durations of TI in Experiment 1 were analyzed using general linear mixed-effects models with complexity (HC/LC), stocking density (HD/LD), day (bird age), day $\times$ complexity, and day $\times$ stocking density as fixed effects, with bird ID and pen as random factors. Tukey's HSD test was used for post-hoc analysis when main factors or their interaction were significant at $p < 0.05$. Occurrence of vigilance behaviors were summed to give a total score, which ranged between 0 (no vigilance behavior observed) and 4 (all vigilance behaviors observed at least once), then were analyzed with complexity and stocking density as fixed effects, and pen as a random factor. Data are presented as LSmeans $\pm$ SEM unless otherwise noted.

Table 3. Summary of data analyses for Experiments 1 and 2.

Fear/Anxiety Test	Response Variable (Unit)	Distribution of Data Residuals	Statistical Approach
Attention bias	Latency to first vocalization (s) [1]	Normal	General linear mixed-effects model
	Latency to first step (s)	Normal	General linear mixed-effects model
	Latency to begin feeding (s)	Other	Chi-square [1] and general linear mixed-effects model [2]
	Latency to resume feeding (s)	Normal	General linear mixed-effects model
	Frequency to resume feeding (% of tested birds) [2]	Other	Chi-square
	Vigilance behavior scores (0–4)	Normal	General linear mixed-effects model
	Frequency of vigilance behaviors	Other	Chi-square
Tonic immobility	Duration (s)	Normal	General linear mixed-effects model

[1] In Experiment 1; [2] In Experiment 2.

3. Results

3.1. Experiment 1

3.1.1. Attention Bias Test

Out of the 60 birds tested, 10 birds (4 from LC/LD, 3 from HC/HD, and 3 from HC/LD) began feeding after the first alarm call was played. No differences in latencies to begin feeding were found between either complexity ($\chi^2 = 0.915$; $p = 0.339$) or stocking density ($\chi^2 = 1.715$; $p = 0.190$) treatments (Table 4). Seven birds (2 from LC/LD, 2 from HC/HD, and 3 from HC/LD) resumed feeding after the second alarm call was played. No differences in latencies to resume feeding were found between either complexity ($F_{1,6} = 0.528$; $p = 0.544$) or stocking density ($F_{1,6} = 0.892$; $p = 0.444$) treatments (Table 4). No differences in latency to first vocalization were found between either complexity ($F_{1,59} = 0.169$; $p = 0.691$) or stocking density ($F_{1,59} = 0.554$; $p = 0.476$) treatments (Table 4). Latency to step did not differ between either complexity ($F_{1,44} = 0.016$; $p = 0.904$) or stocking density ($F_{1,44} = 1.925$; $p = 0.215$) treatments (Table 4). Looking around tended to be observed more frequently for birds from LD pens compared to birds from HD pens ($\chi^2 = 3.298$; $p = 0.069$; Table 5), with no other differences in frequency of observed individual vigilance behaviors between treatments. Vigilance behavior scores did not differ between either complexity ($F_{1,59} = 0.062$; $p = 0.809$) or stocking density ($F_{1,59} = 1.552$; $p = 0.244$) treatments (Table 5).

Table 4. Least squares mean estimates (s $\pm$ SEM) for latency to first vocalization ($n = 60$), first step ($n = 45$), and begin feeding ($n = 60$), as well as raw means (s $\pm$ SEM) for latency to resume feeding ($n = 7$) for broiler chickens kept in high-complexity (HC), low-complexity (LC), high-density (HD), and low-density (LD) treatments in Experiment 1 at 4 weeks of age (days 30, 32, and 33).

Latencies (s)	Complexity Treatment		Stocking Density Treatment	
	HC	LC	HD	LD
First vocalization (s)	16.40 $\pm$ 5.35	19.51 $\pm$ 5.35	20.77 $\pm$ 5.35	15.14 $\pm$ 5.35
First step (s)	39.21 $\pm$ 9.58	37.09 $\pm$ 13.81	49.21 $\pm$ 10.62	27.09 $\pm$ 12.46
Begin feeding (s)	265.28 $\pm$ 13.95	296.80 $\pm$ 1.66	287.69 $\pm$ 7.13	274.39 $\pm$ 12.67
Resume feeding (s)	56.39 $\pm$ 42.24	22.73 $\pm$ 20.44	16.22 $\pm$ 9.87	58.99 $\pm$ 42.03

Table 5. Least squares mean estimates ($\pm$ SEM) for vigilance behavior scores and % of total observations that each type of vigilance behavior was observed for broiler chickens kept in high-complexity (HC), low-complexity (LC), high-density (HD), and low-density (LD) treatments ($n = 60$) in Experiment 1 at 4 weeks of age (days 30, 32, and 33). Birds were scored either 0 (not observed) or 1 (observed) for each of four vigilance behavior characteristics (erect posture, neck stretching, looking around, and freezing), giving a vigilance score between 0 (no vigilance behavior observed) and 4 (all vigilance behaviors observed).

Indicators	Complexity Treatment		Stocking Density Treatment	
	HC	LC	HD	LD
Vigilance behavior score (0–4)	2.53 ± 0.19	2.47 ± 0.19	2.33 ± 0.19	2.67 ± 0.19
Erect posture (% of birds)	43.33	30.00	36.67	36.67
Neck stretching (% of birds)	50.00	53.33	46.67	56.67
Looking around (% of birds)	76.67	46.67	66.67 [B]	86.67 [A]
Freezing (% of birds)	83.33	56.67	83.33	86.67

[A–B] Proportions with uncommon superscripts differ at $p < 0.1$.

3.1.2. Tonic Immobility Test

An interaction effect of environmental complexity and age was found for TI durations ($F_{1,35} = 6.264$; $p = 0.015$), with longer TI durations for birds from HC pens on day 12 compared to day 26 ($p = 0.004$; Table 6). No other pairwise differences were found ($p > 0.12$). Stocking density and age tended to impact TI durations ($F_{1,35} = 3.15$; $p = 0.081$), with birds from HD pens showing longer TI durations on day 12 than on day 26 ($p = 0.016$; Table 6). No other pairwise differences were found ($p > 0.17$). Attempts to induce TI did not differ on day 12 between either complexity ($F_{1,35} = 1.03$; $p = 0.318$) or stocking density ($F_{1,35} = 0.041$; $p = 0.84$) treatments, or on day 26 between either complexity ($F_{1,35} = 1.287$; $p = 0.265$) or stocking density ($F_{1,35} = 0.463$; $p = 0.501$) treatments (Table 6).

Table 6. Least squares mean estimates for tonic immobility duration (s $\pm$ SEM; 0–300 s) and induction attempts (1–3) for broiler chickens kept in high-complexity (HC), low-complexity (LC), high-density (HD), and low-density (LD) treatments in Experiment 1 on days 12 and 26 ($n = 36$).

Measures	Bird Age (Day)	Complexity Treatment		Stocking Density Treatment	
		HC	LC	HD	LD
Tonic immobility duration (s)	12	109.43 ± 18.65 [a]	51.24 ± 18.65 [a,b]	101.42 ± 18.655 [a]	59.25 ± 18.65 [a,b]
	26	31.12 ± 18.65 [b]	49.94 ± 18.65 [a,b]	34.31 ± 18.65 [b]	46.75 ± 18.65 [a,b]
Tonic immobility induction attempt (1–3)	12	2.17 ± 0.19	1.89 ± 0.19	2.06 ± 0.19	2.00 ± 0.19
	26	2.39 ± 0.17	2.11 ± 0.17	2.17 ± 0.17	2.33 ± 0.17

[a,b] Means with uncommon superscripts differ at $p < 0.05$.

3.2. Experiment 2

3.2.1. Attention Bias Test

Out of the 144 birds tested, 92 began feeding following the first alarm call (19 from LC/LD, 21 from LC/HD, 24 from HC/HD, and 28 from HC/LD). Birds from HC pens began feeding faster than birds from LC pens ($F_{1,143} = 4.430$; $p = 0.043$; Figure 4). No differences in latency to begin feeding were found between stocking density treatments ($F_{1,143} = 0.081$; $p = 0.777$). Seventy-eight birds resumed feeding after the second alarm call was played (13 from LC/LD, 15 from LC/HD, 22 from HC/HD, and 28 from HC/LD). No differences in latency to resume feeding were found between either complexity ($F_{1,77} = 2.658$; $p = 0.149$) or stocking density ($F_{1,77} = 2.413$; $p = 0.182$) treatments (Figure 4). More birds from HC pens resumed feeding than birds from LC pens (50 from HC, 28 from LC; $\chi^2 = 4.863$; $p = 0.027$). No differences between stocking density treatments were found ($\chi^2 = 2.109$; $p = 0.146$; Figure 4). No differences in latency to first step were found between either complexity ($F_{1,99} = 0.005$; $p = 0.946$) or stocking density ($F_{1,99} = 0.834$; $p = 0.368$) treatments (HC: 101.55 ± 20.89 s; LC: 101.01 ± 20.82 s; HD: 114.51 ± 20.89 s; LD: 88.05 ± 20.82 s). Neck stretching behavior was observed more frequently in birds from LD pens than HD pens (χ^2 = 4.559; p = 0.033), with no other differences in frequency of observed vigilance behavior between treatments. Vigilance behaviors scores did not differ between either complexity ($F_{1,98}$ = 0.079; p = 0.780) or stocking density ($F_{1,98}$ = 1.233; p = 0.275) treatment (Table 7).

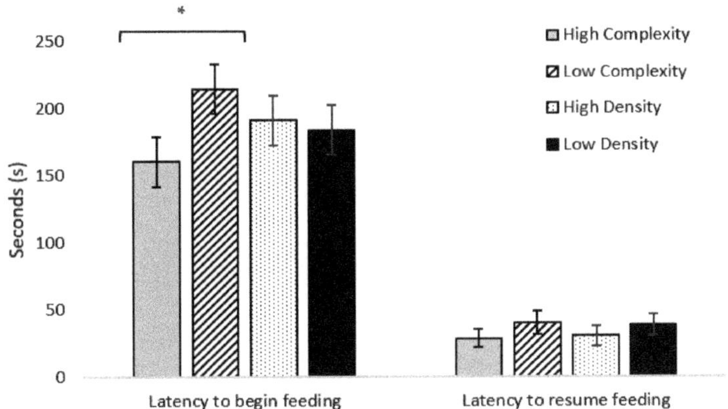

Figure 4. Least squares mean estimates (s $\pm$ SEM) for latency to begin feeding (n = 144) and resume feeding (n = 78) for broiler chickens kept in high-complexity, low-complexity, high-density, and low-density treatments in Experiment 2 at 4 and 5 weeks of age (days 32, 33, and 38). The timer was reset to zero after the second alarm call was played to record latency to resume feeding. * p < 0.05.

Table 7. Least squares mean estimates ($\pm$SEM) for vigilance behavior scores and % of each type of vigilance behavior observed for broiler chickens kept in high-complexity (HC), low-complexity (LC), high-density (HD), and low-density (LD) treatments (n = 99) in Experiment 2 at 4 and 5 weeks of age (days 32, 33, and 38). Birds were scored either 0 (not observed) or 1 (observed) for each of four vigilance behavior characteristics (erect posture, neck stretching, freezing, and looking around), giving a vigilance score between 0 (no vigilance behavior observed) and 4 (all vigilance behaviors observed).

Indicators	Complexity Treatment		Stocking Density Treatment	
	HC	**LC**	**HD**	**LD**
Vigilance behavior score (0–4)	2.72 ± 0.15	2.66 ± 0.15	2.57 ± 0.15	2.80 ± 0.15
Erect poster (% of birds)	52.08	45.10	48.98	48.00
Neck stretching (% of birds)	66.67	56.87	51.02 [b]	72.00 [a]
Freezing (% of birds)	62.50	76.47	69.34	70.00
Looking around (% of birds)	89.58	88.24	87.76	90.00

[a,b] Percentages with uncommon superscripts differ at p < 0.05.

3.2.2. Tonic Immobility Test

There was no difference in TI duration between complexity treatments ($F_{1,70}$ = 0.091; p = 0.770). Birds from HD pens had shorter TI durations than birds from LD pens ($F_{1,70}$ = 12.610; p = 0.006; Figure 5). No differences in attempts to induce TI were found between either complexity ($F_{1,70}$ = 1.016; p = 0.341) or stocking density ($F_{1,70}$ = 0.074; p = 0.793) treatments. Mean TI induction attempts were 2.08 for HC, 1.86 for LC, 2.00 for HD, and 1.94 for LD pens (SEM of 0.15).

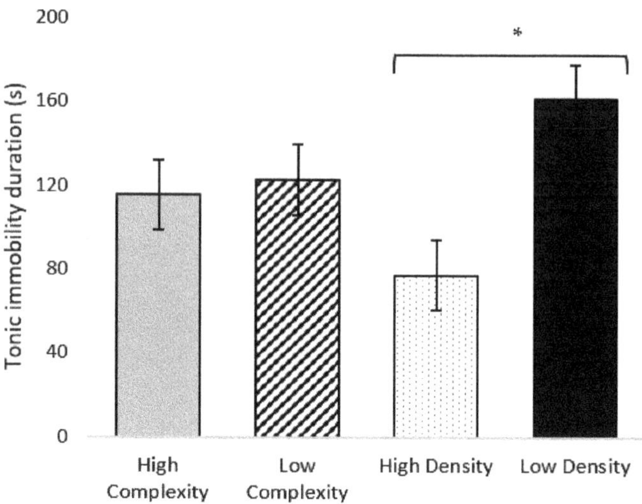

Figure 5. Least squares mean estimates (s ± SEM) for tonic immobility duration (0–300 s) for broiler chickens (*n* = 71) kept in high-complexity, low-complexity, high-density, and low-density treatments in Experiment 2 on day 19 of age. * *p* < 0.05.

4. Discussion

This study investigated fear and anxiety in broiler chickens housed in either high or low environmental complexities and stocking densities. During the AB test in Experiment 1, birds from LD pens tended to look around more frequently than birds from HD pens, with no differences between the complexity treatments. Birds from HC and HD pens had longer TI durations on day 12 compared to day 26, whereas there was no difference for LC and LD birds. During the AB test in Experiment 2, birds from HC pens began feeding faster than birds from LC pens following the first alarm call playback, more birds from HC pens resumed feeding than birds from LC pens following the second alarm call playback, and birds from LD pens stretched their necks more frequently than birds from HD pens. These results suggest reduced anxiety in birds from HC pens compared to LC pens. Furthermore, birds from HD pens had shorter TI durations than birds from LD pens, indicating reduced fearfulness in birds from HD pens compared to LD pens.

4.1. Environmental Complexity

For the AB test, environmental complexity impacted latencies to begin feeding in Experiment 2, but not in Experiment 1. Longer latencies to begin feeding during a threatening situation suggests greater attention allocated towards the threat (alarm call), which indicates a higher level of anxiety. In Experiment 2, birds from HC pens were faster to begin feeding following an alarm call playback than birds from LC pens. This finding suggested reduced anxiousness in broilers housed in complex environments, which was in line with our hypothesis. Conversely, our results suggest that broilers housed in low-complexity environments biased their attention towards a perceived threat compared to a reward (feed). Therefore, these results link low-complexity environments to greater anxiety in broilers. By alleviating these negative states, high-complexity environments appear to improve broiler welfare. Attention bias tests performed with starlings [46] and laying hens [43] showed differences in level of anxiety in relation to environmental conditions or preference. Laying hens that preferred to remain indoors during the day responded more anxiously in an AB test compared to hens that preferred to go outside, observed through a small number of indoor-preferring hens eating during the test (only 7% of indoor-preferring hens resumed feeding after the alarm call playback compared to 36% of outdoor-preferring hens) [43].

Latencies to begin feeding in that study were comparable to those in the present study (indoor hens = 160 s vs. outdoor hens = 85 s compared to broilers from HC pens = 160 s vs. birds from LC pens = 214 s). Furthermore, our results do align with previous work in rodents that shows environmental complexity can reduce anxiety, although different behavioral tests were used in those studies, such as open field or elevated plus maze tests [47–49]. Ultimately, our AB results indicate that broilers housed in a complex environment are less anxious than those housed in a low-complexity environment.

Environmental complexity can decrease fear in broiler chickens, although some previous studies found no relationship. Access to elevated platforms resulted in shorter TI durations (238 s vs. 311 s) compared to access to manipulated standard resources (greater distance between feeders and water lines), suggesting reduced fearfulness in broilers housed with platforms [50]. These TI durations are longer than those observed in the current study, even though test approaches were comparable (LC: 123 s vs. HC: 116 s in Experiment 1). Broilers housed with perches and dust baths had shorter flight distances in an avoidance test, suggesting they were less fearful towards humans than control birds [28]. In Experiment 1, we found a difference in fearfulness within complexity treatments at different ages, but found no difference between complexity treatments. This is in agreement with other studies that did not report an impact of complexity on fear. For example, broilers housed with barrier perches did not have different TI durations compared to control birds [35,51]. Furthermore, responses during a novel object test to assess fearfulness did not differ between broilers housed with or without string enrichments [36]. Our results indicate that providing multiple enrichments concurrently did not impact fearfulness in broilers.

4.2. Stocking Density

Contrary to our predictions, stocking density did not affect birds' responses during the AB test. In line with this finding, one previous study suggested that other housing conditions impact broiler welfare more than stocking density [52]. Stocking density can be especially influential later in life, with broiler welfare compromised when stocking densities are higher than 34–38 kg/m^2, depending on final body weights [53]. Therefore, the potential detrimental effect of high stocking density could have been absent at the age that AB testing was performed, with high densities ranging between 19–21 kg/m^2 (days 30, 32, and 33) in Experiment 1 and 25–30 kg/m^2 (days 32, 33, and 38) in Experiment 2. We recommend that future research investigating the effect of stocking density on AB in broilers should perform the test later in life, when densities are at least 34 kg/m^2.

We hypothesized that birds from HD pens would have longer TI durations and require fewer attempts to induce TI than birds from LD pens, indicating greater fear. However, in Experiment 1, we did not establish a difference between HD or LD treatments on TI durations, but there was a difference depending on age. This decrease in TI duration with age could reflect habituation to the test and handling, as the same birds were tested on both days. In Experiment 2, we found that birds in HD pens had shorter TI durations than birds in LD pens (HD: 72 s versus LD: 161 s in Experiment 2), suggesting birds from HD pens were less fearful compared to birds from LD pens. Past research suggests housing broilers at high stocking densities can contribute to increased fearfulness, which is contrary to our result. For example, broilers housed at a density of more than 18 to 22 birds/m^2 had longer TI durations than broilers housed at lower densities [33,54,55]. Another study found that broilers housed at a high stocking density of 56 kg/m^2 showed longer TI durations (more fearful) than broilers housed at lower densities [33]. Two of these lower stocking densities were comparable to the high and low densities at the time of TI testing in our study (6 kg/m^2 and 15 kg/m^2 compared to 8–16 kg/m^2 and 4–8 kg/m^2 at testing age in the present study), yet they did not find differences in TI duration between those two density levels (112 s for birds housed at 6 kg/m^2 versus 101 s for birds housed at 15 kg/m^2), whereas the present study found that birds from HD pens had shorter TI durations compared to birds from LD pens. The difference in results could be attributed to an age effect, as birds were tested for TI at 6 weeks of age in the previous study compared to 2- and 3 weeks of age in the present study. Broilers may be more fearful early in life, as young, small birds may perceive "safety in numbers" of greater importance than older, large birds. Domestic fowl have maintained pronounced anti-predator behavior, and so the value of being surrounded by many conspecifics is the reduced risk of predation and increased predator detection [56–60]. This could explain why birds in HD pens were less fearful than birds in LD pens at a young age. Contrary to our predictions and previous findings, birds from HD pens were less fearful than birds from LD pens. We recommend further research on this relationship.

4.3. Attention Bias Test Methodology

The AB test was modified after Experiment 1 to increase sample size and apply a group approach (three birds tested simultaneously) rather than testing individual birds. Broilers in Experiment 1 might have attempted to escape the testing arena faster due to social isolation, while in Experiment 2, broilers experienced social support from flock mates present, reducing their motivation to escape. In line, anecdotal observations did suggest social isolation distress based on the volume and pitch frequency of bird vocalizations and attempts to jump over arena walls in Experiment 1, but not 2. Broilers have a strong motivation for social reinstatement and chickens in natural settings live in relatively small, highly social groups [61–64]. Additionally, pairs of chicks placed in a novel open field test exhibited less fear-related behaviors than individual chicks in the same test [44]. Treatments did not impact latency to first vocalization in Experiment 1, latency to begin or resume feeding in Experiment 1, or vigilance behavior scores and latency to first step in both experiments. However, a large numeric difference between latencies to first step in Experiment 1 and 2 was found, with shorter latencies in Experiment 1 (27–49 s vs. 88–114 s). Therefore, latency to first step when birds are tested individually in a novel testing arena may indicate the birds' motivation for social reinstatement rather than a measure of anxiousness.

The effects of environmental complexity and stocking density on attention bias in broiler chickens were previously unknown. AB tests were pharmacologically validated in laying hens—hens given anxiogenic drugs were slower to feed and faster to vocalize than hens receiving a saline injection, suggesting increased anxiousness in the former [25]. In our study, broilers' latency to first vocalization (15–20 s) was much shorter than reported for laying hens, which vocalized after 114 s (control) and 317 s (hens that received an anxiogenic drug in Experiment 2 [25]). Similarly, latencies to first step in broilers was much shorter than (27–114 s) or comparable to previously reported results for laying hens (between 42–52 s and between 211–355 s [25]). Disparities in AB between broilers and laying hens could be due to different ages at the time of AB testing or genetic strain differences associated with selection for production traits [65–67]. Broilers have been genetically selected for fast growth rate [68], while laying hens were selected for traits associated with increased egg production [69]. Generally, it is accepted that different strains and breeds of domestic fowl possess different temperaments, most apparent in terms of fear or flightiness, which can be defined as rapid movement away from a stimulus [70–73]. Therefore, the temperamental differences between broilers and laying hens could explain the difference in responses seen in the AB test.

5. Conclusions

We investigated the effects of housing broiler chickens in a high- or low-complexity environment under high or low stocking densities on their level of fear and anxiety. The group approach to AB testing in Experiment 2 produced a difference in broiler responses between the complexity treatments, compared to the individual testing approach in Experiment 1. Broilers from high-complexity pens exhibited responses in the AB test suggestive of reduced anxiety compared to broilers from low-complexity pens, with no differences between the stocking density treatments. These results suggest that the environmental complexity provided in the present study improved welfare of broilers through reduced anxiety. To our knowledge, this is the first AB test successfully assessing anxiety in broiler chickens. Additionally, birds housed at higher stocking densities showed reduced TI durations, suggesting reduced fearfulness compared to birds housed at lower stocking densities. This finding counterintuitively indicates that, for broilers around 3 weeks old, housing at higher densities may reduce fearfulness.

Author Contributions: Conceptualization and methodology, M.G.A., A.M.C., R.C.N., and L.J.; formal analysis and data curation, M.G.A. and L.J.; writing—original draft preparation, M.G.A. and L.J.; writing, review, and editing, M.G.A., A.M.C., A.C., G.A., R.C.N., and L.J.; project administration, M.G.A., A.M.C., and L.J. All authors have read and agreed to the published version of the manuscript.

Funding: Experiment 2 was funded by the Poultry Science Association Foundation (Arthur W. Perdue Fellowship).

Institutional Review Board Statement: Both experiments in this study were approved by the Virginia Tech's Institutional Animal Care and Use Committee on 30 September 2019 (approval #19-175). The study was conducted in accordance with the IACUC's guidelines.

Data Availability Statement: Data are available upon request from the corresponding author.

Acknowledgments: Thank you to Virginia Tech Poultry farm staff and Jacobs Lab volunteers for assistance with this study. Thank you to Proteka, Inc. for the donation of peck stones used during the study.

Conflicts of Interest: The authors declare no conflict of interest.

References

1. Newberry, R.C. Environmental enrichment: Increasing the biological relevance of captive environments. *Appl. Anim. Behav. Sci.* **1995**, *44*, 229–243. [CrossRef]
2. BenSassi, N.; Vas, J.; Vasdal, G.; Averós, X.; Estévez, I.; Newberry, R.C. On-farm broiler chicken welfare assessment using transect sampling reflects environmental inputs and production outcomes. *PLoS ONE* **2019**, *14*, e0214070. [CrossRef]
3. Tahamtani, F.M.; Pedersen, I.J.; Riber, A.B. Effects of environmental complexity on welfare indicators of fast-growing broiler chickens. *Poult. Sci.* **2020**, *99*, 21–29. [CrossRef]
4. Vasdal, G.; Vas, J.; Newberry, R.C.; Moe, R.O. Effects of environmental enrichment on activity and lameness in commercial broiler production. *J. Appl. Anim. Welf. Sci.* **2019**, *22*, 197–205. [CrossRef] [PubMed]
5. Steimer, T. The biology of fear- and anxiety-related behaviors. *Dialogues Clin. Neurosci.* **2002**, *4*, 231–249. [CrossRef]
6. Monk, J.E.; Doyle, R.E.; Colditz, I.G.; Belson, S.; Cronin, G.M.; Lee, C. Towards a more practical attention bias test to assess affective state in sheep. *PLoS ONE* **2018**, *13*, e0190404. [CrossRef] [PubMed]
7. Campbell, D.L.M.; De Haas, E.N.; Lee, C. A review of environmental enrichment for laying hens during rearing in relation to their behavioral and physiological development. *Poult. Sci.* **2019**, *98*, 9–28. [CrossRef]
8. Hedlund, L.; Whittle, R.; Jensen, P. Effects of commercial hatchery processing on short- and long-term stress responses in laying hens. *Sci. Rep.* **2019**, *9*, 2367. [CrossRef] [PubMed]
9. Bas Rodenburg, T.; de Haas, E.N. Of nature and nurture: The role of genetics and environment in behavioural development of laying hens. *Curr. Opin Behav. Sci.* **2016**, *7*, 91–94. [CrossRef]
10. Sanyal, T.; Kumar, V.; Nag, T.C.; Jain, S.; Sreenivas, V.; Wadhwa, S. Prenatal loud music and noise: Differential impact on physiological arousal, hippocampal synaptogenesis and spatial behavior in one day-old chicks. *PLoS ONE* **2013**, *8*, e67347. [CrossRef] [PubMed]
11. Jones, R.B. Fear and adaptability in poultry: Insights, implications and imperatives. *Worlds Poult. Sci. J.* **1996**, *52*, 131–174. [CrossRef]
12. Altan, Ö.; Pabuçcuoğlu, A.; Altan, A.; Konyalioğlu, S.; Bayraktar, H. Effect of heat stress on oxidative stress, lipid peroxidation and some stress parameters in broilers. *Br. Poult. Sci.* **2003**, *44*, 545–550. [CrossRef] [PubMed]
13. Hemsworth, P.H.; Coleman, G.J.; Barnett, J.L.; Jones, R.B. Behavioural responses to humans and the productivity of commercial broiler chickens. *Appl. Anim. Behav. Sci.* **1994**, *41*, 101–114. [CrossRef]
14. Fogelholm, J.; Inkabi, S.; Höglund, A.; Abbey-Lee, R.; Johnsson, M.; Jensen, P.; Henriksen, R.; Wright, D. Genetical genomics of tonic immobility in the chicken. *Genes* **2019**, *10*, 341. [CrossRef] [PubMed]
15. Jones, R.B. The nature of handling immediately prior to test affects tonic immobility fear reactions in laying hens and broilers. *Appl. Anim. Behav. Sci.* **1992**, *34*, 247–254. [CrossRef]
16. Delezie, E.; Lips, D.; Lips, R.; Decuypere, E. Is the mechanisation of catching broilers a welfare improvement? *Anim. Welf.* **2006**, *15*, 141–147.
17. Hughes, R.A. Shock-potentiated tonic immobility in chickens as a function of posthatch age. *Anim. Behav.* **1979**, *27*, 782–785. [CrossRef]

18. Crump, A.; Arnott, F.; Bethell, E.J. Affect-driven biases as animal welfare indicators: Review and methods. *Animals* **2018**, *8*, 136. [CrossRef]

19. Bradley, B.P.; Mogg, K.; Lee, S.C. Attentional biases for negative information in induced and naturally occurring dysphoria. *Behav. Res. Ther.* **1997**, *35*, 911–927. [CrossRef]

20. Bradley, B.P.; Mogg, K.; Millar, N.; White, J. Selective processing of negative information: Effects of clinical anxiety, concurrent depression, and awareness. *J. Abnorm. Psychol.* **1995**, *104*, 532–536. [CrossRef]

21. Bar-Haim, Y.; Lamy, D.; Pergamin, L.; Bakermans-Kranenburg, M.J.; Van Ijzendoorn, M.H. Threat-related attentional bias in anxious and nonanxious individuals: A meta-analytic study. *Psychol. Bull.* **2007**, *133*, 1–24. [CrossRef] [PubMed]

22. Bethell, E.J.; Holmes, A.; MacLarnon, A.; Semple, S. Evidence that emotion mediates social attention in rhesus macaques. *PLoS ONE* **2012**, *7*, e44387. [CrossRef]

23. Lee, C.; Verbeek, E.; Doyle, R.; Bateson, M. Attention bias to threat indicates anxiety differences in sheep. *Biol. Lett.* **2016**, *12*, 20150977. [CrossRef]

24. Lee, C.; Café, L.M.; Robinson, S.L.; Doyle, R.E.; Lea, J.M.; Small, A.H.; Colditz, I.G. Anxiety influences attention bias but not flight speed and crush score in beef cattle. *Appl. Anim. Behav. Sci.* **2018**, *205*, 210–215. [CrossRef]

25. Campbell, D.L.M.; Taylor, P.S.; Hernandez, C.E.; Stewart, M.; Belson, S.; Lee, C. An attention bias test to assess anxiety states in laying hens. *PeerJ* **2019**, *2019*, e7303. [CrossRef]

26. Monk, J.E.; Belson, S.; Lee, C. Pharmacologically-induced stress has minimal impact on judgement and attention biases in sheep. *Sci. Rep.* **2019**, *9*, 11446. [CrossRef]

27. Luo, L.; Reimert, I.; de Haas, E.N.; Kemp, B.; Bolhuis, J.E. Effects of early and later life environmental enrichment and personality on attention bias in pigs (*Sus scrofa domesticus*). *Anim. Cogn.* **2019**, *22*, 959–972. [CrossRef] [PubMed]

28. Monk, J.E.; Lee, C.; Belson, S.; Colditz, I.G.; Campbell, D.L.M. The influence of pharmacologically-induced affective states on attention bias in sheep. *PeerJ* **2019**, *7*, e7033. [CrossRef]

29. Ventura, B.A.; Siewerdt, F.; Estevez, I. Access to barrier perches improves behavior repertoire in broilers. *PLoS ONE* **2012**, *7*, e29826. [CrossRef] [PubMed]

30. Estevez, I.; Newberry, R.C.; de Reyna, L.A. Broiler chickens: A tolerant social system? *Etologia* **1997**, *5*, 19–29.

31. Mallapur, A.; Miller, C.; Christman, M.; Estevez, I. Short-term and long-term movement patterns in confined environments by domestic fowl: Influence of group size and enclosure size. *Appl. Anim. Behav. Sci.* **2009**, *117*, 28–34. [CrossRef]

32. Bach, M.H.; Tahamtani, F.M.; Pedersen, I.J.; Riber, A.B. Effects of environmental complexity on behaviour in fast-growing broiler chickens. *Appl. Anim. Behav. Sci.* **2019**, *219*, 104840. [CrossRef]

33. Buijs, S.; Keeling, L.J.; Rettenbacher, S.; Van Poucke, E.; Tuyttens, F.A.M. Stocking density effects on broiler welfare: Identifying sensitive ranges for different indicators. *Poult. Sci.* **2009**, *88*, 1536–1543. [CrossRef]

34. Sanotra, G.S.; Damkjer Lund, J.; Vestergaard, K.S. Influence of light-dark schedules and stocking density on behaviour, risk of leg problems and occurrence of chronic fear in broilers. *Br. Poult. Sci.* **2002**, *43*, 344–354. [CrossRef]

35. Ventura, B.A.; Siewerdt, F.; Estevez, I. Effects of barrier perches and density on broiler leg health, fear, and performance. *Poult. Sci.* **2010**, *89*, 1574–1583. [CrossRef]

36. Bailie, C.L.; Ijichi, N.E.; O'Connell, N.E. Effects of stocking density and string provision on welfare-related measures in commercial broiler chickens in windowed houses. *Poult. Sci.* **2018**, *97*, 1503–1510. [CrossRef] [PubMed]

37. Norring, M.; Kaukonen, E.; Valros, A. The use of perches and platforms by broiler chickens. *Appl. Anim. Behav. Sci.* **2016**, *184*, 91–96. [CrossRef]

38. Baxter, M.; Richmond, A.; Lavery, U.; O'Connell, N.E. Investigating optimal levels of platform perch provision for windowed broiler housing. *Appl. Anim. Behav. Sci.* **2020**, *225*, 104967. [CrossRef]

39. Shields, S.J.; Garner, J.P.; Mench, J.A. Dustbathing by broiler chickens: A comparison of preference for four different substrates. *Appl. Anim. Behav. Sci.* **2004**, *87*, 69–82. [CrossRef]

40. Baxter, M.; Bailie, C.L.; O'Connell, N.E. Play behaviour, fear responses and activity levels in commercial broiler chickens provided with preferred environmental enrichments. *Animal* **2019**, *13*, 171–179. [CrossRef]

41. National Research Council. *Nutrient Requirements of Poultry*, 9th ed.; National Academies Press: Washington, DC, USA, 1994. [CrossRef]

42. Evans, C.S.; Evans, L.; Marler, P. On the meaning of alarm calls: Functional reference in an avian vocal system. *Anim. Behav.* **1993**, *46*, 23–38. [CrossRef]

43. Campbell, D.L.M.; Dickson, E.J.; Lee, C. Application of open field, tonic immobility, and attention bias tests to hens with different ranging patterns. *PeerJ* **2019**, *7*, e8122. [CrossRef]

44. Jones, R.B. Open-field responses of domestic chicks in the presence of a cagemate or a strange chick. *IRCS Med. Sci. Psychol. Psychiatry* **1984**, *12*, 482–483.

45. Pichova, K.; Nordgreen, J.; Leterrier, C.; Kostal, L.; Moe, R.O. The effects of food-related environmental complexity on litter directed behaviour, fear and exploration of novel stimuli in young broiler chickens. *Appl. Anim. Behav. Sci.* **2016**, *174*, 83–89. [CrossRef]

46. Brilot, B.O.; Bateson, M. Water bathing alters threat perception in starlings. *Anim. Behav.* **2012**, *8*, 379–381. [CrossRef] [PubMed]

47. Gong, X.; Chen, Y.; Chang, J.; Huang, Y.; Cai, M.; Zhang, M. Environmental enrichment reduces adolescent anxiety- and depression-like behaviors of rats subjected to infant nerve injury. *J. Neuroinflamm.* **2018**, *15*, 262. [CrossRef]

48. Benaroya-Milshtein, N.; Hollander, N.; Apter, A.; Kukulansky, T.; Raz, N.; Wilf, A.; Yaniv, I.; Pick, C.G. Environmental enrichment in mice decreases anxiety, attenuates stress responses and enhances natural killer cell activity. *Eur. J. Neurosci.* **2004**, *20*, 1341–1347. [CrossRef]
49. Grippo, A.J.; Ihm, E.; Wardwell, J.; McNeal, N.; Scotti, M.A.L.; Moenk, D.A.; Chandler, D.L.; LaRocca, M.A.; Preihs, K. The effects of environmental enrichment on depressive- and anxiety-relevant behaviors in socially isolated prairie voles. *Psychosom. Med.* **2014**, *76*, 277. [CrossRef]
50. Tahamtani, F.M.; Pedersen, I.J.; Toinon, C.; Riber, A.B. Effects of environmental complexity on fearfulness and learning ability in fast growing broiler chickens. *Appl. Anim. Behav. Sci.* **2018**, *207*, 49–56. [CrossRef]
51. Bizeray, D.; Estevez, I.; Leterrier, C.; Faure, J.M. Influence of increased environmental complexity on leg condition, performance, and level of fearfulness in broilers. *Poult. Sci.* **2002**, *81*, 767–773. [CrossRef]
52. Dawkins, M.S.; Donnelly, C.A.; Jones, T.A. Chicken welfare is influenced more by housing conditions than by stocking density. *Nature* **2004**, *427*, 342–344. [CrossRef] [PubMed]
53. Estevez, I. Density allowances for broilers: Where to set the limits? In *Poultry Science*; Poultry Science Association: Champaign, IL, USA, 2007; Volume 86, pp. 1265–1272. [CrossRef]
54. Onbasilar, E.E.; Pyraz, O.; Erdem, E.; Ozturk, H. Influence of lighting periods and stocking densities on performance, carcass characteristics and some stress parameters in broilers. *Arch. Fur Geflugelkd.* **2008**, *72*, 193–200.
55. De Jong, I.; Berg, C.; Butterworth, A.; Estevéz, I. Scientific report updating the EFSA opinions on the welfare of broilers and broiler breeders. *EFSA Support. Publ.* **2012**, *9*, 295E. [CrossRef]
56. Newberry, R.C.; Estevez, I.; Keeling, L.J. Group size and perching behaviour in young domestic fowl. *Appl. Anim. Behav. Sci.* **2001**, *73*, 117–129. [CrossRef]
57. Roberts, G. Why individual vigilance declines as group size increases. *Anim. Behav.* **1996**, *51*, 1077–1086. [CrossRef]
58. Beauchamp, G. Group-size effects on vigilance: A search for mechanisms. *Behav. Process.* **2003**, *63*, 111–121. [CrossRef]
59. Pulliam, H.R. On the advantages of flocking. *J. Theor. Biol.* **1973**, *38*, 419–422. [CrossRef]
60. Krause, J.; Ruxton, G.D. Living in Groups. In *Oxford Series in Ecology and Evolution*; The University of Chicago Press: Chicago, IL, USA, 2003; Volume 78, pp. 509–510. [CrossRef]
61. Bayram, A.; Ozkan, S. Effects of a 16-hour light, 8-hour dark lighting schedule on behavioral traits and performance in male broiler chickens. *J. Appl. Poult. Res.* **2010**, *19*, 263–273. [CrossRef]
62. Guhl, A.M. Social behavior of the domestic fowl. *Trans. Kans. Acad. Sci.* **1968**, *71*, 379–384. [CrossRef]
63. Collias, N.E.; Collias, E.C. Social organization of a red junglefowl, gallus gallus, population related to evolution theory. *Anim. Behav.* **1996**, *51*, 1337–1354. [CrossRef]
64. Katajamaa, R.; Larsson, L.H.; Lundberg, P.; Sörensen, I.; Jensen, P. Activity, social and sexual behaviour in Red Junglefowl selected for divergent levels of fear of humans. *PLoS ONE* **2018**, *13*, e0204303. [CrossRef] [PubMed]
65. Jackson, S.; Diamond, J. Metabolic and digestive responses to artificial selection in chickens. *Evolution (NY)* **1996**, *50*, 1638–1650. [CrossRef] [PubMed]
66. Mahagna, M.; Nir, I. Comparative development of digestive organs, intestinal disaccharidases and some blood metabolites in broiler and layer-type chicks after hatching. *Br. Poult. Sci.* **1996**, *37*, 359–371. [CrossRef]
67. Koenen, M.E.; Boonstra-Blom, A.G.; Jeurissen, S.H.M. Immunological differences between layer- and broiler-type chickens. *Vet. Immunol. Immunopathol.* **2002**, *89*, 47–56. [CrossRef]
68. Havenstein, G.B.; Ferket, P.R.; Qureshi, M.A. Growth, livability, and feed conversion of 1957 versus 2001 broilers when fed representative 1957 and 2001 broiler diets. *Poult. Sci.* **2003**, *82*, 1500–1508. [CrossRef]
69. Appleby, M.C.; Mench, J.A.; Hughes, B.O. *Poultry Behaviour and Welfare*; CABI Publication: Wallingford, UK, 2004.
70. Keer-Keer, S.; Hughes, B.O.; Hocking, P.M.; Jones, R.B. Behavioural comparison of layer and broiler fowl: Measuring fear responses. *Appl. Anim. Behav. Sci.* **1996**, *49*, 321–333. [CrossRef]
71. Murphy, L.B. Responses of domestic fowl to novel food and objects. *Appl. Anim. Ethol.* **1977**, *3*, 335–349. [CrossRef]
72. Jones, R.B. Social and environmental aspects of fear in the domestic fowl. In *Cognitive Aspects of Social Behaviour in the Domestic Fowl*; Zayan, R., Duncan, I.J.H., Eds.; Elsevier: Amsterdam, The Netherlands, 1987; pp. 82–149.
73. Duncan, I.J.H. How do fearful birds respond? In *Second European Symposium on Poultry Welfare*; Wegner, R., Ed.; German Branch of the World's Poultry Science Association: Celle, Germany, 1985; pp. 96–106.

Article

From the Point of View of the Chickens: What Difference Does a Window Make?

Elaine Cristina de Oliveira Sans [1,*], Frank André Maurice Tuyttens [2,3], Cesar Augusto Taconeli [4], Ana Silvia Pedrazzani [1], Marcos Martinez Vale [5] and Carla Forte Maiolino Molento [1]

[1]   Animal Welfare Laboratory, Federal University of Paraná, Rua dos Funcionários, 1540, Curitiba 80035-050, Paraná, Brazil; anasilviap@yahoo.com.br (A.S.P.); carlamolento@ufpr.br (C.F.M.M.)
[2]   Flanders Research Institute for Agriculture, Fisheries and Food (ILVO), Scheldeweg 68, 9090 Melle, Belgium; frank.tuyttens@ilvo.vlaanderen.be
[3]   Faculty of Veterinary Medicine, Ghent University, Salisburylaan 133, 9820 Merelbeke, Belgium
[4]   Department of Statistics, Federal University of Paraná, Rua Evaristo F. Ferreira da Costa, 408, Curitiba 81531-990, Paraná, Brazil; cetaconeli@gmail.com
[5]   Department of Animal Science, Federal University of Paraná, Rua dos Funcionários, 1540, Curitiba 80035-050, Paraná, Brazil; marcos.vale@ufpr.br
*   Correspondence: elainesans@gmail.com

Citation: Sans, E.C.d.O.; Tuyttens, F.A.M.; Taconeli, C.A.; Pedrazzani, A.S.; Vale, M.M.; Molento, C.F.M. From the Point of View of the Chickens: What Difference Does a Window Make? *Animals* **2021**, *11*, 3397. https://doi.org/10.3390/ani11123397

Academic Editors: Victoria Sandilands and Tina Widowski

Received: 21 September 2021
Accepted: 25 November 2021
Published: 28 November 2021

Publisher's Note: MDPI stays neutral with regard to jurisdictional claims in published maps and institutional affiliations.

Simple Summary: Light is an important environmental factor in many aspects for broiler chickens, such as behaviour and physiology, and welfare may be compromised when they are reared under low illuminance. We aimed to investigate what broiler chickens prefer when given free choice between a barn side with artificial lighting only as opposed to the other barn side with natural light through glass windows and artificial light. Environmental indicators and external conditions were monitored inside and outside the experimental barn, as well as chickens' preference regarding location in each side of the barn and their behavioural repertoire. Chickens preferred the barn side with natural and artificial light from 18 days onwards, after the heating light was removed. Chickens' behavioural repertoire changed according to barn side and their ages, expressing more natural behaviours and activity in the barn side with natural light. In summary, the birds indicated that natural light from windows makes a relevant difference in their lives, as it is what they choose when the only other option is the same in-barn environment with only artificial lighting.

Abstract: We aimed to investigate what broiler chickens prefer when given free choice between a barn side with artificial lighting only as opposed to the other barn side with natural light through glass windows and artificial light. Eighty-five 1 day-old male Cobb 500 broiler chickens were divided into 10 pens; half of each pen area was provided with only artificial light (OAL) and the other half with natural and artificial light (NAL), and birds were free to move across sides. Environmental indicators and external conditions such as temperature, relative humidity, air velocity, ammonia and illuminance were monitored inside and outside the barn. Chickens' preference was registered each three days, divided in categories: I (at 9, 12, and 15 days), II (at 18, 21, 24, and 27 days), and III (at 30, 33 and 36 days). The effect of the interaction between environmental indicators and week was statistically different only for illuminance. Chickens preferred NAL to OAL from 18 days onwards (II $p < 0.001$; III $p = 0.016$). Drinking ($p = 0.034$) and exploration or locomotion ($p = 0.042$) behaviours were more frequent, and "not visible" behaviours ($p < 0.001$) were less frequent, in NAL. Foraging was the only behaviour with an interaction effect between age category and light treatment, as birds during period II expressed this behaviour more frequently in NAL than OAL ($p = 0.003$). For our experimental conditions, the chickens preferred NAL from 18 days of age onwards, when the confounding effect of the heating light was removed, and their behavioural repertoire was also different according to each side of the barn and to their ages.

Keywords: artificial light; behaviour; dark side; environment; glass window; natural light; poultry; preference test

1. Introduction

In general, broiler chickens are intensively reared worldwide in large flocks confined in indoor houses where food, water and environmental control are available to provide for their basic physiological needs [1]. However, considering bird evolutionary history, conditions provided by the production chain are far apart from that found by chickens in a natural life. In nature, they are exposed to a variety of circumstances and environmental conditions which include the day length and photoperiod [1,2].

Broiler chickens subjected to commercial management are typically housed in dim lighting because it is presumed to improve productivity and feed conversion efficiency, reducing overall activity and injurious pecking [3,4]. Such inactivity caused by low illuminance is likely related to an apathetic state, as responsiveness to many stimuli seems reduced, even though it is commonly confounded with a calm state [5]. In fact, light is an important environmental factor for the animals [4,6]. More specifically for broiler chickens, lighting quality and intensity affect their behaviour and physiology [3,6–10]. Natural lighting as a positive factor for bird welfare is a common assumption. However, it is not clear whether this assumption holds when natural light is offered through glass windows and, thus, in a different constitution as compared to outdoor natural lighting. It remains true, though, that natural lighting through windows may provide a dynamic range of illuminance levels in different areas within the house, with considerably higher intensities as compared to the regular artificial lighting recommended for birds. Thus, the potential for enrichment of the perceived environment and, consequently, for improving bird welfare through barn windows [11] seems to warrant further investigation. The birds do express more natural behaviour and are more active compared to birds not exposed to natural light [11]. Although there are types of lamps that can offer the same characteristics as natural illuminance, such as bulbs supplemented with ultraviolet (UV) light fixtures [12], these technologies are not widely used in Brazilian chicken barns, for which a variety of lamp types is observed, such as incandescent and fluorescent lamps [13,14]. In this case, according to the light source type, artificial illuminance may differ from natural light in terms of light colour, intensity, photoperiod, and flicker [9], and these characteristics may influence bird preferences [15]. Moreover, worldwide recommendations for illuminance inside the barns accept extremely low levels of 20 lux (lx) [16,17] and this seems to represent an important subject to be discussed regarding broiler chicken welfare.

Vision is probably the dominant sense in domestic poultry, and the evolution of vision was determined, in part, by the natural light available [18]. The photoreceptive pigments in the retina allow birds to perceive colours in a more detailed way than humans [19]. Birds also have the ability to perceive ultraviolet (UV) light, with the spectral sensitivity below 350 nm [12,19], and may experience a better quality of vision in brighter environments [20]. In a natural scenario, UV light is important for birds in relation to orientation, foraging, calibration of their circadian clock, and sexual selection [21]. In intensive systems, according to glass types, the full passage of UV light is blocked, but windows may be an alternative for providing some UV wavelengths to chickens [11,22,23].

If birds perceive natural and artificial light in different ways, this may influence their behaviour. Manser [7] suggested that light intensities between 5 and 22 lx, currently used for broiler chickens and turkeys, may contribute to the decrease of their engaging in exploratory behaviour and social interaction, high prevalence of leg abnormalities, mortality, eye abnormalities, breast blisters in growing birds, and fearfulness. Surely, the study of behaviour is an important tool for the identification of relevant environments and devices to the animals, justifying the provision of adequate resources to the animals [24]. Preference tests suggest that most broiler chickens make consistent and rational choices associated with the environments that are associated with lower fear and stress responses [25,26]. However, there is a lack of studies about lighting preferences by the birds, and this is especially relevant nowadays, when there is an increase in the number of closed-houses [27]. There is an increasing number of companies replacing natural by artificial lighting, in systems that apply the minimal illuminance recommended for broiler chickens houses (20 lx) [16], or even less than the recommended minimum.

Although there are no public data regarding the proportion of each type of poultry house type in Brazil, broiler chickens in intensive systems are mostly raised in two main barn types [13,14]. The conventional system employs open-sided poultry houses, where the natural daylight may enter without passing through glass when their movable curtains are open; they are called conventional because they used to predominate in the Brazilian poultry meat industry. Lately, the closed-poultry house type is rapidly becoming more popular in Brazil, and it uses only artificial light. Open- and closed-sided poultry houses have positive and negative welfare aspects, which may also vary according to season [13,14]. However, the quantity and quality of the light available to the birds may be considered a major factor that differentiates these two barn types in terms of their animal welfare potential.

Our objective was to investigate the importance of the existence of windows in the barns by studying what the chickens prefer when given free choice between an area with only artificial lighting (OAL) and an area with natural and artificial lighting (NAL). Our hypothesis was that the NAL has a significant effect on animal behaviour and that it would be preferred by birds.

2. Materials and Methods

The study was conducted between January and February 2021, in an experimental broiler house measuring 10 m × 6 m × 2.5 m (Figure 1), of the Federal University of Paraná farm, Pinhais, Brazil (25°23′36.2″ S, 49°08′2.9″ W) at an altitude of 935 m. The house was built in a North-South orientation with 10 pens, each one with a total area of 3.36 m^2 (0.80 m × 2.10 m). Eighty-five one-day-old male Cobb 500 broiler chickens were randomly distributed into ten pens, as groups of eight birds in five pens and of nine birds in the other five pens. The experimental design was planned for eight birds per pen and the additional birds were included to cover for eventual mortality throughout the experimental period. The experimental barn was longitudinally divided into a barn side with no windows and only artificial light (OAL), and the other side was built with a window throughout its lateral wall and received both natural and artificial light (NAL); pens were built transversally, so that half of each pen was in the OAL and the other half in NAL side (Figure 2a), resulting in 1.68 m^2 per pen in each barn side. Ten LED lights were evenly spread across the entire pen areas, in both the OAL and NAL sides. The passage between NAL and OAL barn sides was always open, and the birds were allowed to move freely across the sides as they chose.

Artificial light was provided by Light Emitting Diodes (LED) white lamps of 9 W, 6500 K (correlated colour temperature), dimmable, with no UV or infrared emission, distributed along each side of the barn, suspended from the ceiling at a height of 1.50 m from the floor. In NAL sides, in addition to the same quantity and quality of artificial light as in OAL side, natural daylight was provided through eight windows along the west lateral wall of the barn, measuring 1.25 m × 0.95 m each, equipped with 8 mm colorless tempered glass. The use of glass is opposed to the more common open-sided barns in Brazil. According to the glass type, some UV wavelength may be blocked [22,23]; however, the glass was a necessary resource to maintain the control on internal environmental conditions other than lighting between barn sides, to ensure that bird preference was based exclusively on illuminance, without any interference of other factors such as temperature or relative humidity. Approximately $\frac{3}{4}$ of the window areas on the wall of the NAL side were shut by black curtains between 06:00 PM and 07:00 AM, and for the OAL side, the windows were totally closed by black curtains throughout the experimental period.

A black curtain was used in the center of the barn to separate the OAL and NAL sides (Figure 2a), installed from the ceiling down to 60 cm from the floor. Wooden separators filled this 60 cm close to the floor, and this wooden separation contained passages of 0.50 cm, which allowed for the birds to have free access to both sides of the pen (Figure 2b).

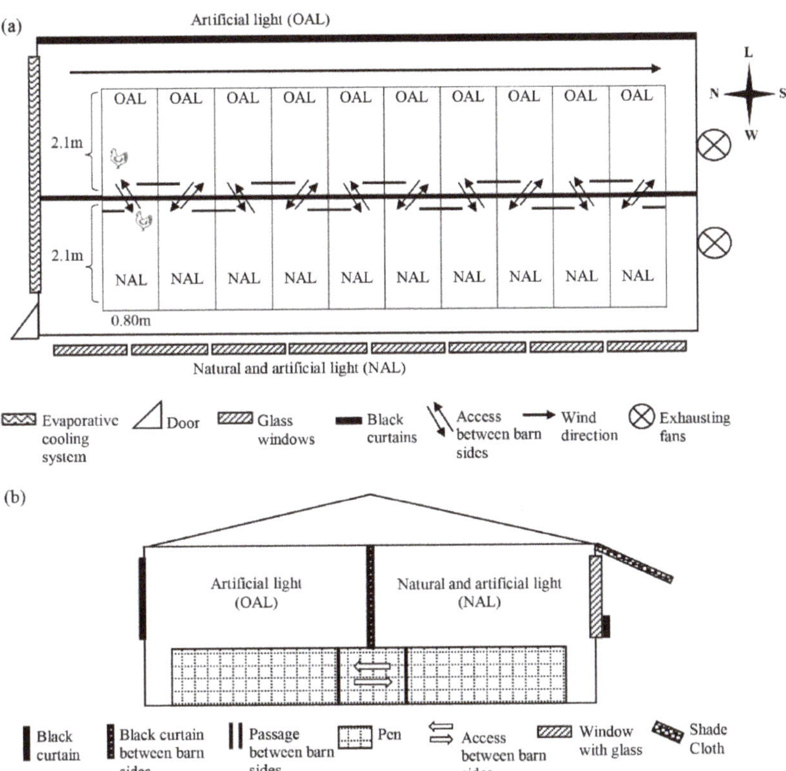

Figure 1. Experimental design of preference test seen from above (**a**) and from the barn entry side (**b**). The house was divided in two sides, one with Only Artificial Light (OAL) provide by LED lamps, and the other side, with Natural Light provided by glass windows and artificial light provided by the same lamp type and quantity (NAL), from January to February 2021, in the State of Paraná, South of Brazil.

(a)   (b)

Figure 2. Overview of the inside of the house (**a**) both sides shown, Only Artificial Light (OAL) on the left, and Natural and Artificial Light (NAL) on the right; and (**b**) separation between sides constructed of black curtain and wooden panel, in a preference test performed from January to February 2021, in the State of Paraná, South of Brazil.

All the pens were equipped with the same quantity and quality of feed, litter, heaters, manual feeders and drinkers, and, from 10 d old onwards, nipple with cups drinkers. Infrared lamps of 240 V and 175 W, for both barn sides and all pens, were used to heat the birds. The heating lamps added, on average, up to 25 lx more in each pen. The pens were made by plastic mesh fence, to facilitate the passage of air. Two exhaust fans, one for each side of the barn, and one evaporative cooling system ensured appropriate temperatures in the entire poultry house. A polyethylene shade cloth was installed on the West side to decrease the direct solar incidence through the glass windows that was observed after 03:00 PM. Natural shadow was provided by trees on the East side of the house (OAL).

2.1. Environment Measurements

Environmental indicators were measured twice every day, at 10:00 a.m. and 03:00 p.m., for the duration of experiment. According to tests carried out before the start of the experiment, the environmental conditions were similar across the house. For this reason, we took measurements at two places indoors, in the middle of the house, on each barn side. The outdoor conditions were monitored for approximately 3 m in front of the barn main entrance, in a place with no coverage. The indoor environmental indicators were monitored at bird level, in the center of pen five in each barn side (OAL and NAL). Temperature, relative humidity, air velocity, and illuminance were measured using Lutron LM 8000A (Akso, São Leopoldo, Brazil). Ammonia concentration (NH_3) was measured by SP2nd Portable Single-Gas Detector (Senko, Osan-si, Korea).

2.2. Experimental Design

On the first day of birds' lives, five groups of birds were initially housed in the OAL side, and the other five groups in the NAL side. Birds had six days of adaptation, for learning between the barn sides offered within each pen, and avoiding any potential confounding effects due to fear of novelty or other factors related to the new environment initially faced by the animals. From day 7 on, each bird group was relocated every three days to the next pen located to their right, allowing for all the groups to stay for three days in each of the 10 pens available in the experiment. If a bird or a group of birds was in the OAL side, they were relocated to the next pen also in the OAL side; the same was done for the birds in the NAL side. Birds were only relocated after emptying the destination pen, thus avoiding contact between birds from different groups; birds in the last pen of the barn were relocated to the first pen, considering the barn door. This management allowed for testing whether there was a pen effect by separating it from group effects. The beginning of assessments only started after two days of the group change, allowing the birds to get used to their new pen. In case of mortality, birds were relocated as needed to maintain a minimum of eight birds per pen. Until 18 d, in both sides, the birds were exposed to 24 h of light and no dark periods, i.e., 24L:0D, on both side. After this period, the birds received a 16L:8D continuous lighting regimen. The switch was done in the following manner: after 14 d of age, heating lamps were turned on only during the night period; after 18 d of age, all heating lamps were removed and birds became exposed to complete darkness from 09:30 PM until 05:30 AM.

2.3. Bird Preference and Behaviour

We video-recorded both sides of two different pens per day, the number of birds in either OAL or NAL sides, and their behavioural repertoire, by fitting four video cameras, Canon Vixia HF R800 (Canon Inc., Zhuhai, China), one installed in front of each side of each two pens. Recordings started on day 9 and ended on day 36, always from 07:30 a.m. to 05:30 p.m., and were conducted every third day, totaling 10 d of observations with 100 h of video-recordings, with 20 pen observations. All the pens recorded were chosen at random, allowing for different pens and groups of birds to be recorded during the experimental period.

Birds' preference was measured by the count of birds present in each side of the barn. Their behaviour was analyzed according to a predefined ethogram (Table 1), using the same video-recording. Both count of birds and behaviours were observed by scanning methodology, with instantaneous sampling every 1 h [28,29].

Table 1. Ethogram with definition of the behaviours recorded for broiler chickens during the preference test, performed from January to February 2021, in the State of Paraná, South of Brazil.

Behaviour	Definition
Feeding	Head in the feeder or pecking at the feed within the feeder
Drinking	Beak touching the drinker
Foraging	Pecking or scratching on the floor or both
Exploration or locomotion	Interacting with pen walls or locomotion behaviour, such as running, walking or jumping
Comfort	Preening, wing flapping, wing stretching, feather ruffling or shaking, and elements of dustbathing behaviour
Inactive	Sitting, lying or standing while not engaged in any activity, eyes open or closed
Not visible	Any behaviour that was not identified, due to birds standing very close or in front of each other or in the shielded part of passage ways between barn sides, resulting in an unsatisfactory recording angle

Bird health condition and mortality were checked daily. Birds with severe lameness that compromised their ability to drink and feed, i.e., scores 4 and 5 [30], were culled by cervical dislocation.

2.4. Statistical Analyses

Mortality and outdoor environmental conditions such as temperature, relative humidity, air velocity, NH_3 concentration, and illuminance were analyzed by descriptive statistics. For the same environmental indicators, measured indoor and in both barn sides, linear regression models were fitted to test the main effects of house side (OAL or NAL) and age (from 1 to 6 weeks), in addition to the interaction effect. The Tukey's test for multiple comparison was used to ensure a global significance level of $p < 0.05$, and the goodness of the fitted models was assessed through residual analysis using half-normal plots with simulated bands.

Bird preference and behaviour data were analysed by mixed regression models. Total counts of birds in OAL and NAL barn sides, which means the total of birds verified in each barn side, throughout the day for each pen, and the recorded counts, were considered as the response variable. The fixed effect of chicken age and the random effects of group of birds and pens were considered. Age was categorized according to period: I (at 9, 12 and 15 d old), II (at 18, 21, 24 and 27 d old), and III (at 30, 33 and 36 d old). A binomial generalized linear mixed model was initially fitted, but the residual diagnostics clearly indicated that it was inadequate. Then, to account for overdispersion verified in this experimental data, a beta-binomial mixed regression model [31] was adopted, which are useful for analyzing discrete rates, such as the proportions of birds verified in NAL and OAL barn sides throughout the day for each pen.

For each of the remaining behavioural variables, a beta-binomial mixed regression model was also fitted. In such cases, the fixed effects of age categories, side of the barn (OAL or NAL), and the corresponding interaction effect were evaluated. The variables group of birds (birds that were reared together during all experimental period), pen (10 boxes distributed throughout the barn sides), and pen/day (the exact group of birds in each pen for a specific day of behavioural observation) were considered as random effects; this last random effect was needed as the design included the rotation of bird groups across pens, thus allowing for the study of any pen effect without the confounding effects of bird group. The fitted models were successively simplified by removing the non-significant fixed effects, starting with the interaction effect, then the main effects of age class and barn side, and considered birds rates recorded in OAL and NAL barn sides, taking into account only the total number of live birds.

The model results were summarized through the estimated probabilities and corresponding confidence intervals (CI; 95%). The estimates and standard errors for the variance components of random effects are also presented. The age categories, when statistically significant, were compared using a multiple comparison procedure with properly adjusted *p*-values.

Statistical analyses were performed using the R software 4.0.2 [32] and conclusions were based on a significance level of $p < 0.05$. The contrasts of means for environmental indicators were estimated using the emmeans package [33]. The hnp package [34] was used for the residual analysis, and the plots were produced through the ggplot2 package [35]. The PROreg package [36] was used to fit beta-binomial mixed regression models for preference and behaviour analysis.

3. Results

From 2 d, it was observed that some chickens started to move spontaneously between OAL and NAL barn sides, and from 4 d old, at least one bird in each pen had already accessed both sides of the barn. Soon afterwards, from day 6, the number of birds crossing between barn sides became high. Thus, it was not necessary to intercede or teach the birds how to move between the barn sides.

The total mortality was 9.4% (8 of 85 birds). The main cause of death, for four of the eight birds, was associated with culling due to severe lameness. Other mortality causes indicated one bird with ascites and another bird with avian infectious bronchitis; the other two birds did not have their deaths investigated, and died at 7 and 13 d. Two birds, one per pen, were relocated to maintain eight birds per pen, and this procedure occurred before the birds were 15 d old.

3.1. Environmental Measurements

The average (min to max) values for outdoor environmental conditions during data collection periods were: temperature 25.5 °C (17.0 to 31.5 °C), relative humidity 72.6% (51.0 to 99.9%), air velocity 0.7 m s^{-1} (0.0 to 3.6 m s^{-1}), illuminance 11,716 lx (2500 to >20,000 lx), and NH$_3$ concentration 1.0 ppm (0.0 to 2.0 ppm). Results for indoor environmental measures showed minimal difference, and did not differ statistically between the OAL and NAL barn sides for temperature, relative humidity, air velocity, and NH3 concentration; however, overall differences across experimental weeks were observed (Figure 3).

Illuminance was the only indoor environmental indicator with a significant effect of the interaction between barn sides and weeks. Even though overall illuminance was significantly higher in the NAL side, it is clear that it significantly increased as weeks went by in the NAL side, while it remained constant throughout the period of six weeks for the OAL side (Figure 4). This increase in illuminance occurred due to a continuous period of rain, especially in the first three weeks of our experimental period. The average (min to max) values for illuminance during all weeks were 32.4 lx (22 to 44 lx) in OAL and 545.5 lx (280 to 900 lx) in NAL.

3.2. Bird Preference and Behaviour

After the heating light was removed, from 18 d of age onwards, results showed in Figure 5 suggest that broiler chickens preferred NAL to OAL. This preference was significant for age categories II and III (Table 2). Results regarding birds' preference by age categories, not included in the tables, show that birds in period II expressed higher preference for NAL when compared with period III ($p = 0.007$). Averaged for all ages, 32.9% of the birds were seen in OAL and 67.1% in NAL.

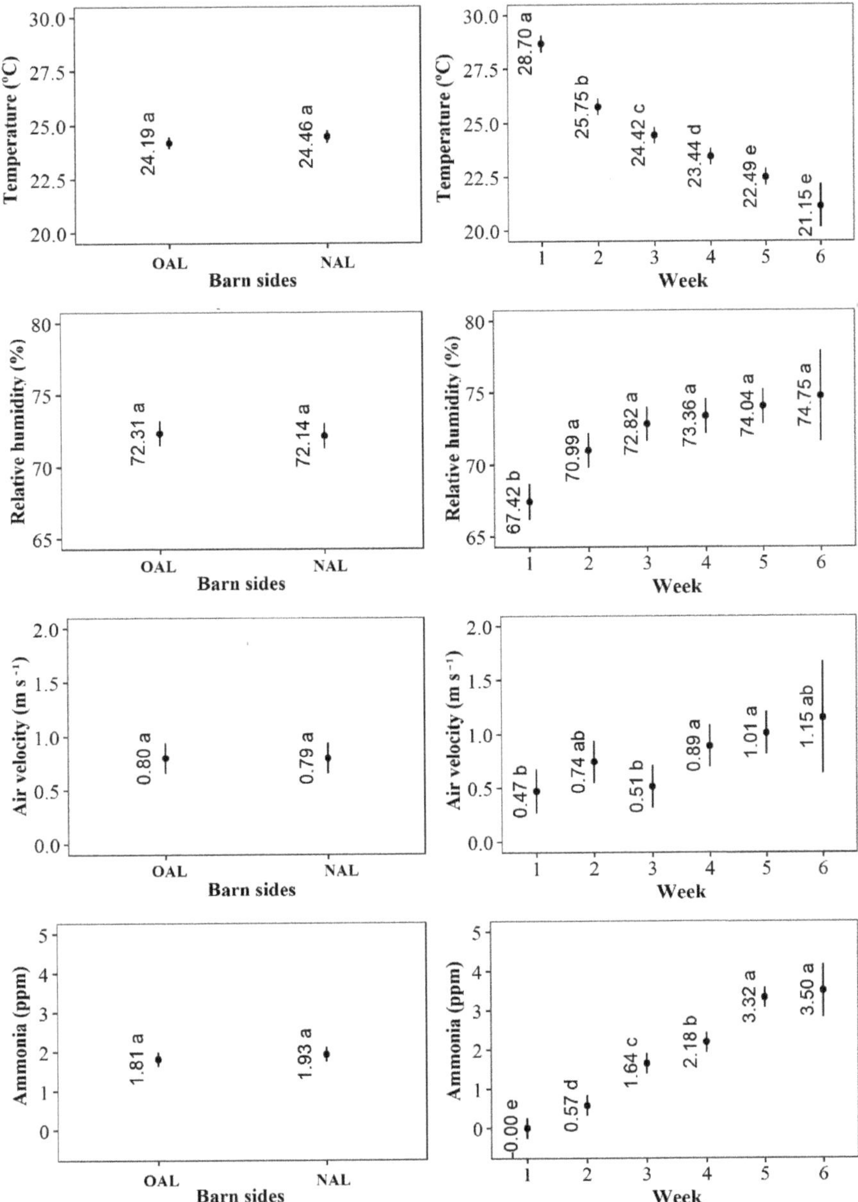

Figure 3. Estimated means and confidence intervals for indoor temperature (°C), relative humidity (%), air velocity (m s^{-1}), and ammonia concentration (ppm), for Only Artificial Light (OAL) vs Natural and Artificial Light (NAL; left column of panels), and across weeks (right column panels). Data were collected twice daily from week 1 to 6, at 10:00 a.m. and 03:00 p.m., in a preference test performed from January to February 2021, in the State of Paraná, South of Brazil; means followed by the same letters do not differ (Tukey test, $p < 0.05$).

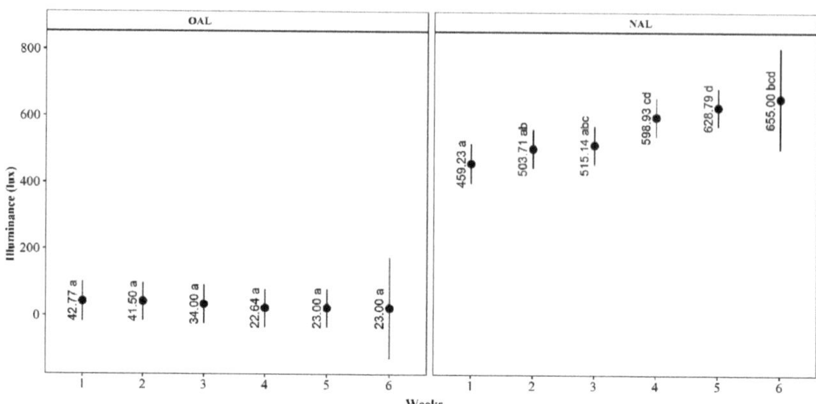

Figure 4. Means and confidence intervals for illuminance (lux) according to barn sides (OAL and NAL) across age from 1 to 6 weeks, in a preference test performed from January to February 2021, in the state of Paraná, South of Brazil; averages followed by equal letters do not differ statistically (Tukey test, $p < 0.05$).

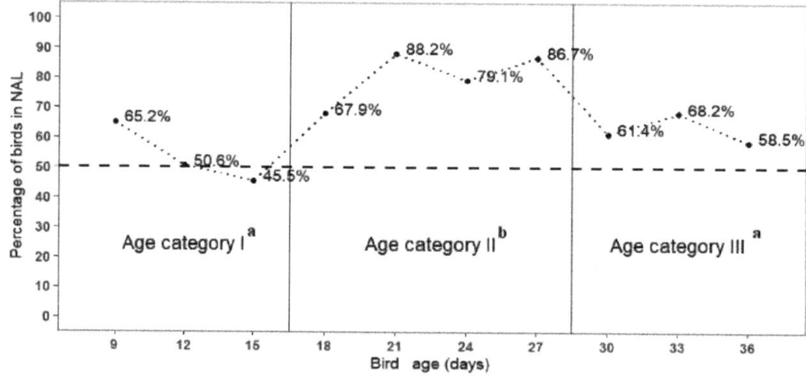

Figure 5. Percentage of broiler chickens observed in Natural and Artificial Light (NAL) barn side, according to bird age category (I at 9, 12, 15 d old; II at 18, 21, 24, 27 d old; III at 30, 33, 36 d old), in a preference test performed from January to February 2021, in the State of Paraná, South of Brazil; age categories followed by the same letters do not differ ($p < 0.05$); and dashed line indicates weekly values within age categories.

Table 2. Estimated preference probabilities and confidence intervals (CI) for Natural and Artificial Light (NAL) barn side according to bird age category, in a preference test performed from January to February 2021, in the State of Paraná, South of Brazil.

Bird Age Category		Preference		
Period	Observation Days	Estimates [1]	CI (95%)	*p*-Value [2]
I	9, 12, 15	0.538 [a]	(0.435; 0.637)	0.470
II	18, 21, 24, 27	0.803 [b]	(0.724; 0.864)	<0.001
III	30, 33, 36	0.627 [a]	(0.523; 0.719)	0.016
		∂^2group = 0.169 (0.096); ∂^2pen = 0.191 (0.090);		

[1] Different letters mean different probabilities ($p < 0.05$); [2] *p*-value for testing null hypothesis that choice of barn side is random.

Results regarding feeding and comfort behaviours showed no window effect (Table 3), but a significant effect of the age categories. The difference in frequency of feeding behaviour was significant between period I vs. period III ($p = 0.020$). The frequencies for comfort behaviour were different across all the three age categories: period I vs. period II ($p = 0.002$), period I vs. period III ($p < 0.001$), and period III vs. period II ($p = 0.036$). The presence of the window was a significant factor for drinking ($p = 0.034$) and exploration or locomotion behaviours ($p = 0.042$), which were more frequent in NAL. The category "not visible" showed higher counts in OAL ($p < 0.001$), and the only behaviour observed was "any behavior that was not identified, due to birds standing in the shielded part of passage ways between barn sides due to unsatisfactory recording angle". There was no significant effect for inactive behaviour ($p > 0.05$) and this was the most common behaviour in both OAL (47.0%) and NAL (44.6%) barn sides.

Table 3. Estimated probabilities for behaviours, according to the presence of windows (OAL vs. NAL) and broiler chicken age category (period I, II, and III), in a preference test performed from January to February 2021, in the State of Paraná, South of Brazil.

Behaviour	Effect		Window	Estimates [3]	CI (95%)
	Bird Age Category				
	Period	**Observation Days**			
Feeding	I	9, 12, 15		0.343 [a]	(0.279–0.413)
	II	18, 21, 24, 27	ns [1]	0.275 [ab]	(0.219–0.337)
	III	30, 33, 36		0.217 [b]	(0.163–0.281)
	$\hat{\sigma}^2$group = 0.065 (0.091); $\hat{\sigma}^2$pen = 0.073 (0.083); $\hat{\sigma}^2$pen/day = 0.113 (0.070)				
Comfort	I	9, 12, 15		0.034 [a]	(0.023–0.052)
	II	18, 21, 24, 27	ns [1]	0.086 [b]	(0.068–0.109)
	III	30, 33, 36		0.123 [c]	(0.097–0.156)
	$\hat{\sigma}^2$group = 0.089 (0.077); $\hat{\sigma}^2$pen = 0.367 (0.098); $\hat{\sigma}^2$pen/day = 0.057 (0.067)				
Drinking	ns [1]		OAL [2]	0.026 [a]	(0.016–0.041)
			NAL [2]	0.045 [b]	(0.035–0.059)
	$\hat{\sigma}^2$group = 0.208 (0.142); $\hat{\sigma}^2$pen = 0.379 (0.146); $\hat{\sigma}^2$pen/day = 0.535 (0.125)				
Exploration or locomotion	ns [1]		OAL [2]	0.031 [a]	(0.020–0.049)
			NAL [2]	0.053 [b]	(0.042–0.068)
	$\hat{\sigma}^2$group = 0.493 (0.158); $\hat{\sigma}^2$pen = 0.083 (0.099); $\hat{\sigma}^2$pen/day = 0.191 (0.083)				
Not visible	ns [1]		OAL [2]	0.118 [a]	(0.088–0.156)
			NAL [2]	0.035 [b]	(0.024–0.053)
	$\hat{\sigma}^2$group = 0.174 (0.125); $\hat{\sigma}^2$pen = 0.327 (0.127); $\hat{\sigma}^2$pen/day = 0.440 (0.110)				
Inactive	ns [1]		ns [1]	0.455	(0.416–0.494)
	$\hat{\sigma}^2$group = 0.053 (0.069); $\hat{\sigma}^2$pen = 0.086 (0.059); $\hat{\sigma}^2$pen/day = 0.122 (0.055)				

[1] ns = not significant; [2] OAL = Only Artificial Light; NAL = Natural and Artificial Light; [3] Different letters mean different probabilities ($p < 0.05$).

There was a significant effect for the interaction between windows and age categories for foraging behaviour (Table 4): when chickens were younger, in period I, they foraged more frequently in NAL than OAL ($p = 0.003$), while for the other two age categories, there was no difference. Considering the behaviour observed when the chickens were on the NAL side, birds in period I foraged more frequently than when they were in age category II ($p < 0.001$); the difference remained significant when birds in period I were compared with the same birds in period III ($p = 0.009$). There were no differences across the age categories when the chickens were observed in the OAL barn side.

Table 4. Estimated probabilities for foraging behaviour, according to the presence of windows (OAL vs. NAL) and broiler chicken age category, in a preference test performed from January to February 2021, in the State of Paraná, South of Brazil.

Bird Age Category		Window Presence	
Period	Observation Days	OAL [1]	NAL [1]
I	9, 12, 15	0.014 [Aa] (0.005; 0.037) [2]	0.067 [Ba] (0.045; 0.097) [2]
II	18, 21, 24, 27	0.009 [Aa] (0.002; 0.036) [2]	0.007 [Ab] (0.004; 0.016) [2]
III	30, 33, 36	0.019 [Aa] (0.006; 0.058) [2]	0.009 [Ab] (0.003; 0.027) [2]
∂^2group = 0.343 (0.242); ∂^2pen = 0.853 (0.271); ∂^2pen/day = 0.453 (0.207)			

[1] OAL = only artificial light; NAL = natural and artificial light; [2] Different capital letters refer to significant differences between barn sides ($p < 0.05$), and different low case letters indicate significant differences amongst birds' age ($p < 0.05$).

4. Discussion

In general, our results showed that, after the heating light was removed, from 18 d of age onwards, broiler chickens preferred NAL to OAL. This preference was significant for age categories II and III. The chickens spent more time drinking, exploring and moving, and foraging in NAL than OAL. Inactive (the most commonly observed behaviour), feeding, and comfort behaviour did not differ significantly between OAL vs. NAL, only according to bird age category.

Regarding birds' preference, our results are in agreement with other studies which showed that birds chose environments with higher illuminance and also expressed other changes in their behavioural repertoire due to differences in light intensity [37–40], and in our study we observed average of 32.4 lx in OAL and 545.5 lx in NAL. According to Lima and Silva [41], the absence of natural light, especially in closed-sided houses, may limit the expression of natural behaviours, with negative impacts on chicken welfare. Prescott et al. [8] strongly recommend a combination of natural daylight and artificial light for poultry barns. These considerations regarding the use of natural light are dependent on the importance of this choice for the birds themselves, with a potential to improve their welfare which tends to be proportional to the importance of natural light from the point of view of the birds. Our results especially contribute to the understanding of the birds preference, as the only internal environmental indicator that showed significant difference between OAL and NAL barn sides was illuminance. This represents an overall response of the birds to light conditions which warrants further studies, to understand the importance of other light characteristics, such as wavelength or spectrum variances. The light intensity is one of the most studied light characteristics for broiler chickens [20,37,39,40], and the bird preference for higher illuminance encouraged behaviours such as drinking, exploration or locomotion, and foraging in our study. Regarding other internal environmental measures, our results for relative humidity were not ideal, especially from the third week onwards. Even though values were close to the acceptable range between 45–70% [16], this non-compliance may be a welfare problem for the animals. However, this situation most likely did not influence the choice of birds (NAL or OAL), because relative humidity was the same on both barn sides.

Solar radiation reaching the earth surface is divided into infrared radiation, visible light, and UV; the latter is divided into three types according to wavelength: UVA (315–400 nm), UVB (280–315 nm), and UVC (100–280 nm), but 99% of the UV that reaches earth is UVA [42]. The solar radiation types that effectively reach individuals vary according to existence and type of eventual physical barriers. Tempered glass of 4 mm may block up to 28.4% of UV light from reaching the individuals [22], and 8 mm, 54.5% [23]. The glass type may also block at least 90% of wavelengths under 350 nm ([22,42]. However, windows with glass allow both visible wavelengths and a small amount of UV to pass to inside the houses [11] and, thus, alter chicken behaviour [43]. In the NAL barn side, birds may have received UV light that was not available in the OAL side. This may have motivated their preference, as poultry have a fourth retinal cone photoreceptor that allows them to see in the UVA wavelength (315–400 nm) [19]. Birds exposed to some UV light may have decreased stress susceptibility and fear responses than those raised without UV [12], showing that the illuminance of poultry houses can be improved in several aspects.

Regarding lamp types, LED bulbs with colour temperatures over 5000 K, called cold [44], contain more blue than warm white light [45], and in our study 6500 K lamps were used in both NAL and OAL barn sides. Understanding light temperature is relevant to our study as, in addition to light intensity, birds may also choose specific colour temperature [9,18]. In our study, the OAL light source was similar in terms of colour temperature, as measured in degrees of Kelvin (K), to the average daylight that birds were searching for by moving to the NAL side of the barn, suggesting that the light intensity, as measured in Lux (lx), may have been the main driver for the preference.

New lighting technologies are currently being developed as potential replacements for incandescent light sources, and some sources may be better to the welfare of broiler chickens [46]. However, our results suggest that the exposure to natural lighting may be an ideal solution according to the preference of the birds. This warrants further preference studies with different types of artificial light bulbs, as well as asking the birds how strong their preference is, through motivation tests. Considering the higher visual perception capacities that birds have as compared to humans, it seems relevant to explore light characteristics in addition to intensity to better understand what the birds are responding to when they express their preferences. In future research, the real perception of birds in relation to illuminance may be further studied. Although the differences in perceived light intensity by birds, known as Clux or Gallilux, may be estimated by adding between 20–25% in relation to lx, i.e., 25 Clux = 17.4 lx [47–49], it is important to study light from a bird perspective with more precision technology.

Bird preferences may be influenced not only by barn sides and their characteristics regarding light, but by their natural behaviours [24]. A special consideration is that chickens are social animals, and bird preferences may be influenced not only by individual choices, but also by their social nature and its effects, such as social facilitation [8,29,50]. Because of social facilitation, the birds tend to behave as a social unit, where most members exhibit the same behaviour at the same point in time [5]. Thus, the higher number of chickens in NAL side may have acted as an additional force for more birds to migrate to this side.

Bateson and Seanurne-Way [51] suggested that when birds were exposed to constant light, the elicitation of social behaviour became more likely. Our results seem to reinforce the statement that a place with higher illuminance fosters group formation that may be positive for the animals. Recognition between individuals is also part of the social interaction process, and this characteristic may be affected when birds are reared in very low illuminance [3,7,18]. Although in our study we have not observed any aggressive behaviour among birds, according to Porter et al. [52], chicks that had been housed in pairs in the dark showed no evidence that they discriminated between familiar and unfamiliar test partners. Thus, the NAL side may also have provided a better recognition of individual birds and, consequently, this may be potentially considered an additional factor explaining bird choice. Collins et al. [53] reinforce the importance of vision in key behaviours such as feeding and social behaviour in poultry, and suggest that the birds may experience lower welfare as a result of their lack of sight. Therefore, when birds choose the NAL barn side, they may be making choices to favour their natural social interaction behaviours.

The birds spent a considerable proportion of their time in the OAL barn side, and this choice should also be considered. Although higher light intensity has been associated with increases in activity levels and improvement in leg health of commercial broiler chickens [11], birds should have access to different types of illuminance, so that they can choose according to their preferences. As it is recognized that enrichment strategies should be provided for chickens [16], illuminance may follow the same principle. An adequate density associated with the availability of different types of light environments may reduce bird crowding, preserving their safety and health. Thus, the illuminance distribution must be adequate, avoiding a contrast between the lightest and darkest points of more than 20% [54]. The provision of areas with reduced light intensity for resting and other activities has been suggested before [40]. On farms where windows are provided, resting behaviour occurs more often in areas with lower light intensities, whereas active behaviours occur in areas with higher light intensities, but it is up to the birds to choose [55]. This pattern of light intensity choices is expected for diurnal animal species. Vergneau-Grosset and Peron [56] recommended that when exposing an animal to UV light, it is important to provide a hiding place or shade; in our study, both the passage way between each barn side and the OAL side may have fulfilled this function. Other negative effects of excess UV radiation intensity may be observed in both natural and artificial light sources, and revolve around the occurrence of burns in animals and behavioral changes, such as increased stress or incidence of severe feather pecking [56,57]. During the experimental period, none of these characteristics were observed in our birds, which agrees with the probably low UV exposure through the glass window.

In our study, the birds showed preference for the NAL barn side only from 18 d onwards. The association of this preference with bird age was also observed during other preference test, when chicks spent most time in the brightest light (200 lx) at 2 weeks, and at 6 week the birds preferred the environment with dimmest light (6 lx) [37]. The age for birds to begin expressing light preferences coherently coincides with the total removal of the heating lamps. Although this type of lamp is not suitable for lighting, it was responsible for adding up to 25 lx in each pen, which may have acted as an important confounding effect for birds to detect the lighting differences between barn sides. In addition, according to Gunnarsson et al. [29], early exposure to natural or artificial light might have an effect on later preference for light type and on the behaviour of the birds, even after a house transition. Therefore, the birds may have grown habituated with the illuminance from the heating lamps and, after their removal, they may have been obliged to make new choices, as their early life light experience became absent. In addition, the heating lamps may have provided an early imprinted association between light intensity and heat, reinforcing a positive perception of light by the birds. Even though it was not possible to identify the exact reason for bird preference for the NAL barn side, most possible explanations seem coherent with the more natural characteristic of the lighting on this side of the barn. Our hypothesis is that the windows tend to be closer to meeting the birds' basic needs in relation to light and, thus, tend to increase animal welfare. Examples of such needs include the establishment and maintenance of social hierarchies, social encounters, group aggregation and peer recognition.

Results regarding chicken behaviours showed that the frequencies varied according to barn sides (drinking, exploration and locomotion, and not visible), and bird age categories (feeding and comfort). The behaviours of drinking and exploration and locomotion showed higher frequencies in the NAL side, and the category of not visible birds was more frequent in the OAL side. Davis et al. [37] observed that broiler chicks performed more feeding, drinking, and locomotion behaviours in the brighter environments. However, for Deep et al. [58] light intensity had no effect on expression of drinking behaviour. Adding further evidence to this discussion, our results indicate that providing windows increases the behaviour repertoire, a fact observed in previous studies. Sans et al. [13,14] observed that broiler chickens reared in open-sided houses, with natural light provided by no-glass windows, but with curtains during summer/autumn, showed higher relative frequencies for exploration behaviour when compared with birds in closed-sided houses; during the winter, there was a higher frequency for drinking and a lower inactivity. Thus, our results suggest that, even with eventual changes in natural daylight characteristics due presence of glass in the windows, it remains possible to observe a potential improve in bird welfare as the increase of activities considered important for the birds, i.e., social activities in the NAL barn side. Furthermore, windowed industry barns in Brazil do not fit glass barriers, and this was an experimental resource to control for other in-barn environmental conditions such as temperature and relative humidity, in order to study the specific effect of lighting. These results reinforce that, when given the opportunity, birds prefer to perform their behaviours in an environment with natural daylight or, minimally, higher levels of illuminance than those provided inside the barns with only artificial lighting.

The exploratory or locomotion behaviours, observed in higher frequency in NAL side, tend to be viewed as positive behaviours, because may increase the birds' activity and improve the interactions between the birds and their environment [1]. However, if the house is not stimulating, as birds age they may get bored and reduce exploratory behaviour [1]. It seems important that broiler chickens are reared in stimulating poultry houses, with adequate lighting characteristics that allow for the birds to perform activities which are essential for their welfare.

Foraging was the only behaviour for which a significant interaction effect between window and age categories was present, indicating that birds foraged more, when were younger, in the NAL than the OAL barn side. One of the reasons for the interaction with bird age may be the appearance of locomotor problems which tend to become more severe as birds age, in addition to increasing body weights and non-stimulating environments [16,51]. Alvino et al. [4] also observed that foraging was affected by light intensity, and broiler chickens in the 5 lx treatment spent significantly less time performing this behaviour than when the light intensity was 50 and 200 lx. Foraging, exploration, or locomotion are important behaviours, since they involve actions related to knowing the environment and searching for feed [16]. According to Manser [7], newly hatched birds, both domestic poultry and turkeys, may die of malnutrition if they have difficulty in seeing the feeders due low light intensity, which may reduce overall activity, reducing the chances of foraging, finding a feeder, and learning how to feed. Although this describes an extreme situation, it demonstrates the importance of adequate lighting from the first days of birds' life, so that they can enjoy the opportunity to explore the environment, the other birds, and the resources available.

Inactive behaviour was not different between barn sides or across different age categories. According to some studies, this behaviour may be associated with increased bird age, walking ability deterioration, body weight, and fast growth rates [11,51,59]. Although our study did not test the birds' walking ability, the number of culls regarding leg problems suggests that this problem was prevalent, causing suffering and pain to the birds, as well as limiting their behavioural repertoire.

Although light is an important element for birds, when provided in isolation, it may not be enough to reduce inactive behaviour. According to El-Deek and El-Sabrout [60], most of intensive production systems that are currently used do not usually support the natural behavioural needs of poultry. Therefore, farm animals may be reared in an environmental with enrichment and light which more closely resembles their natural characteristics. These options, acting together, may increase activity, improve leg health [11,61], and stimulate behaviours such as foraging and exploration [43]. However, selection for fast growth may lead to several welfare problems, such as metabolic disorders, decrease locomotor activity, and extend time spent sitting or lying [51]. For European Food Safety Authority (EFSA) [62], the risk assessment regarding poor welfare effects showed that fast growth is one of the major risk scores, including unbalanced body conformation, high stocking density, wet litter, and light intensity. Including slower-growing genetic strains may be a way to decrease welfare restrictions [51,63], adding to important environmental changes to indoor houses to meet the birds' needs in the current poultry industry.

In general, animals engaged in pleasant activities, such as exploring, feeding, and interacting with other animals in a social group, may experience positive feelings, and without this engagement, the animal will not experience the full range of positive welfare states that are potentially available [64]. Although in our study, a qualitative behavioral assessment [30] was not used, it is likely that birds were more likely to experience positive feelings while they were in the NAL barn side, due to higher opportunities to increase behaviours that are more active, such as exploring, foraging, moving, and interacting with other birds.

Some behaviours were only also associated with age, such as comfort and feeding. Comfort behaviours were associated with increases in bird age categories. In the literature, this behaviour is associated to increases in chicken welfare, as the activities may related to the maintenance of bird health [65]. Alvino et al. [4] observed that broiler chickens reared in 5 lx spent less time in preening behaviour, as compared to those in 50 and 200 lx. The increase of comfort may be understood as positive results, indicating a possibility to encourage higher expression of behaviours associated to increases in chicken welfare. Although we only observed difference in feeding frequency regarding bird age categories, some authors observed a clear preference of laying hens and broiler chickens to eat in brighter lightings, from 20 to 200 lx [38,40], and that they ate more under 30 lx than 1 lx [66]. Birds may also find it aversive to eat in very dim light, because this behaviour is normally guided visually, and they see better in brighter environments [20,38]. Although feeding behaviour decreased with age in our study, no emaciated chicken was observed during the experimental period and feeding showed the second highest frequency, only behind inactive behaviour. It is also important to consider that, when the birds are foraging, they may also searching for feed [16], which may explain the lower number of visits to the feeder in the NAL side.

As for the "not visible" behavioural category, when the birds were younger, some of them stayed together in the passage between the barn sides, which may have given an enhanced sense of social interaction or protection. As birds aged, they may also have been looking for a different lighting, according to specific momentaneous needs. Birds observed in OAL spent less time in exploration, moving, and foraging, and when observed in this barn side, stayed lying very close or in front of each other, which also prevented appropriate behavioural identification. Thus, a potential reason for finding more birds in the not visible behaviour category in the OAL side may be an association between seeking an environment with lower light intensities and pen areas associated with a feeling of protection, provided by staying either close to wall angles in the passageways or close to another bird. Such potential reasons seem to indicate that the OAL side was chosen by the birds when they were searching for a cozy place to either rest or sleep.

5. Conclusions

For our experimental conditions, the chickens preferred natural and artificial lighting from 18 d of age onwards, when the confounding effect of the heating light was removed, and their behavioural repertoire was also different according to each side of the barn and to their ages. As the chickens also used the lower lit pen areas, barns with light gradient options seem important for them. In summary, the birds indicated that windows make a relevant difference in their indoor lives, as it is what they choose when the only other option is the same in-barn environment with only artificial lighting. Further preference studies are warranted to understand the potential effects of geographical, seasonal, climatic and genetic variations, amongst others.

Author Contributions: Conceptualization E.C.d.O.S., F.A.M.T. and C.F.M.M.; methodology, E.C.d.O.S. and C.F.M.M.; investigation, E.C.d.O.S., M.M.V. and C.F.M.M.; formal analysis, C.A.T.; writing—original draft preparation, E.C.d.O.S., F.A.M.T. and C.F.M.M.; writing—review and editing, E.C.d.O.S., F.A.M.T., C.A.T., A.S.P., M.M.V. and C.F.M.M.; supervision, C.F.M.M.; project administration, E.C.d.O.S. and C.F.M.M.; funding acquisition, E.C.d.O.S. and C.F.M.M. All authors have read and agreed to the published version of the manuscript.

Funding: This work was supported by the World Animal Protection. Also, this study was financed in part by Coordenação de Aperfeiçoamento de Pessoal de Nível Superior—Brasil (CAPES) (funding number: 001).

Institutional Review Board Statement: The experiment in this study was approved by the Animal Use Ethics Committee of the Agricultural Campus, N° 104/2017, Federal University of Paraná.

Data Availability Statement: Data are available upon request from the corresponding author.

Acknowledgments: We are grateful for all employees of the Federal University of Paraná farm. We are grateful for the valuable suggestions from the reviewers and the academic editor.

Conflicts of Interest: The authors declare no conflict of interest.

References

1. Newberry, R.C. Exploratory behaviour of young domestic fowl. *Appl. Anim. Behav. Sci.* **1999**, *63*, 311–321. [CrossRef]
2. Collias, N.E.; Collias, E.C. Social organization of a red junglefowl, *Gallus gallus*, population related to evolution theory. *Anim. Behav.* **1996**, *51*, 1337–1354. [CrossRef]
3. Prescott, N.B.; Wathes, C.M. Reflective properties of domestic fowl (*Gallus g. domesticus*), the fabric of their housing and the characteristics of the light environment in environmentally controlled poultry houses. *Br. Poult. Sci.* **1999**, *40*, 185–193. [CrossRef] [PubMed]
4. Alvino, G.M.; Archer, G.S.; Mench, J.A. Behavioural time budgets of broiler chickens reared in varying light intensities. *Appl. Anim. Behav. Sci.* **2009**, *1*, 54–61. [CrossRef]
5. Paranhos da Costa, M.J.R.; Lima, V.A.; Sant'anna, A.C. Comportamento e Bem-Estar Animal. In *Fisiologia Das Aves Comerciais*; Macari, M., Maiorka, A., Eds.; Funep: Jaboticabal, Brazil, 2017; pp. 605–646.
6. Kristensen, H.H.; Prescott, N.B.; Perry, G.C.; Ladewig, J.; Ersbøll, A.K.; Overvad, K.C.; Wathes, C.M. The behaviour of broiler chickens in different light sources and illuminances. *Appl. Anim. Behav. Sci.* **2007**, *103*, 75–89. [CrossRef]
7. Manser, C.E. Effects of lighting on the welfare of domestic poultry: A review. *Anim. Welf.* **1996**, *5*, 341–360.
8. Prescott, N.B.; Kristensen, H.H.; Wathes, C.M. Light. In *Measuring and Auditing Broiler Welfare*; Weeks, S., Butterworth, A., Eds.; CABI Publishing Int.: Wallingford, UK, 2004; pp. 101–116.
9. Kristensen, H.H.; Perry, G.C.; Prescott, N.B.; Ladewig, J.; Ersbøll, A.K.; Wathes, C.M. Leg health and performance of broiler chickens reared in different light environments. *Br. Poult. Sci.* **2006**, *47*, 257–263. [CrossRef]
10. Kumar, V.S. Avian photoreceptors and their role in the regulation of daily and seasonal physiology. *Gen. Comp. Endocrinol.* **2015**, *220*, 13–22. [CrossRef]
11. Bailie, C.L.; Ball, M.E.E.; O'Connell, N.E. Influence of the provision of natural light and straw bales on activity levels and leg health in commercial broiler chickens. *Animal* **2013**, *7*, 618–626. [CrossRef]
12. House, G.M.; Sobotik, E.B.; Nelson, J.R.; Archer, G.S. Effect of the addition of ultraviolet light on broiler growth, fear, and stress response. *J. Appl. Poult. Res.* **2020**, *29*, 402–408. [CrossRef]
13. Sans, E.C.O.; Tuyttens, F.A.M.; Taconeli, C.A.; Rueda, P.M.; Ciocca, J.R.; Molento, C.F.M. Welfare of broiler chickens reared under two different types of housing. *Anim. Welf.* **2021**, *30*, 341–353. [CrossRef]
14. Sans, E.C.O.; Tuyttens, F.A.M.; Taconeli, C.A.; Rueda, P.M.; Ciocca, J.R.; Molento, C.F.M. Welfare of broiler chickens reared in two different industrial house types during the winter season in Southern Brazil. *Br. Poult. Sci.* **2021**, *62*, 621–631. [CrossRef]
15. James, C.; Asher, L.; Herborn, K.; Wiseman, J. The effect of supplementary ultraviolet wavelengths on broiler chicken welfare indicators. *Appl. Anim. Behav. Sci.* **2018**, *209*, 55–64. [CrossRef]
16. De Jong, I.; Berg, C.; Butterworth, A.; Estevéz, I. External Scientific Report—Scientific Report Updating the European Foof Safety Authority Opinions on the Welfare of Broilers and BROILER breeders. Supporting Publications 2012, EN-295, 116p. Available online: www.efsa.europa.eu/publications (accessed on 20 February 2021).
17. RSPCA. *RSPCA Welfare Standards for Meat Chickens*; Royal Society for the Prevention of Cruelty to Animals: Horsham, UK, 2017; 106p.
18. Prescott, N.B.; Wathes, C.M.; Jarvis, J.R. Light, vision and the welfare of poultry. *Anim. Welf.* **2003**, *12*, 269–288.
19. Prescott, N.B.; Wathes, C.M. Spectral sensitivity of the domestic fowl (*Gallus gallus domesticus*). *Br. Poult. Sci.* **1999**, *40*, 332–339. [CrossRef] [PubMed]
20. Blatchford, R.A.; Klasing, K.C.; Shivaprasad, H.L.; Wakenell, P.S.; Archer, G.S.; Mench, J.A. The effect of light intensity on the behavior, eye and health, and immune function of broiler chicken. *Poult. Sci.* **2009**, *88*, 20–28. [CrossRef] [PubMed]
21. Bennett, A.T.D.; Cuthill, I.C. Ultraviolet vision in birds: What is its function? *Vis. Res.* **1994**, *34*, 1471–1478. [CrossRef]
22. Duarte, I.; Rotter, A.; Malvestiti, A.; Silva, M. The role of glass as a barrier against the transmission of ultraviolet radiation: An experimental study. *Photodermatol. Photoimmunol. Photomed.* **2009**, *25*, 181–184. [CrossRef] [PubMed]
23. Silva, S.K.; Batista, M.L.; Porfiro, L.D. Study of blocking efficiency of ultraviolet radiation by the main types of glass used in civil construction. *Rev. Anápolis Dig.* **2020**, *11*, 126–145.
24. Fraser, D.; Matthews, L.R. Preference and motivation testing. In *Animal Welfare*; Appleby, M.C., Hughes, B.O., Eds.; CAB International: New York, NY, USA, 1997; pp. 159–173.
25. Nicol, C.J.; Caplen, G.; Edgar, J.; Browne, W.J. Associations between welfare indicators and environmental choice in laying hens. *Anim. Beh.* **2009**, *78*, 423–424. [CrossRef]
26. Browne, W.J.; Caplen, G.; Edgar, J.; Wilson, L.R.; Nicol, C.J. Consistency, transitivity and inter-relationships between measures of choice in environmental preference tests with chickens. *Behav. Process.* **2010**, *83*, 72–78. [CrossRef] [PubMed]
27. Souza, A.P.O.; Molento, C.F.M. Good agricultural practices in broiler chicken production in the state of Paraná: Focus on animal welfare. *Cienc. Rural* **2015**, *45*, 2239–2244. [CrossRef]
28. Martin, P.; Bateson, P. *Measuring Behaviour: An Introductory Guide*, 2nd ed.; Cambridge University Press: Cambridge, UK, 1993.

29. Gunnarsson, S.; Heikkilä, M.; Hultgren, J.; Valros, A. A note on light preference in layer pullets reared in incandescent or natural light. *Appl. Anim. Beh. Sci.* **2008**, *112*, 395–399. [CrossRef]

30. Welfare Quality. *Welfare Quality® Assessment Protocol for Poultry (Broilers, Laying Hens)*; Welfare Quality® Consortium: Lalystad, The Netherlands, 2009.

31. Najera-Zuloaga, J.; Lee, D.J.; Arostegui, I. A beta-binomial mixed-effects model approach for analysing longitudinal discrete and bounded outcomes. *Biom. J.* **2019**, *61*, 600–615. [CrossRef] [PubMed]

32. R Core Team. *R: A Language and Environment for Statistical Computing*; R Foundation for Statistical Computing: Vienna, Austria, 2020; Available online: https://www.R-project.org/ (accessed on 14 March 2021).

33. Lenth, R. Emmeans: Estimated Marginal mMeans, aka Least-Squares Means. 2020. Available online: https://cran.r-project.org/web/packages/emmeans/index.html (accessed on 10 March 2021).

34. Moral, R.A.; Hinde, J.; Demétrio, C.G.B. Half-normal plots and overdispersed models in R: The hnp package. *J. Stat. Softw.* **2017**, *81*, 1–23. [CrossRef]

35. Wickham, H. *Ggplot2: Elegant Graphics for Data Analysis*; Springer: New York, NY, USA, 2016.

36. Najera-Zuloaga, J.; Lee, D.J.; Arostegui, I. PROreg: Patient Reported Outcomes Regression Analysis. R Package Version 1.1. 2020. Available online: https://CRAN.R-project.org/package=PROreg (accessed on 20 March 2021).

37. Davis, N.J.; Prescott, N.B.; Savory, C.J.; Wathes, C.M. Preference of growing fowls for different light intensities in relation to age, strain and behaviour. *Anim. Welf.* **1999**, *8*, 193–203.

38. Prescott, N.B.; Wathes, C.M. Preference and motivation of laying hens to eat under different illuminances and the effect of illuminance on eating behaviour. *Br. Poult. Sci.* **2002**, *43*, 190–195. [CrossRef]

39. Rault, J.L.; Clark, K.; Groves, P.J.; Cronin, G.M. Light intensity of 5 or 20 lux on broiler behavior, welfare and productivity. *Poult. Sci.* **2017**, *96*, 779–787. [CrossRef]

40. Raccoursier, M.; Thaxton, Y.V.; Christensen, K.; Aldridge, D.J.; Scanes, C.G. Light intensity preferences of broiler chickens: Implications for welfare. *Animal* **2019**, *13*, 2857–2863. [CrossRef]

41. Lima, V.A.; Silva, I.J.O. Aves. In *O Bem-Estar Animal no Brasil e na Alemanha: Responsabilidade e Sustentabilidade*; Hartung, J., Paranhos da Costa, M.J.R., Perez, C., Eds.; GRAFTEC Gráfica e Editora Ltda.: São Paulo, Brazil, 2019; pp. 117–124.

42. Lewis, P.D.; Gous, R.M. Responses of poultry to ultraviolet radiation. *World's Poult. Sci. J.* **2009**, *65*, 499–510. [CrossRef]

43. De Jong, I.C.; Gunnink, H. Effects of a commercial broiler enrichment programme with or without natural light on behaviour and other welfare indicators. *Animal* **2018**, *13*, 384–391. [CrossRef] [PubMed]

44. Archer, G.S. Color temperature of light-emitting diode lighting matters for optimum growth and welfare of broiler chickens. *Animal* **2018**, *12*, 1015–1021. [CrossRef]

45. Sultana, S.; Hassan, M.R.; Choe, H.S.; Ryu, K.S. The effect of monochromatic and mixed LED light colour on the behaviour and fear responses of broiler chicken. *Avian Biol. Res.* **2013**, *6*, 207–214. [CrossRef]

46. Olanrewaju, H.A.; Miller, W.W.; Maslin, W.R.; Collier, S.D.; Purswell, J.L.; Branton, S.L. Effects of light sources and intensity on broilers grown to heavy weights. Part 1: Growth performance, carcass characteristics, and welfare indices. *Poult. Sci.* **2016**, *95*, 727–735. [CrossRef] [PubMed]

47. Lewis, P.D.; Morris, T.R. Poultry and coloured light. *World's Poult. Sci. J.* **2000**, *56*, 189–207. [CrossRef]

48. Olanrewaju, H.A.; Miller, W.W.; Maslin, W.R.; Collier, S.D.; Purswell, J.L.; Branton, S.L. Influence of light sources and photoperiod on growth performance, carcass characteristics, and health indices of broilers grown to heavy weights. *Poult. Sci.* **2018**, *97*, 1109–1116. [CrossRef] [PubMed]

49. Olanrewaju, H.A.; Purswell, J.L.; Collier, S.D.; Branton, S.L. Effect of light intensity adjusted for species-specific spectral sensitivity on blood physiological variables of male broiler chickens. *Poult. Sci.* **2019**, *98*, 1090–1095. [CrossRef] [PubMed]

50. Bessei, W. Welfare of broilers: A review. *World's Poult. Sci. J.* **2006**, *62*, 455–466. [CrossRef]

51. Bateson, P.P.G.; Seaburne-May, G. Effects of prior exposure to light on chicks' behaviour in the imprinting situation. *Anim. Behav.* **1973**, *21*, 720–725. [CrossRef]

52. Porter, R.H.; Roelofsen, R.; Picard, M.; Arnould, C. The temporal development and sensory mediation of social discrimination in domestic chicks. *Anim. Behav.* **2005**, *70*, 359–364. [CrossRef]

53. Collins, S.; Forkman, B.; Kristensen, H.H.; Sandøe, P.; Hocking, P.M. Investigating the importance of vision in poultry: Comparing the behaviour of blind and sighted chickens. *Appl. Anim. Beh. Sci.* **2011**, *133*, 60–69. [CrossRef]

54. Rutz, F.; Silva, F.H.A.; Nunes, J.K. Fundamentos de um programa de luz para frangos de corte. In *Produção de Frangos de Corte*; Macari, M., Mendes, A.A., Menten, J.F., Nääs, I.A., Eds.; FACTA: Campinas, Brazil, 2014; pp. 225–250.

55. Souza da Silva, C.; De Jong, I. *Literature Update on Effective Environmental Enrichment and Light Provision in Broiler Chickens*; Report 1204; Wageningen Livestock Research: Wageningen, The Netherlands, 2019.

56. Vergneau-Grosset, C.; Péron, F. Effect of ultraviolet radiation on vertebrate animals: Update from ethological and medical Perspectives. *Photochem. Photobiol. Sci.* **2020**, *19*, 752–762. [CrossRef]

57. Ruis, M.A.W.; Reuvekamp, B.F.; Gunnink, H.; Binnendijk, G.P. *The Effect of Optimized Lighting Conditions on Feather Pecking and Production of Laying Hens*; Report 335; Wageningen Livestock Research: Wageningen, The Netherlands, 2010. Available online: https://library.wur.nl/WebQuery/wurpubs/fulltext/137033 (accessed on 8 March 2021).

58. Deep, A.; Schwean-Lardner, K.; Crowe, T.G.; Fancher, B.I.; Classen, H.L. Effect of light intensity on broiler behaviour and diurnal rhythms. *Appl. Anim. Behav. Sci.* **2012**, *136*, 50–56. [CrossRef]

59. Weeks, C.A.; Danbury, T.D.; Davies, H.C.; Hunt, P.; Kestin, S.C. The behaviour of broiler chickens and its modification by lameness. *Appl. Anim. Behav. Sci.* **2000**, *67*, 111–125. [CrossRef]
60. El-Deek, A.; El-Sabrout, K. Behaviour and meat quality of chicken under different housing systems. *World's Poult. Sci. J.* **2019**, *75*, 1–9. [CrossRef]
61. Sherwin, C.M.; Lewis, P.D.; Perry, G.C. The effects of environmental enrichment and intermittent lighting on the behaviour and welfare of male domestic turkeys. *Appl. Anim. Behav. Sci.* **1999**, *62*, 319–333. [CrossRef]
62. EFSA—European Food Safety Authority. Panel on Animal Health and Welfare (AHAW): Scientific Opinion on the influence of genetic parameters on the welfare and the resistance to stress of commercial broilers. *EFSA J.* **2010**, *8*, 38–48. [CrossRef]
63. Hartcher, K.M.; Lum, H.K. Genetic selection of broiler and welfare consequences: A review. *World's Poult. Sci. J.* **2019**, *76*, 154–167. [CrossRef]
64. Mellor, D.J. Positive animal welfare states and reference standards for welfare assessment. *N. Z. Vet. J.* **2014**, *63*, 17–23. [CrossRef]
65. Appleby, M.C.; Mench, J.A.; Hughes, B.O. *Poultry Behaviour and Welfare*; CABI: Wallingford, UK, 2004.
66. Li, G.; Li, B.; Shi, Z.; Zhao, Y.; Tong, Q.; Liu, Y. Diurnal rhythms of group-housed layer pullets with free choices between light and dim environments. *Can. J. Anim. Sci.* **2020**, *100*, 37–46. [CrossRef]

Article

Structuring Broiler Barns: How a Perforated Flooring System Affects Animal Behavior

Franziska May [1,*], Jenny Stracke [2], Sophia Heitmann [3], Carolin Adler [4], Alica Krasny [1], Nicole Kemper [1] and Birgit Spindler [1]

[1] Institute for Animal Hygiene, Animal Welfare and Farm Animal Behavior, University of Veterinary Medicine Hannover, Foundation, 30173 Hannover, Germany; alica.krasny@tiho-hannover.de (A.K.); nicole.kemper@tiho-hannover.de (N.K.); birgit.spindler@tiho-hannover.de (B.S.)
[2] Institute of Animal Science, Ethology, University of Bonn, 53115 Bonn, Germany; jenny.stracke@itw.uni-bonn.de
[3] Public Veterinary Service, Animal Health and Animal Husbandry, Schleswig-Holstein, 23795 Bad Segeberg, Germany; s.heitmann@segeberg.de
[4] Department of Animal and Poultry Science, University of Saskatchewan, Saskatoon, SK S7N 5A8, Canada; carolin.adler@usask.ca
* Correspondence: franziska.may@tiho-hannover.de

Citation: May, F.; Stracke, J.; Heitmann, S.; Adler, C.; Krasny, A.; Kemper, N.; Spindler, B. Structuring Broiler Barns: How a Perforated Flooring System Affects Animal Behavior. *Animals* **2022**, *12*, 735. https://doi.org/10.3390/ani12060735

Academic Editors: Victoria Sandilands and Tina Widowski

Received: 4 February 2022
Accepted: 14 March 2022
Published: 15 March 2022

Publisher's Note: MDPI stays neutral with regard to jurisdictional claims in published maps and institutional affiliations.

Simple Summary: Broiler chickens in Europe are usually raised in littered barns without structuring elements. Previous studies have found a positive influence on the health and welfare of broiler chickens when they have access to elevated platforms. This study aimed to evaluate the effect of an elevated perforated floor on the behavior of broiler chickens. Therefore, one of two barns was equipped with a perforated floor under the food and water supply. The second barn was used as a control. In total, three fattening periods were observed, with 500 broiler chickens kept in each barn. To compare the behavior of the birds, cameras were installed in both barns. The videos were analyzed by counting the number of birds in a defined area and observing focal animals continuously while recording their behavior. More animals were observed on the perforated floor than in the littered control area, but, in total, the focal animals spent less time on the perforated floor compared to the observed littered area in the control barn. There were no differences in the length of the recorded behaviors between the treatments. These findings suggest that, in general, the elevated perforated floor is attractive for the animals. However, it does not promote one of the recorded behavior patterns. Our results show that an elevated perforated floor could be an option for structuring broiler barns.

Abstract: Broiler chickens in Europe are usually raised in a barren environment. Elevated perforated platforms address this problem and can positively influence animal health and welfare. To evaluate the effect of an elevated perforated floor on the behavior of broiler chickens, one of two barns was equipped with a perforated flooring system under the food and water supply. The second barn was used as a control. In total, three fattening periods were observed, with 500 broiler chickens (Ross 308 breed) kept in each barn. To compare the behavior of the birds in these groups, cameras were installed in the two barns. The videos were analyzed by counting the number of birds and observing focal animals while recording their behavior. More animals were observed on the perforated floor than in the littered control area ($p < 0.001$), but focal animals spent less time on the perforated floor compared to the observed littered area in the control barn ($p < 0.05$). There were no differences in the length of the recorded behaviors between the treatments. These findings suggest that, in general, the elevated perforated floor is attractive for the animals. However, it does not promote one of the recorded behavior patterns. Our results show that an elevated perforated floor could be an option for structuring broiler barns.

Keywords: broiler; behavior; perforated floor; elevated platform; animal welfare

---

1. Introduction

According to council directive 2007/43/EC, fast-growing broiler chickens in Europe are usually kept on littered concrete floors. This results in barren environments on many farms, providing no structural elements besides feeding and water lines. This lack of environmental complexity is a concern of animal welfare [1], since broilers are not able to perform highly motivated behavior patterns such as perching, which is presumed to lead to frustration and a negative emotional state [2,3]. Further, a barren environment provides no stimuli for activity, which can be one factor for health problems such as leg abnormalities [4]. Providing elevated platforms can be an option to address this problem. They can encourage higher activity [5] and positively impact leg health [6]. Moreover, separating the animals from manure using elevated platforms with a perforated surface has been reported to improve foot pad health [7–9]. A pioneer in this field is Switzerland, providing access to elevated platforms to over 90% of broiler chickens by raising them under the animal welfare label 'BTS' [10]. The Dutch 'Better Life' label also considers elevated structures, for example, straw bales, as crucial [11].

Like their ancestor, the red jungle fowl, domestic chickens are highly motivated to perch on elevated structures such as branches [12]. In husbandry systems for laying hens, elevated structures are an integral part. It is known that the lack of perches leads to frustration in laying hens and has been suggested to reduce animal welfare [2]. Previous research has demonstrated that broilers are also motivated to perch when given the opportunity. However, they prefer elevated platforms to perches [13–15], which might be due to their relatively greater body weight [14,15]. Laying hens always choose the highest structure available for perching [16]. Interestingly, Malchow et al. [15] found that mixed-sex fast-growing-broiler chickens at the end of the fattening period also preferred the highest platform tested in their study, which was 50 cm high. However, in the following study, Malchow et al. [5] reported that fast-growing male broilers were observed more often on the lowest level of the platforms (10 cm), presumably because they were heavier than the chickens in their earlier study. A higher body weight, lower leg health, and therefore a decline in activity are suggested to cause a decrease in perching [5,13,14]. It has been observed that installing ramps at an angle of about 15°–35° is a possibility to adapt elevated structures to the needs of broiler chickens [13–15,17]. Younger birds also benefit from ramps that allow them to perch [14]. Broiler chickens are motivated to use elevated structures from the first week on [14,15]. An increase in the use of elevated platforms by broiler chickens over the fattening period was found by Bailie et al. [14], with a peak during week five. Afterward, the use of platforms declines presumably because the broilers are larger and need more space and due to their decreasing physical abilities [13,14].

An approach for implementing elevated perforated floors into broiler husbandry is a partially perforated flooring system. The installation of a perforated floor beneath the feeding and the water lines could be useful to control the drainage of water lines and could, therefore, improve litter quality [18]. At the same time, the system allows the animals to sit elevated and use the littered area for pecking, scratching, and dustbathing. Previous studies evaluated the effect of a partially perforated flooring system on animal welfare [8,9], production performance [8,9], litter quality [18,19], ammonia emission [20], and antimicrobial resistance [21]. However, the behavior of broiler chickens in this system has not yet been investigated [9,18].

This study aimed to evaluate the use of an elevated perforated floor equipped with food and water supply by broiler chickens. We assessed the number of animals on the floor and their time spent on it in the different phases of the fattening period. We predicted that there would be a higher number of animals on the perforated floor and that they would spend more time on it than in a littered control area. We evaluated whether the animals used the elevated floor apart from using the resources. More specifically, we expected that the birds would spend more time either with locomotion or sitting inactive on the perforated floor than in a littered control area. Further, we assessed whether the animals sat inactive for longer periods on the perforated floor. We predicted that the number of animals in both observed areas would decrease and that the duration of sitting inactive would increase over the fattening period.

2. Materials and Methods

2.1. Birds and Housing

The study was conducted in two similarly constructed broiler barns at the research farm of the University of Bonn (Königswinter, Germany). Five-hundred fast-growing broiler chickens of the breed Ross 308 (BWE-Brüterei Weser-Ems GmbH & Co. KG, Rechterfeld, Germany) were kept in each barn for 32 days over three non-consecutive trials (1000 animals per trial). The barns had a floor space of 5.3 m × 4.7 m each, which led to a stocking density of 39 kg/m^2 at the end of the fattening period. This corresponds to 19.5 animals per square meter at a final body weight of about 2 kg in both barns. In both barns, the concrete floor was littered with wood shavings (600 g/m^2) for one trial and straw granulate (1000 g/m^2) for two trials. Fresh wood shavings/straw granulate were distributed if necessary. In addition, Miscanthus briquettes (Campus Klein-Altendorf, Universität Bonn, Rheinbach, Germany) were added on both sides of the barns on days 14, 21, and 28 to accommodate pecking behavior. Both barns were equipped with two water lines with drinking nipples, which were shared by nine birds each and four round feeding troughs (1.07 cm space per bird).

In one of the barns (experimental barn, Exp), about 50% of the floor space was equipped with an elevated perforated floor (5.3 m × 2.1 m), which was installed in the middle of the barn, under the water lines and feeding troughs at a height of 15 cm (see Figure 1a). The perforated floor was constructed out of plastic elements (Golden Broiler Floor, FIT Farm Innovation Team GmbH, Steinfurt, Germany), measuring 196 cm × 56 cm. The mesh size of the grid was 1.5 cm × 1.5 cm, and the slats were 0.5 cm thick. The residual floor space on the left and the right side of the perforated floor was littered. Adjusted elements connected the littered floor space and the elevated perforated floor to be used as ramps for accessing the floor. The angle between the barn floor and the ramps was about 30°. In the control barn (Con), the animals were kept under approximately commercial conditions on a littered concrete floor (see Figure 1b).

(a)

(b)

Figure 1. (a) Experimental barn with elevated perforated floor and littered areas. (b) Control barn with the littered floor.

The broiler chickens of both barns were fed standard broiler feed (Deuka, Deutsche Tiernahrung Cremer GmbH & Co. KG, Düsseldorf, Germany) and water, both were provided ad libitum. In both barns, a negative pressure ventilation system and a climate computer (PL9400, Stienen Bedrijfselektronica B.V., Nederweert, the Netherlands) were used to provide temperature and humidity within the commercial standard for broilers [22]. There was natural light through windows (the size of the windows corresponding to 3% of the floor) and an artificial lighting program with a transition time of 1 h between light and dark periods. The dark period was increased by 2 h per day, from 2 h on the first day of life to 6 h on the third day. It was kept constant at 6 h per day from day 3 to day 29. After day 29, the dark period was decreased by 2 h per day until day 32. The mean light intensity was about 100 Lux in both barns. The broiler chickens were vaccinated against Newcastle disease (day 13), Gumboro (day 18), and infectious bronchitis (day 18).

2.2. Behavioral Observations

For behavioral observations, a camera (EQ900F eZ.Hd Series, EverFocus Electronics Corporation, New Taipei City, Taiwan, China) was installed at the ceiling (height: two meters) in the middle of each barn. It was used to film a certain area on the perforated floor and the corresponding area in the control barn. A recorder (AXR-108, Monacor International GmbH & Co. KG, Bremen, Germany) was programmed to record 2 days per week for 24 h. Only daytime videos were analyzed due to the inadequate quality of the videos in the nighttime. One observer conducted the video analysis. To ensure consistent measurements, a subset of 60 scans and ten focal animals was analyzed by a second observer, not involved in the main data analysis, to calculate the inter-observer reliability. Furthermore, intra-observer reliability was calculated, scoring a sample of 60 scans and ten focal animals twice with an offset in time.

2.2.1. Use of the Perforated Floor

To quantify the use of the perforated floor and compare it to the corresponding area in the control barn, the footage was analyzed on two observation days at the beginning (days 2 and 5), the middle (days 16 and 19), and the end (days 26 and 30) of the fattening period of the three trials. Therefore, a screenshot of the videos was taken every 30 min between 6:00 am and 10:30 pm (in total, 1017 pictures for all trials) (scan sampling). A representative area of the perforated floor was defined and marked on the pictures using a digital frame produced with the program GoldenRatio (Markus Welz, Vs. 3.1.4, Krailling, Germany). The width of the area (210 cm) resulted from the width of the perforated floor with ramps in the experimental barn and the length (115 cm) from the distance of five nipples (23 cm between two nipples) of the waterline (in total 2.42 m^2). The distance between the nipples was also used to adapt the frame to the corresponding area on the pictures of the control barn. Both areas contained two water lines with five nipples each and one round feeding trough with a diameter of 40 cm, so the usable space was 2.29 m^2. Birds with more than half of their bodies (including head and tail) within the frame were counted using ImageJ (Wayne Rasband, Vs. 1.51q, National Institutes of Health, Maryland, USA), and the location of each animal was categorized following the definitions in the ethogram in Table 1. All categories were considered as mutually exclusive. The videos were consulted if it was unclear in which direction an animal was holding its head.

Table 1. Ethogram used in the scan sampling and in the focal animal observation.

Method	Behavior/Location	Description
Scan sampling and focal animals	Located near the feeding trough (NF)	The bird is located with more than a half of its body within a radius of one animal's length around the feeding trough and is holding its head in the direction of the feeding trough (regardless of whether upright position or sitting)
Scan sampling and focal animals	Located near the waterline (NW)	The bird is located with more than a half of its body within a radius of one animal's length around the water line and is holding its head in the direction of a nipple (regardless of whether upright position or sitting)
Scan sampling	Other (Oth)	All animals not 'located near the feeding trough' or 'located near the waterline'
Focal animals	Locomotion [1] (Loc)	The bird is standing or walking (upright position) and is not 'located near the feeding trough' or 'located near the waterline'
Focal animals	Sitting inactive[1] (Sit)	The bird is resting (sitting with head under the wing or resting on the ground) or lying (the bird is lying on one side with a leg and/or wing stretched out) and is not 'located near the feeding trough' or 'located near the waterline'

[1] Adapted from Baxter et al. [23].

2.2.2. Focal Animal Observations

The duration of behavior patterns was evaluated by observing focal animals. Again, a frame was added to the footage by using the distance between the nipples of the waterline. The area within the frame measured about 210 cm × 210 cm (in total 4.41 m^2, usable space 4.28 m^2). The experimental barn contained the perforated floor with ramps on both sides, two water lines with ten nipples each and one feeding trough. The corresponding littered area in the control barn included the same resources. On three observation days per fattening period (beginning, day two; middle, day 16 or 19; and end, day 30), ten focal birds per day and barn were selected pseudo-randomly (in total 180 animals for Exp and Con). The first animal, which entered the frame from the left side after 2:00 pm, was observed continuously for two hours or until it left the frame. Then, the video was rewound to the moment the animal entered the frame, and the next animal, which entered the frame from the right side, was chosen for observation. This was repeated until ten birds were observed. To avoid observing an animal twice, the side for choosing the next animal was alternated. The observation started at 2:00 pm because no external disturbances occurred for at least 2 h at this time of the day. While observing the focal birds, their behavior patterns were categorized using INTERACT (Mangold International GmbH, Vs. 17.1, Arnstorf, Germany). The activities defined in the ethogram in Table 1 were differentiated. All categories were considered as mutually exclusive. Based on Norring et al. (2016), the activity was considered to have ended if the animal stopped the activity for 3 s.

2.3. Statistical Analyses

Statistical analysis was conducted using the SAS software (Statistical Analysis Institute, Vs. 9.4, Cary, NC, USA). The observer reliability was calculated for each location separately for the scan sampling and for each behavior in summary for the focal animal observation using Krippendorff's alpha [24]. The data type was set to metric for each parameter; the number of bootstraps was set to 2000. The classifications suggested by Landis and Koch [25] were used to evaluate reliability (<0.00 = poor, 0.00–0.20 = slight, 0.21–0.40 = fair, 0.41–0.60 = moderate, 0.61–0.8 = substantial, 0.81–1.00 = almost perfect). For descriptive analysis, the mean values and standard deviations (±) were calculated for all parameters.

Residuals and data were checked for distribution based on normality plots created with the univariate procedure in SAS. Generalized linear mixed models were calculated using the GLIMMIX procedure, defining a normal distribution for the data of the usage of the perforated floor, while the distribution for the parameters of focal bird observations was specified as lognormal. For the usage of the perforated floor, a total number of animals, as well as the percentage of those located at 'Oth', 'NF', and 'NW' in relation to all animals on the respective area, were analyzed. All parameters were analyzed separately, including treatment (experimental group, control group), phase of the fattening period (beginning, middle, end), and the interaction between both as fixed effects. The hierarchical structure of the data was considered in the random statement by nesting each screenshot on the day of observation, while the day of observation was nested in the respective trial.

Data analyzed for the focal bird observation were 'Loc', 'Sit', 'NF', and 'NW'. Here, again, all parameters were analyzed separately, including the fixed factors as described for the usage of the perforated floor. Again, the hierarchical structure of data was accounted for in the random statement, nesting the observed animal on the observation day; these, in turn, were nested in the respective trial. Furthermore, the effects of the abovementioned fixed factors on the duration of each behavioral event of the behavior 'Sitting inactive' was analyzed, again including the hierarchical structure in the random statement. Moreover, repeated measures in different animals were accounted for in the random statement. Pairwise comparisons were made using Tukey–Kramer tests. The level of significance was set for $p < 0.05$. A level of $p < 0.1$ was regarded as a tendency.

3. Results

For the scan sampling, intra-observer reliability was found to be 'almost perfect' for all locations. Inter-observer reliability was 'moderate' for the location Oth, 'substantial' for the location NF, and 'almost perfect' for the location NW. For the focal animal observation, the measurement resulted in 'almost perfect' for the intra-observer and 'substantial' for the inter-observer reliability. The values of Krippendorff's alpha coefficients for the different behaviors/locations are presented in Table 2.

Table 2. Observer reliabilities for scan sampling ($n = 60$ scans) and the focal animal observation ($n = 10$ animals).

Parameter		Krippendorff's Alpha Intra-Observer	Krippendorff's Alpha Inter-Observer
Scan sampling:	Oth	0.84	0.42
	NF	0.83	0.75
	NW	0.90	0.85
Focal animal observation:		0.99	0.75

3.1. Use of the Perforated Floor

The treatment had a significant effect on the total number of animals in the observed area ($F_{(1, 505)} = 232.76$, $p < 0.001$), with more birds per square meter in the Exp (12.83 ± 4.24) than in the Con (9.67 ± 3.58). A tendency was found for the phase of the fattening period to affect the total number of birds ($F_{(2, 505)} = 2.55$, $p < 0.1$), with more animals counted at the beginning of the fattening period than in the middle ($t = |2.04|$, $p < 0.1$) and the end ($t = |1.77|$, $p < 0.1$). Further, the effect of the interaction between treatment and phase was found not to be significant but revealed a tendency ($F_{(2, 505)} = 2.63$, $p < 0.1$). As shown in Figure 2, there were significantly more birds per square meter in the Exp than in the Con at the beginning (Exp: 14.41 ± 4.69, Con: 10.91 ± 4.37($t = |10.89|$, $p < 0.001$)), the middle (Exp: 11.76 ± 2.70, Con: 8.28 ± 3.25 ($t = |8.67|$, $p < 0.001$)), and at the end (Exp: 11.77 ± 4.06, Con: 9.26 ± 1.96 ($t = |7.17|$, $p < 0.001$)). There were no significant differences between the phases within the treatment groups. However, a tendency was found for the Exp group, with a higher total number of animals per square meter at the beginning than in the middle ($t = |2.02|$, $p < 0.1$) and at the end ($t = |2.15|$, $p < 0.1$). In the Con group, we detected a tendency for more animals per square meter at the beginning than in the middle of the fattening period ($t = |1.99|$, $p < 0.1$).

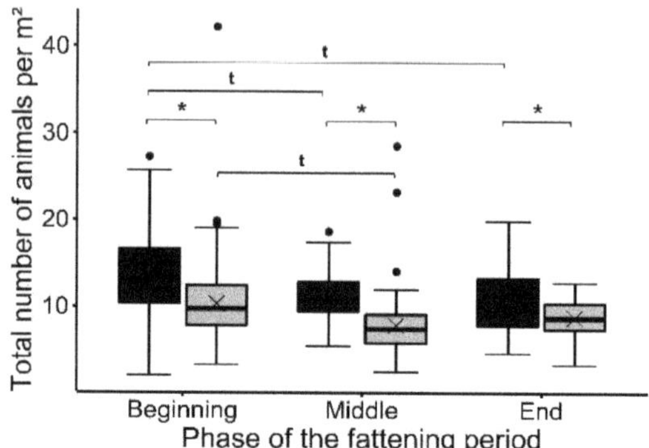

Figure 2. The total number of animals per m^2 in the observed area at the different phases of the fattening period (n = three trials each). Results are presented as boxplots (data range, median, and lower and upper quartile; outliers are included in the graph as dots and means as a cross). Results of the experimental group (Exp) are presented in dark gray; the control group (Con) results are presented in light gray. * Significant difference at $p < 0.05$, ^{t} tendency at $p < 0.1$.

The proportion of observed birds in the Exp and Con group located in the different areas and the effects of treatment and phase of the fattening period are summarized in Table 3. The percentage of animals located at Oth differed significantly between the treatment groups ($F_{(1, 505)} = 463.99$, $p < 0.001$). In general, a higher percentage of animals were located at Oth in the Exp group than in the Con group ($43.54 \pm 14.82\%$ vs. $28.56 \pm 15.53\%$). Further, the phase had a significant effect on the proportion of animals being located at Oth ($F_{(2, 505)} = 3.81$, $p < 0.05$), with animals being located at Oth more often at the beginning ($42.90 \pm 16.05\%$) than at the end of the fattening period ($28.94 \pm 15.02\%$; $t = |2.73|$, $p < 0.05$). The interaction between treatment and phase also had a significant effect on the proportion of animals being located at Oth ($F_{(2, 505)} = 32.04$, $p < 0.001$). In the Exp group, we detected a higher percentage of animals being located at Oth than in the Con group during all phases (beginning $t = |7.59|$, $p < 0.001$; middle $t = |16.30|$, $p < 0.001$; end $t = |12.62|$, $p < 0.001$). Within the Exp group, there was a tendency for a higher proportion of animals being located at Oth at the beginning than at the end of the fattening period ($t = |1.93|$, $p < 0.1$). In the Con group, the proportion of animals being located at Oth was significantly different between the beginning and middle of the fattening period ($t = |2.98|$, $p < 0.05$) and beginning and end ($t = |3.43|$, $p < 0.01$). In both cases, a higher percentage of broilers were located at Oth at the beginning.

Treatment had a significant effect on the proportion of animals located NF ($F_{(1, 505)} = 105.01$, $p < 0.001$). NF was found in a higher proportion of animals in the Con group ($32.77 \pm 12.21\%$) than in the Exp group ($26.63 \pm 11.12\%$). A tendency was found for the effect of phase ($F_{(2, 505)} = 2.82$, $p < 0.1$), with a higher amount of animals located NF at the beginning than at the end of the fattening period ($t = |1.93|$, $p < 0.1$).

NW was found to be affected by the treatment ($F_{(1, 505)} = 196.36$, $p < 0.001$), the phase ($F_{(2, 505)} = 4.09$, $p < 0.05$), and the interaction between both ($F_{(2, 505)} = 25.36$, $p < 0.001$). This resulted in a lower percentage of animals located NW at the beginning of the fattening period ($31.10 \pm 11.77\%$) than at the middle ($35.53 \pm 13.62\%$; $t = |1.95|$, $p < 0.1$) and the end ($37.00 \pm 11.99\%$; $t = |2.74|$, $p < 0.05$). Further, a higher percentage of broilers was observed NW in the Con group ($38.67 \pm 12.82\%$) than in the Exp group ($29.83 \pm 10.73\%$). This pattern was found within every phase of the fattening period (beginning $t = |3.74|$, $p < 0.01$; middle $t = |12.14|$, $p < 0.001$; end $t = |7.56|$, $p < 0.001$). In the Con group, the percentage of animals located NW was lower at the beginning of the fattening period than in the middle ($t = |4.35|$, $p < 0.001$) and the end ($t = |3.65|$, $p < 0.01$). A tendency was found within the Exp group, with a higher proportion of animals located NW at the end than at the beginning ($t = |1.93|$, $p < 0.1$).

Table 3. Mean proportion (%) of animals categorized as located at 'other' (Oth), 'located near the feeding trough' (NF), and 'located near the water line' (NW) at the different phases of the fattening period on the perforated floor (Exp) and in the control area (Con) (n = three trials each). Standard deviations are presented in parentheses, n.s. = result is not significant at $p < 0.05$.

Location	Treatment	Phase of the Fattening Period			Treatment p	Phase p	Treatment × Phase p
		Beginning	Middle	End			
Oth	Exp	47.29 (16.66)	46.52 (12.24)	36.86 (11.77)	<0.001	<0.05	<0.001
	Con	38.51 (14.16)	23.06 (10.89)	21.01 (13.69)			
NF	Exp	23.62 (12.71)	26.12 (7.06)	30.57 (10.46)	<0.001	<0.1	n.s.
	Con	28.37 (11.27)	33.29 (10.20)	37.56 (12.83)			
NW	Exp	29.08 (11.63)	27.36 (10.09)	32.57 (9.47)	<0.001	<0.05	<0.001
	Con	33.11 (11.59)	43.65 (11.70)	41.42 (12.62)			

3.2. Focal Animal Observations

Since the observation area did not cover the entire perforated floor, focal birds could leave the frame while not leaving the perforated floor. Therefore, these birds were named 'runaways' based on Norring et al. (2016) and were excluded from further analysis in both treatment groups (Exp n = 56, Con n = 44). This resulted in 34 completely observed focal birds on the perforated floor and 46 in the littered control area. The duration of the runaways observed in the defined area compared to the completely observed animals is shown in Figure 3.

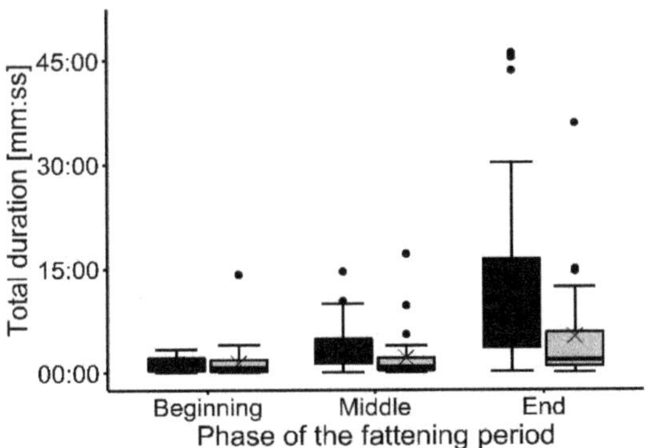

Figure 3. Total time completely observed focal animals (n = 80) spent in the observed area compared to runaways (n = 100) at the different phases of the fattening period. Results are presented as boxplots (data range, median, and lower and upper quartile; outliers are included in the graph as dots and means as a cross). Completely observed animals are presented in dark gray; runaways in light gray.

The treatment had a significant effect on the time the focal animals spent in total in the observed area ($F_{(1, 15)}$ = 6.85, $p < 0.05$). More specifically, in the Con group, the animals stayed longer in the observed area than in the Exp group (Exp: 03:09 ± 03:27, Con: 08:39 ± 11:41 (mm:ss)) (see Figure 4a). The total time animals spent in the observed area was found to be affected by the phase of the fattening period ($F_{(2, 15)}$ = 11.47, $p < 0.001$). The duration increased during the fattening period, with a higher total time at the end than in the middle (t = |3.06|, $p < 0.05$) and at the beginning (t = |4.64|, $p < 0.001$) and a higher total in the middle than at the beginning (t = |1.94|, $p < 0.1$) (Figure 4b).

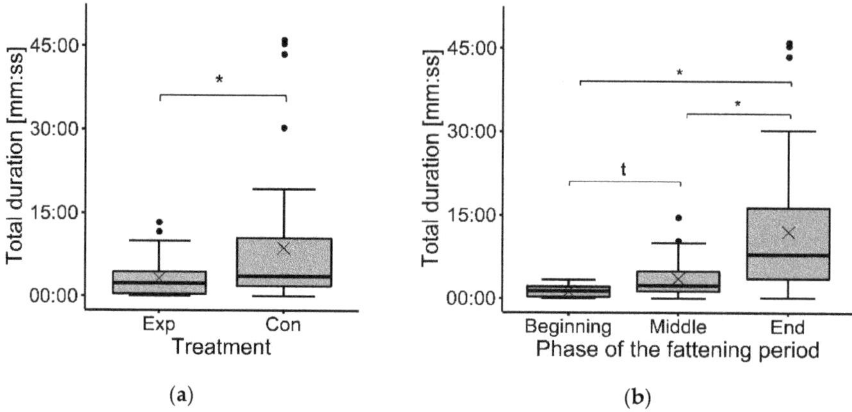

Figure 4. Total time focal animals spent in the observed area: (**a**) in the experimental group (Exp) and the control group (Con); (**b**) in the different phases of the fattening period. Results are presented as boxplots (data range, median, and lower and upper quartile; outliers are included in the graph as dots and means as a cross). * Significant difference at $p < 0.05$, ^{t} tendency at $p < 0.1$.

While 16.3% of the observed animals never showed the behavior Loc, 32.5% never showed Sit. Further, 63.8% of the focal animals had never been located NF, and 26.3% were never observed NW. The treatment had no significant effect on the mean durations of Loc, Sit, NF, and NW (all $F > 0.00$, all $p > 0.05$) (see Figure 5). While Loc, NF, and NW were not affected by the phase (all $F > 0.81$, all $p > 0.05$), an effect of phase was found on the behavior Sit ($F_{(2, 7)}$ = 2.28, $p < 0.05$). Pairwise comparisons revealed a longer duration of the behavior Sit at the end (07:25 ± 09:56 (mm:ss)) of the fattening period than at the beginning (00:06 ± 00:10 (mm:ss); t = |4.09|, $p < 0.05$) and in the middle (01:25 ± 02:23, (mm:ss); t = |2.26|, $p < 0.1$) and a longer duration in the middle than at the beginning (t = |2.47|, $p < 0.05$). There was no significant interaction between treatment and phase for the behaviors Loc, Sit, NF, and NW (all $F > 0.13$, all $p > 0.05$).

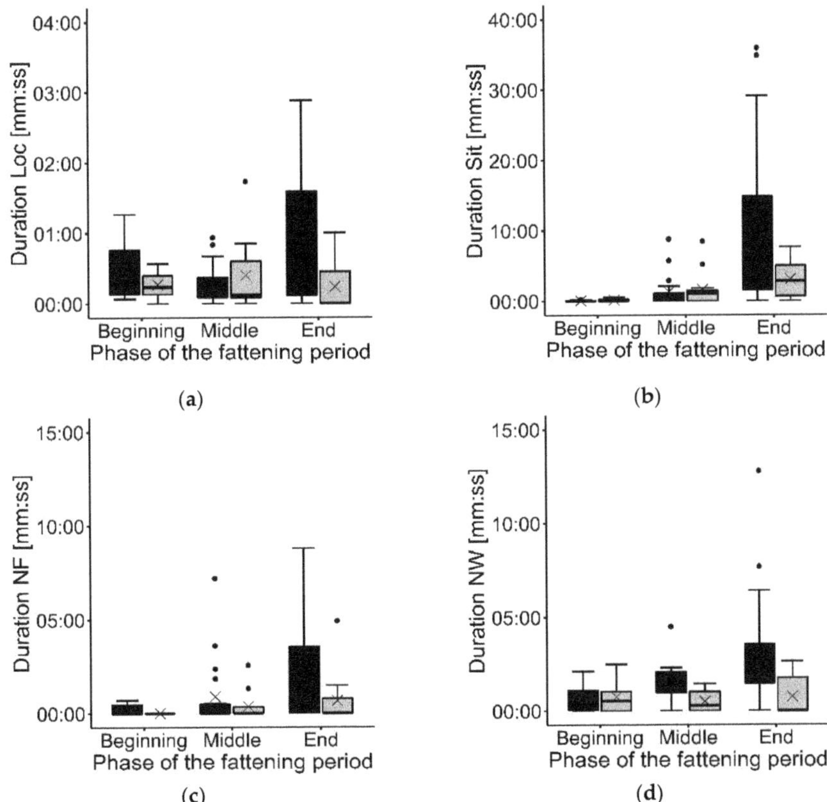

Figure 5. Duration of the different behaviors at the different phases of the fattening period: (**a**) locomotion (Loc); (**b**) sitting inactive (Sit); (**c**) located near the feeding trough (NF); and (**d**) located near the waterline (NW). Results are presented as boxplots (data range, median, and lower and upper quartile; outliers are included in the graph as dots and means as a cross). Results of the experimental group (Exp) are presented in dark gray; the control group (Con) results are presented in light gray.

The mean proportions of the different behaviors of the total time spent in the observed area at the different phases of the fattening period in Exp and Con are shown in Figure 6. Generally, the proportion of the behavior Loc decreased over the fattening period, while the proportion of Sit increased in both treatment groups.

Further, the duration of each behavioral event of the behavior 'Sitting inactive' (Sitting inactive bout, n = 184) of 53 different focal animals was analyzed. While the treatment had no significant effect on the duration of the single sitting inactive bouts ($F_{(1, 137)}$ = 0.34, $p > 0.05$), the phase had a significant effect ($F_{(2, 137)}$ = 5.10, $p < 0.01$). This resulted in longer sitting inactive bouts in the middle (01:15 $\pm$ 01:43 (mm:ss); t = |2.81|, $p < 0.05$) and at the end of the fattening period (01:38 $\pm$ 02:16 (mm:ss); t = |3.15|, $p < 0.01$) than at the beginning (00:11 $\pm$ 00:10 (mm:ss)).

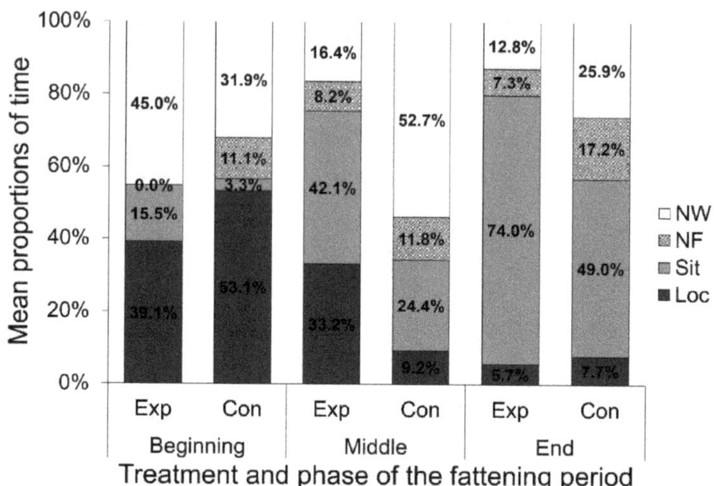

Figure 6. Mean proportions (%) of the behaviors Loc (locomotion), Sit (sitting inactive), NF (located near the feeding trough), and NW (located near the waterline) of the total time focal animals spent in the observed area at the different phases of the fattening period on the perforated floor (Exp) and in the control area (Con).

4. Discussion

The aim of the study was to evaluate the use of an elevated perforated floor equipped with food and water supply by broiler chickens. More animals were observed on the perforated floor than in the littered control area, but focal animals spent less time on the perforated floor compared to the observed littered control area. The animals on the elevated perforated floor were located to a higher proportion at Oth than in the Con. However, there were no differences in the duration of the recorded behaviors. The number of animals in both observed areas decreased, and the duration of the behavior sitting inactive increased over the fattening period. These results show that the broilers' behavior was not negatively affected by the perforated floor to a high degree.

As expected, there were more animals on the perforated floor than in the littered control area. The 15 cm high floor might be attractive for the animals to perch. Prior studies found that fast-growing broiler chickens are motivated to perch if they are given the opportunity [13–15,17]. They prefer elevated platforms to perches, probably due to an increasing body weight, which leads to difficulties finding balance on a perch [14,15]. In accordance with the present results, previous studies have demonstrated that broilers had no difficulties using the elevated floor at the height of 10–50 cm and ramps at an angle of 15°–35° [13–15,17].

The mean number of animals on the elevated floor was 12.8 birds/m², which is in accordance with Norring et al. [13] and Bailie et al. [14]. In contrast to these prior studies, in this study, the food and water supply was located on the perforated floor, which forced the animals to use the floor. The fact that the animals in the Exp barn did not have the choice of using the perforated floor should always be taken into account when interpreting the results of this study. In the scan sampling, a higher proportion of animals was categorized as located at 'other' on the perforated floor than in the control area. In addition, a higher percentage of broilers was 'located near the feeding trough' and 'located near the water line' in the littered control area. These findings indicate a usage of the elevated platform apart from using the resources.

Contrary to expectations, the focal birds in the Con stayed longer in the observed area than the animals on the perforated floor. A possible explanation could be that the climate on the perforated floor was uncomfortable for the animals. The manure under the perforated floor led to a higher amount of ammonia in the Exp barn, evaluated in the same experimental setting by Adler et al. [20]. This may have resulted in shorter times spent on the perforated floor. To avoid the problem of high ammonia concentration in barns with perforated floors, installing a manure belt under the elevated perforated floor could be an option and should be further investigated [20].

Prior studies found increased activity when elevated structures were added to experimental pens [4,5]. In addition, we suggested the perforated floor would also be attractive for resting [26]. Therefore, we expected longer durations of either Loc or Sit in the Exp than in the Con. However, we found no differences in the duration of the recorded behavior patterns of the focal animals in both treatment groups.

In the duration of the single sitting inactive bouts, no differences occurred between Exp and Con. This is contrary to studies by Yngvesson et al. [27] and Forslind et al. [26]. Both studies found that elevated structures increased the duration of resting bouts during the daytime because there was less disturbance by other animals. The results in the current study may be explained by the location of the resources in the observed area. Broilers searching for food and water may have disturbed sitting inactive animals in our study.

In general, the results of the focal animal observations need to be interpreted with caution due to the small sample size after excluding the 'runaways'. Further, the observation always began at 2 pm and ended at 4 pm at the latest. The behavior of broiler chickens is highly related to the light intensity and differs depending on the time of day [28,29]. In addition, the use of elevated structures may also vary depending on the time of day [5,15]. The results of the focal animal observation only represent the behavior in a narrow time slot. Another limitation of the focal animal observation is that animals categorized as 'located near the feeding trough' or 'located near the waterline' could also merely walk or sit near the feeding trough or waterline; accordingly, such animals would not be recorded as 'locomotion' or 'sitting inactive'. These variables are not completely independent. Nevertheless, the measured proportions of the different behavior patterns at the total time the animals spent in the observed area were, in general, consistent with prior studies [29,30].

Regarding the phase of the fattening period, the results of the focal animal observations showed that the animals spent in total more time in the observed area, spent more time 'sitting inactive', and showed longer sitting inactive bouts at the end of the fattening period than at the beginning. This is consistent with other studies and is suggested to be due to a decrease in the activity of broiler chickens with age [29–31]. This is presumably due to higher body weight and an increasing rate of lameness [32] and, therefore, discomfort or pain [33].

As expected, we found slightly more animals per square meter at the beginning of the fattening period than at the end in both treatment groups. In case of the observed area on the elevated perforated floor, this effect could be due to broiler chickens' decreased walking ability with age and, therefore, difficulties walking up the ramp [5,14]. Another explanation for finding the effect in both treatment groups is an increase in body size with age [34].

5. Conclusions

To conclude, this study has identified differences in the behavior of broilers when comparing an elevated perforated floor equipped with food and water supply to a control area. There were more animals per square meter on the elevated floor, with a higher proportion of animals located at 'other', which implies that the birds used the perforated floor beyond simply using the resources. However, the study did not find differences between the durations of behaviors when comparing the treatment groups. These findings suggest that, in general, the elevated perforated floor is attractive for the animals. However, it does not promote one of the recorded behavior patterns. Further research is necessary to find the optimum design for elevated platforms for fast-growing broiler chickens. A study design with perforated elevated platforms away from feed and water supply would allow a true choice for the animals of opting to be on a certain substrate. It would describe more accurately the physical effort animals invest to require access. Nevertheless, our results show that a partially perforated flooring system could be an option for structuring broiler barns.

Author Contributions: Conceptualization, F.M., J.S., S.H., C.A., N.K. and B.S.; methodology, F.M., J.S. and B.S.; software, F.M., J.S. and B.S.; validation, F.M., A.K. and J.S.; formal analysis, F.M. and J.S.; investigation, F.M., A.K. and J.S.; resources, N.K. and B.S.; data curation, F.M., S.H. and C.A.; writing—original draft preparation, F.M.; writing—review and editing, F.M., J.S., S.H., C.A., A.K., N.K. and B.S.; visualization, F.M. and J.S.; supervision, J.S., N.K. and B.S.; project administration, N.K. and B.S.; funding acquisition, N.K. and B.S. All authors have read and agreed to the published version of the manuscript.

Funding: This project "System for poultry husbandry to reduce the risk of bacteria recontamination during the fattening period, to improve the animal health and to reduce the risk of cross-contamination in the human food chain for bacteria in general and especially for antibiotic-resistant bacteria" (fitAvis) was supported by funds from the Federal Ministry of Food and Agriculture (BMEL, Germany), based on a decision by the Parliament of the Federal Republic of Germany via the Federal Office for Agriculture and Food (BLE, Germany) under the innovation support program (2817700214). This Open Access publication was funded by the Deutsche Forschungsgemeinschaft (DFG, German Research Foundation) within the program LE 824/10-1 "Open Access Publication Costs" and University of Veterinary Medicine Hannover, Foundation.

Institutional Review Board Statement: The study was carried out at the Educational and Research Center Frankenforst of the Faculty of Agriculture, University of Bonn (Königswinter, Germany). The experiments were performed in accordance with German regulations and approved by the relevant authority (Landesamt für Natur-, Umwelt- und Verbraucherschutz Nordrhein-Westfalen, Recklinghausen; 81.02.04.2018.A057).

Data Availability Statement: The data presented in this study are available on request from the corresponding author.

Acknowledgments: The authors wish to thank the Educational and Research Center Frankenforst of the Faculty of Agriculture, University of Bonn (Königswinter, Germany), for the care of the animals and their support.

Conflicts of Interest: The authors declare no conflict of interest.

References

1. European Commission, Committee on Animal Health and Animal Welfare. The Welfare of Chickens Kept for Meat Production (Broilers). 2000. Available online: https://ec.europa.eu/food/horizontal-topics/expert-groups/scientific-committees/scientific-committee-animal-health-and-animal_en (accessed on 10 January 2022).
2. Olsson, I.A.S.; Keeling, L.J. Night-time roosting in laying hens and the effect of thwarting access to perches. *Appl. Anim. Behav. Sci.* **2000**, *68*, 243–256. [CrossRef]
3. Newberry, R.C. Environmental enrichment: Increasing the biological relevance of captive environments. *Appl. Anim. Behav. Sci.* **1995**, *44*, 229–243. [CrossRef]
4. Bizeray, D.; Estevez, I.; Leterrier, C.; Faure, J.M. Effects of increasing environmental complexity on the physical activity of broiler chickens. *Appl. Anim. Behav. Sci.* **2002**, *79*, 27–41. [CrossRef]
5. Malchow, J.; Puppe, B.; Berk, J.; Schrader, L. Effects of elevated grids on growing male chickens differing in growth performance. *Front. Vet. Sci.* **2019**, *6*, 203. [CrossRef] [PubMed]
6. Kaukonen, E.; Norring, M.; Valros, A. Perches and elevated platforms in commercial broiler farms: Use and effect on walking ability, incidence of tibial dyschondroplasia and bone mineral content. *Animal* **2017**, *11*, 864–871. [CrossRef]
7. Cavusoglu, E.; Petek, M. Effects of different floor materials on the welfare and behaviour of slow- and fast-growing broilers. *Arch. Anim. Breed.* **2019**, *62*, 335–344. [CrossRef]
8. Adler, C.; Tiemann, I.; Hillemacher, S.; Schmithausen, A.J.; Muller, U.; Heitmann, S.; Spindler, B.; Kemper, N.; Buscher, W. Effects of a partially perforated flooring system on animal-based welfare indicators in broiler housing. *Poult. Sci.* **2020**, *99*, 3343–3354. [CrossRef]
9. Topal, E.; Petek, M. Effects of fully or partially slatted flooring designs on the performance, welfare and carcass characteristics of broiler chickens. *Br. Poult. Sci.* **2021**, *62*, 804–809. [CrossRef]
10. Gebhardt-Henrich, S.G.; Stratmann, A.; Dawkins, M.S. Groups and individuals: Optical flow patterns of broiler chicken flocks are correlated with the behavior of individual birds. *Animals* **2021**, *11*, 568. [CrossRef]
11. The Better Life Label. Available online: https://beterleven.dierenbescherming.nl/zakelijk/en/participate-2/company-type/livestock-farms/broilers/ (accessed on 10 January 2022).

12. Wood-Gush, D.; Duncan, I. Some behavioural observations on domestic fowl in the wild. *Appl. Anim. Ethol.* **1976**, *2*, 255–260. [CrossRef]
13. Norring, M.; Kaukonen, E.; Valros, A. The use of perches and platforms by broiler chickens. *Appl. Anim. Behav. Sci.* **2016**, *184*, 91–96. [CrossRef]
14. Bailie, C.L.; Baxter, M.; O'Connell, N.E. Exploring perch provision options for commercial broiler chickens. *Appl. Anim. Behav. Sci.* **2018**, *200*, 114–122. [CrossRef]
15. Malchow, J.; Berk, J.; Puppe, B.; Schrader, L. Perches or grids? What do rearing chickens differing in growth performance prefer for roosting? *Poult. Sci.* **2019**, *98*, 29–38. [CrossRef] [PubMed]
16. Newberry, R.C.; Estevez, I.; Keeling, L.J. Group size and perching behaviour in young domestic fowl. *Appl. Anim. Behav. Sci.* **2001**, *73*, 117–129. [CrossRef]
17. Tahamtani, F.M.; Pedersen, I.J.; Riber, A.B. Effects of environmental complexity on welfare indicators of fast-growing broiler chickens. *Poult. Sci.* **2020**, *99*, 21–29. [CrossRef]
18. Heitmann, S.; Stracke, J.; Adler, C.; Ahmed, M.F.E.; Schulz, J.; Buscher, W.; Kemper, N.; Spindler, B. Effects of a slatted floor on bacteria and physical parameters in litter in broiler houses. *Vet. Anim. Sci* **2020**, *9*, 100115. [CrossRef]
19. Chuppava, B.; Visscher, C.; Kamphues, J. Effect of different flooring designs on the performance and foot pad health in broilers and turkeys. *Animals* **2018**, *8*, 70. [CrossRef]
20. Adler, C.; Schmithausen, A.J.; Trimborn, M.; Heitmann, S.; Spindler, B.; Tiemann, I.; Kemper, N.; Buscher, W. Effects of a partially perforated flooring system on ammonia amissions in broiler housing-conflict of objectives between animal welfare and environment? *Animals* **2021**, *11*, 707. [CrossRef]
21. Chuppava, B.; Keller, B.; Abd El-Wahab, A.; Surie, C.; Visscher, C. Resistance reservoirs and multi-drug resistance of commensal escherichia coli from excreta and manure isolated in broiler houses with different flooring designs. *Front. Microbiol.* **2019**, *10*, 2633. [CrossRef]
22. Aviagen. Ross 308 Broiler. Management Handbook . 2018. Available online: http://en.aviagen.com/assets/Tech_Center/Ross_Broiler/Ross-BroilerHandbook2018-EN.pdf (accessed on 24 October 2019).
23. Baxter, M.; Bailie, C.L.; O'Connell, N.E. An evaluation of potential dustbathing substrates for commercial broiler chickens. *Animal* **2018**, *12*, 1933–1941. [CrossRef]
24. Hayes, A.F.; Krippendorff, K. Answering the call for a standard reliability measure for coding data. *Communication Methods and Measures* **2007**, *1*, 77–89. [CrossRef]
25. Landis, J.R.; Koch, G.G. The measurement of observer agreement for categorical data. *Biometrics* **1977**, *33*, 159–174. [CrossRef] [PubMed]
26. Forslind, S.; Blokhuis, H.J.; Riber, A.B. Disturbance of resting behaviour of broilers under different environmental conditions. *Appl. Anim. Behav. Sci.* **2021**, *242*, 105425. [CrossRef]
27. Yngvesson, J.; Wedin, M.; Gunnarsson, S.; Jönsson, L.; Blokhuis, H.; Wallenbeck, A. Let me sleep! Welfare of broilers (*Gallus gallus domesticus*) with disrupted resting behaviour. *Acta Agric. Scand. Sect. A—Anim. Sci.* **2018**, *67*, 123–133. [CrossRef]
28. Schwean-Lardner, K.; Fancher, B.I.; Laarveld, B.; Classen, H.L. Effect of day length on flock behavioural patterns and melatonin rhythms in broilers. *Br. Poult. Sci.* **2014**, *55*, 21–30. [CrossRef]
29. Bessei, W. Das Verhalten von Broilern unter intensiven Haltungsbedingungen. *Arch. Geflügelk.* **1992**, *56*, 1–7.
30. Bizeray, D.; Leterrier, C.; Constantin, P.; Picard, M.; Faure, J.M. Early locomotor behaviour in genetic stocks of chickens with different growth rates. *Appl. Anim. Behav. Sci.* **2000**, *68*, 231–242. [CrossRef]
31. Ventura, B.A.; Siewerdt, F.; Estevez, I. Access to barrier perches improves behavior repertoire in broilers. *PLoS ONE* **2012**, *7*, e29826. [CrossRef]
32. Weeks, C.A.; Danbury, T.D.; Davies, H.C.; Hunt, P.; Kestin, S.C. The behaviour of broiler chickens and its modification by lameness. *Appl. Anim. Behav. Sci.* **2000**, *67*, 111–125. [CrossRef]
33. Caplen, G.; Colborne, G.R.; Hothersall, B.; Nicol, C.J.; Waterman-Pearson, A.E.; Weeks, C.A.; Murrell, J.C. Lame broiler chickens respond to non-steroidal anti-inflammatory drugs with objective changes in gait function: A controlled clinical trial. *Vet. J.* **2013**, *196*, 477–482. [CrossRef]
34. Giersberg, M.F.; Hartung, J.; Kemper, N.; Spindler, B. Floor space covered by broiler chickens kept at stocking densities according to Council Directive 2007/43/EC. *Vet. Rec.* **2016**, *179*, 124. [CrossRef] [PubMed]

Review

The Relationships between Damaging Behaviours and Health in Laying Hens

Virginie Michel [1,*], Jutta Berk [2], Nadya Bozakova [3], Jerine van der Eijk [4], Inma Estevez [5], Teodora Mircheva [6], Renata Relic [7], T. Bas Rodenburg [8], Evangelia N. Sossidou [9] and Maryse Guinebretière [10]

[1] Direction de la Stratégie et des Programmes, French Agency for Food, Environmental and Occupational Health & Safety (ANSES), 94701 Maisons-Alfort, France

[2] Institute for Animal Welfare and Animal Husbandry, Friedrich-Loeffler-Institut, 29223 Celle, Germany; jutta.berk@tiho-hannover.de

[3] Department of General Animal Breeding, Animal Hygiene, Ethology and Animal Protection Section, Faculty of Veterinary Medicine, Student's Campus, Trakia University, 6000 Stara Zagora, Bulgaria; nadiab@abv.bg

[4] Animal Health and Welfare, Wageningen Livestock Research, Wageningen University and Research, De Elst 1, 6708 Wageningen, The Netherlands; jerine.vandereijk@wur.nl

[5] Department of Animal Production, Neiker-Basque Institute for Agricultural Research and Development, 01080 Vitoria-Gasteiz, Spain; iestevez@neiker.eus

[6] Section of Biochemistry, Faculty of Veterinary Medicine, Trakia University, 6000 Stara Zagora, Bulgaria; teodoramirchevag@abv.bg

[7] Faculty of agriculture, University of Belgrade, 11080 Belgrade, Serbia; rrelic@agrif.bg.ac.rs

[8] Animals in Science and Society, Faculty of Veterinary Medicine, Utrecht University, Yalelaan 2, 3584 Utrecht, The Netherlands; t.b.rodenburg@uu.nl

[9] Laboratory of Farm Animal Health and Welfare, Veterinary Research Institute, Ellinikos Georgikos Or-Ganismos-DIMITRA (ELGO-DIMITRA), 57001 Thessaloniki, Greece; sossidou@vri.gr

[10] Epidemiology, Health and Welfare Unit, French Agency for Food, Environmental and Occupational Health & Safety (ANSES), 22440 Ploufragan, France; maryse.guinebretiere@anses.fr

* Correspondence: virginie.michel@anses.fr; Tel.: +33-156-291-881

Citation: Michel, V.; Berk, J.; Bozakova, N.; van der Eijk, J.; Estevez, I.; Mircheva, T.; Relic, R.; Rodenburg, T.B.; Sossidou, E.N.; Guinebretière, M. The Relationships between Damaging Behaviours and Health in Laying Hens. *Animals* **2022**, *12*, 986. https://doi.org/10.3390/ani12080986

Academic Editors: Victoria Sandilands and Tina Widowski

Received: 12 February 2022
Accepted: 4 April 2022
Published: 11 April 2022

Publisher's Note: MDPI stays neutral with regard to jurisdictional claims in published maps and institutional affiliations.

Simple Summary: The design of housing systems and genetic selection of laying hens have in the past focused mainly on productivity, excluding issues around the animals' behavioural needs and welfare. Because of inadequate housing conditions and especially a barren environment, behavioural disorders such as feather and body pecking, as well as cannibalism, occur in the modern layer industry. Since conventional cages for egg production were banned in the European Union in January 2012, alternative systems such as floor, aviary, free-range, and organic systems have become increasingly common and now concern over 50% of hens housed in Europe. Despite the many advantages that come with non-cage systems, the shift to a housing system where laying hens are kept in larger groups and more complex environments has given rise to new challenges related to management, health, and welfare. We have carried out a review showing the close relationships between damaging behaviours and health in modern husbandry systems for laying hens.

Abstract: Since the ban in January 2012 of conventional cages for egg production in the European Union (Council Directive 1999/74/EC), alternative systems such as floor, aviary, free-range, and organic systems have become increasingly common, reaching 50% of housing for hens in 2019. Despite the many advantages associated with non-cage systems, the shift to a housing system where laying hens are kept in larger groups and more complex environments has given rise to new challenges related to management, health, and welfare. This review examines the close relationships between damaging behaviours and health in modern husbandry systems for laying hens. These new housing conditions increase social interactions between animals. In cases of suboptimal rearing and/or housing and management conditions, damaging behaviour or infectious diseases are likely to spread to the whole flock. Additionally, health issues, and therefore stimulation of the immune system, may lead to the development of damaging behaviours, which in turn may result in impaired body conditions, leading to health and welfare issues. This raises the need to monitor both behaviour and health of laying hens in order to intervene as quickly as possible to preserve both the welfare and health of the animals.

Keywords: hen health; damaging behaviour; laying hens; housing system

1. From "Productivism" Systems to More "Welfare-Friendly" Approaches: New Challenges to Face

After the Second World War, animal production systems were automatised and rationalised in order to reach higher productivity to be able to feed Europe with cheap animal protein. The spectacular development of poultry husbandry systems until the 1990s led to systems that were optimal in terms of working conditions, productivity, and food safety and animal health, for example, with the separation of animals and eggs from manure, such as in cages systems. Housing system design and the genetic selection of animals focused on productivity, excluding considerations around animal behavioural needs and welfare [1,2]. Due to inadequate housing conditions, and especially the barren environment, behavioural disorders such as feather pecking, toe pecking, vent/cloacal pecking, and cannibalism can occur [3].

Since the ban in January 2012 of battery cages for egg production in the European Union (Council Directive 1999/74/EC), alternative systems such as floor, aviary, free-range, and organic systems have become increasingly common, reaching 50% of hen housing in Europe in 2019 [4]. Non-cage (alternative) systems provide the birds much more behavioural freedom as well as ample access to litter, nests, and perches, which improves their welfare. Additionally, free-range and aviary systems allow higher bird activity, which may result in increased bone density and strength [5]. For instance, free-ranging laying hens have been shown to have better plumage conditions and higher final body weights, which leads to higher egg weight than hens in indoor systems [5–7].

Despite the many advantages associated with non-cage systems, the shift to a housing system where laying hens are kept in larger groups and more complex environments has given rise to new challenges related to management, health, and welfare. Considerable research has been performed to study environmental conditions and management practices in non-cage systems in different climatic conditions [7–10]. For instance, several studies have found that aviaries can have a negative impact on indoor air quality, with higher concentrations of suspended dust than in cage systems, resulting from the presence of floor litter (higher ammonia levels) and hens' activities (higher particulate matter levels) in it [11]. Dust is composed of inorganic and organic compounds from the birds themselves as well as from feed, litter, and building materials [12]. Dust may be a vector of microorganisms and toxins. High dust levels may compromise the health and welfare of both birds and their caretakers [13,14]. Bird health can also be negatively affected in non-cage systems by a higher risk of bacterial and fungal infections spreading among the birds [15,16].

Finally, welfare challenges persist, even after the switch to non-cage systems, including keel bone damage (reviewed by Riber et al. [17]), feather pecking, toe pecking, vent/cloacal pecking, and cannibalism [16,18]. Damaging pecking may occur during rearing periods of pullets as well as during the laying period [19–21], even though it is more prevalent in the laying period. Severe feather pecking, leading to feather loss, can result in economic losses as a result of increased food consumption in defeathered birds [22,23] and increased mortality [24,25], as well as in reduced animal welfare since feather pecking is painful for the birds being pecked [26]. Additionally, hens with feather damage are more susceptible to cannibalistic pecking [27]. Free-range systems are also associated with a higher risk of exposure to parasites, pathogens, and predation [28].

2. Description and Definition of the Different Concepts Linked to Health and Damaging Behaviours

2.1. Health of Laying Hen Flocks

There are many ways to define the health of productive animals. Considering the World Health Organisation definitions of 1946 and 2006, health is "a state of complete physical, mental and social well-being and not just the absence of disease or infirmity". Animal health can be defined as "a lack of disease or normal functioning of the animal body and normal behaviour" [28]. Here, we see that the development of abnormal behaviour is considered an impairment of animal health. In the production sector, Gunnarson [29] defines health as "the state of the animal organism that allows highest productivity based on a balance between animals and their environment, as well as the animal's physical well-being". More recently, animal health has been considered one of the pillars of the "One Health" concept, developed with the aim of protecting public health [30]. The vision of One Health is that human health can be better protected through policies that ensure the health of animals and of ecosystems since human, animal, and environmental health are all interconnected [30]. Within the One Health framework, animal welfare offers opportunities to define the conditions for animals to grow healthily and to be able to cope with pathogens while reducing the need for the use of antibiotics. Such conditions are defined by the animals´ behavioural needs that have been shaped by their own evolutionary history and are deeply imbedded in their genetic makeup. Understanding the factors that affect the social behaviour of laying hens [31–33], or their responses to the features of their surrounding environment [6,34,35], provides the scientific information needed to manage flocks according to these biological needs, to avoid sources of potential stressors, and to reduce the risk of damaging behaviour. This type of holistic approach will help to preserve animal health and welfare while allowing optimal animal performance in modern animal production systems. From the definitions of health cited above, we can see that they include the mental state of an animal and that both physical and mental health can be captured in the term "welfare". The French Agency for Food, Environmental and Occupational Health & Safety (ANSES) [36] defines the welfare of an animal as "its positive mental and physical state as related to the fulfilment of its physiological and behavioural needs, in addition to its expectations". Importantly, when the behavioural needs or expectations of animals are not fulfilled, damaging behaviours may develop or increase. The next chapter discusses the most common damaging behaviours that can be encountered in current intensive housing systems for layers and that may compromise animal welfare.

2.2. Damaging Behaviours in Laying Hens

2.2.1. Pecking Behaviours in Chickens

Pecking is a natural behaviour in chickens during foraging and exploration of the environment. When a chicken pecks a conspecific, a distinction is made between pecking arising from aggressive or non-aggressive motivations, as the body parts targeted and risk factors for the behaviour differ. Non-aggressive injurious pecking is considered a redirected form of foraging behaviour, as both pecking during feeding and injurious pecking show similar fixed motor patterns [37]. An association has been found between the high occurrence of litter-directed pecks by individuals when they are young and a high level of severe feather pecking and litter-directed pecks when they are adults [37]. This suggests that severe feather pecking is not a direct substitute for foraging but that some individuals have high pecking motivation overall and are, thereby, more prone to develop injurious pecking in addition to foraging.

In cage-free systems, hens have greater behavioural opportunities and freedom of movement, but these systems may also be associated with a greater risk of damaging behaviours as compared to cages [20,38]. Even though these behaviours can still happen in cages, they are limited to the cage where they develop.

2.2.2. From Pecking Behaviour to Damaging Behaviour

Damaging behaviours represent a collection of unwanted behaviours that develop under certain circumstances at high frequency and intensity in laying hens, other poultry species, and avian species [39] and can cause harm to other group members. They include feather pecking [40–42], aggressive pecking [43] (outside of the frame of hierarchy establishment), different forms of cannibalism [44–46], which include vent/cloacal pecking [42,47], and toe pecking [48,49].

Gentle feather pecking is a frequent behaviour in young birds and is also important in social recognition. In adult birds, stereotyped gentle feather pecking can be observed, where birds, for instance, spend a long time pecking at the tips of the tail feathers of another bird. Although this behaviour indicates a welfare problem for the pecker, it usually does not lead to much feather damage [45,50]. The main problematic behaviour is severe feather pecking, directly affecting the health of the hens—several feathers are lost, or whole stripping of certain areas of the body is observed. This is associated with pain in the affected hen [26] and can cause skin eruption or bleeding. Severe cases of feather pecking can escalate into cannibalism and death of the bird victim, cannibalistic tissue pecking, and vent/cloacal pecking, potentially leading to severely wounded or dead birds [45,46,50].

Cannibalistic behaviour involves beak-inflicted damage followed by the consumption of blood and tissues of conspecifics while they are still alive or after death [51]. Cannibalistic behaviour is learned by individual birds and can spread to others through social learning [51], even through adjacent cages [52]. Severe feather pecking can lead to increased risks of cannibalism [53]. Cannibalism in the cloacal area, also known as "vent pecking", is considered a distinct form of cannibalistic pecking [42] and may negatively affect the welfare and health of the bird by causing considerable pain and even leading to mortality [54]. Serious inflammatory and even infectious processes can follow skin breakage. Toe pecking is another behaviour that is harmful to victims and that negatively affects hen health. It occurs when a bird starts to peck the toes of another bird [48,49]. In severe forms, toe swelling can be attributed to cannibalism, and complications may be lethal [55].

2.2.3. Main Causes of Damaging Behaviour and Control Strategies

Regarding the causal factors leading to feather pecking, a classical hypothesis suggests that it is a redirected form of foraging that develops in the absence of foraging material [56–59]. The hypothesis is that under commercial conditions where chicks are reared in the absence of their mother's guidance, the direction of foraging pecks toward flock mates could result from a chick's failure to learn to direct these pecks toward appropriate substrates and food items. In addition, the absence of suitable manipulable foraging material can lead to injurious pecking in chicks [37]. In a review, De Haas et al. [37] explored how behavioural programming via prenatal conditions (role of maternal stress, egg conditions, incubation settings) and early postnatal conditions (chick brooding conditions) could influence the development of injurious pecking in laying hens. This review argues that it may be possible to prevent injurious pecking in commercial laying hen flocks by adapting the environmental conditions of previous generations, optimising incubation conditions, reducing stress around hatching, and guiding the early learning of chicks.

Damaging behaviour can emerge at different ages in most breeds, although with varying intensity depending on the genetic line [34], and can affect a large number of birds in the flock. Reported percentages of affected flocks at the end of lay can reach values as high as 60% of the flocks, with more than 10% of hens having moderate or severe feather damage in one body region [21], or 86% of the flocks in which severe feather pecking was observed [60].

Although no strategy can guarantee the complete absence of pecking behaviours, optimised management practices, especially concerning feeding, lighting, and climatic conditions [35] and environmental enrichment in pullets and adult birds [61–63], can help to reduce the risk. Access to outdoor free-range areas is associated with plumage preservation [6,7,64] and a reduced risk of injurious pecking [65]. Genetic selection at the commercial scale will help in the control of feather pecking [41,44,66]. For instance, Rodenburg et al. [67] offer various genetic means to limit feather pecking, cannibalism, and vent/cloacal pecking based on the systematic selection of birds with less-pronounced damaging behaviours than other birds.

Another hypothesis suggests that mild feather pecking could be a redirected form of social grooming and may have a social recognition function [68]. Kjaer et al. [69,70] suggest that severe feather pecking is related to neurological changes that cause hyperactivity, although Krause et al. showed that selection for high locomotor activity did not result in an increase in feather pecking [71]. Recent studies found that genes involved in cholinergic signalling, channel activity, synaptic transmission, and immune response are involved in feather-pecking mechanisms [66].

Although Borda-Molina et al. did not find any relations between microbiota and feather pecking [72], there is growing evidence that gut microbiota influence hens behaviour and physiology [73,74]. However, whether microbiota can influence the development of feather pecking is not fully demonstrated [74,75]. This shows the complexity of the situation, involving the modulation by the gut microbes of the immune system, or maybe brain function not modulated through the immune system.

The way neurophysiology, gut physiology, and health in a broader sense impact the development of damaging behaviours in layers is described in the chapter below, immediately followed by the description of how, in return, the consequences of damaging behaviours will impact the health and welfare of animals.

3. Inter-Relationships between Damaging Behaviours and Health Problems in Current Housing Systems for Layers

Some damaging behavioural patterns may be associated with certain diseases in hens.

3.1. Recent Knowledge about the Impact of the Health Condition, Including Immune Status of Animals, on the Occurrence of Damaging Behaviours

3.1.1. Immune System

The immune system plays a critical role in brain development. In particular, microglia (macrophage-like immune cells in the brain) have been shown to be involved in many aspects of brain development, such as synapse formation and neuronal survival [76]. Cytokines, chemokines, major histocompatibility complex (MHC) molecules, and toll-like receptors (TLRs) have been shown to play a critical role in neural development [76–78]. Cytokines can target neurocircuits that are involved in regulating mood, motor activity, motivation, and anxiety [79]. As a result, the immune system could influence behaviours through its role in brain development.

The immune system is also more directly involved in regulating behaviour. Cytokines and chemokines can alter behaviour, for example, in sickness behaviour, where sick animals show reduced feed and water intake, lower activity levels, decreased exploration and social interactions, and increased sleep [80,81]. Cytokines could influence behaviour via their effects on the synthesis, re-uptake, and release of neurotransmitters, such as serotonin, dopamine, and glutamate [79,82,83]. As an example, cytokines can influence the functioning of the hypothalamic-pituitary-adrenal axis (HPA axis). They can activate corticotropin-releasing hormone (CRH) and thereby stimulate the release of adrenocorticotropic hormone (ACTH) or can stimulate ACTH release, directly resulting in glucocorticoid release [79,84]. Cytokines can, in fact, influence behaviour via multiple routes.

In humans, there are similarities between sickness behaviour and behaviour expressed by individuals with certain neuropsychiatric disorders, such as depression [85]. Furthermore, many psychiatric disorders have been linked to immune dysregulation, including schizophrenia, anxiety and stress disorders, autism, and major depressive disorder [77].

Several studies have found relationships between the immune system and feather pecking. Most show genetic associations between feather damage (as an indicator of feather pecking) and the immune system. As mentioned previously, cytokines can influence the serotonergic and dopaminergic systems, and, in turn, these systems seem to be involved in the development of damaging behaviours such as feather pecking (for a review, see de Haas and van der Eijk [86]). In addition, through their effects on HPA axis functioning, cytokines could further influence how animals respond to or cope with stress. Feather pecking has been linked to coping styles and increased stress sensitivity [50,87]. Furthermore, feather pecking has been linked to motor activity [70], motivation, and fearfulness [88], and cytokines target brain areas that are involved in the regulation of these behaviours. The serotonergic and dopaminergic systems also appear to be dysregulated in many of these brain areas when feather pecking occurs [86]. The immune system may, therefore, play a role in the development of feather pecking.

Genetic associations have been found between immune-related genes, such as interleukin (IL4, IL9), nuclear factor NF-kappa-B (NFKB), chemokine (CCL4) genes, and feather damage score, providing evidence of a relationship between feather pecking and immunity at the genetic level [89]. Genetic mutations in the *IL4* and *IL9* genes were also associated with levels of natural antibodies (NAb) IgM and IgG [90]. NAb are antibodies that can bind antigens without prior exposure to the antigen [91]. These associations were mostly associative genetic effects on feather damage scores and not direct genetic effects, suggesting that NAb levels may be related to the propensity to perform feather pecking. This is further supported by the finding that when cage mates had higher NAb IgG levels, the individual had more feather damage [90]. Genetic associations were further found between severe feather pecking and specific antibody responses [92], indicating that there are genes simultaneously involved in both feather pecking and specific antibody response. Interestingly, several genes involved in immune responses, for example, TNF ligand and mitogen-activated protein kinase, were either upregulated or downregulated in the hypothalamus of feather-pecking birds compared to neutrals and victim birds [40]. Furthermore, a chicken line performing more feather damage showed upregulation of genes related to immune system processes in the brain compared to a chicken line showing less feather damage [93,94]. These findings provide additional arguments supporting a relationship between the immune system and feather pecking (see also Brunberg et al. [95]).

Further evidence for a relationship between the immune system and feather pecking comes from lines that were divergently selected on feather pecking and that differ in several immune parameters. High-feather-pecking (HFP) birds showed a higher antibody response to infectious bursal disease virus vaccination, while low-feather-pecking (LFP) birds had a higher number of white blood cells and higher expression of MHC class I molecules on T (CD4, CD8) and B cells [96]. Recently, the FP selection lines were shown to differ in both innate and adaptive immune characteristics, with HFP birds having lower IgM NAb but higher IgG NAb levels, specific antibody levels, and nitric oxide production by monocytes compared to LFP birds [97]. These findings suggest that HFP and LFP birds differ in immune responsiveness and provide further support to a relationship between the immune system and feather pecking. Yet, these relationships could be the result of genes that are simultaneously involved in the immune system and in feather pecking, as also indicated by previous studies [89,98].

It remains to be elucidated whether these relationships between the immune system and feather pecking are causal. Preliminary findings show that the immune system may play a role in feather pecking. Birds that received an immune challenge at a young age showed more feather damage at an adult age [99], suggesting that activation of the specific immune response at a young age may stimulate birds to feather peck. Following this rationale, it can be considered that a health issue in a flock, such as infection implying immune system activation, may increase the risk of feather pecking in the future. More research is needed on this topic.

3.1.2. Other Impacts of Health on Damaging Behaviour

The health and integument status of laying hens are closely related. Plumage presence, persistence, and distribution on the body can be indicative of the nutritional status, health, and behaviour of the birds [25,100]. Close inspection of growing feathers can also provide information about physiological and systemic infectious issues while the feathers are formed.

Other dimensions of health, such as parasitic infestation, may affect the development of damaging behaviour in laying hens. Parasitic infestation, for example, with *Ascaridia galli*, can decrease health, performance production, and plumage coverage in layer flocks [101]. In this study, parasitic infestation was significantly associated with plumage damage, while treated animals showed better plumage conditions. The authors claim that lower worm burdens were associated with improved plumage condition, possibly through reduced parasite-induced stress, without providing a precise explanation of the mechanism. These results are consistent with the previous hypothesis of this review, where immune stimulation might trigger feather pecking.

Concerning external parasites, red mite (*Dermanyssus gallinae*) infestation can cause anaemia, while the presence of red mites can also lead to itching, disturbing the flock, and possibly acting as a trigger for injurious pecking [100]. The poultry red mite is the most common ectoparasite on laying hen farms worldwide, causing considerable economic losses and reduced hen health and welfare. Even in moderate numbers, they can cause considerable stress, agitation, and severe feather pecking in hens. As an example, it was shown in a study undertaken in 47 Belgian aviaries that the plumage condition of the flock is better on farms with no red mite infestations [25]. Temple et al. [102], in an experiment where infested layers were treated with fluralaner (Exzolt®), showed improvements in behavioural variables (less preening, head scratching, head checking, severe feather pecking, and aggressive behaviour), physiological biomarkers, and health parameters following the elimination of red mites on a commercial farm. These results indicate that infestations can reduce hen welfare. The severity of feather pecking associated with red mite infestation may increase in non-beak-trimmed flocks.

Other mites, such as the northern fowl mite (*Ornithonyssus sylviarum*), are also key pest species for caged laying hens. Jacobs et al. [103] showed that mite-infested hens had increased nocturnal activity, including preening, as well as fragmentation of behavioural activities together with decreased dozing, indicating disturbed resting behaviour and suggesting a reduction in the welfare of hens infested by these mites.

Plumage and integument damage can also result from clinical diseases, such as diarrhoea or nutrient deficiency. Hens perform more feather pecking when diets contain mineral, protein, or amino acid (methionine, arginine) levels below recommended levels [104]. Systemic bacterial infections such as *Erysipelas* can be associated with poor feather coverage and skin damage [100].

These findings indicate that health issues may stimulate damaging behaviour, but more research is needed to explain the mechanisms involved and to identify prevention strategies. The following chapter explores the consequences of damaging behaviour on laying hen health outcomes.

3.2. Impact of Damaging Behaviours on Health

When discussing the effects of damaging behaviour on the physical and mental health of laying hens, we are primarily referring to the "victim", i.e., "the recipient". First of all, the feather-pecking activity may degrade feather cover in recipients, which may interfere with the bird's body heat regulation, and hens that have lost parts of their plumage are extremely susceptible to the cold [105]. Chickens are sensitive to touch; their skin contains numerous receptors for temperature, pressure, and pain [106]. In crowded systems, feather loss may give rise to skin damage caused by abrasion from the environment and flock mates [57]. Additionally, skin damage can trigger cannibalism [107], often resulting in the mortality of recipients. It has been shown that the victims of cannibalism have lower body weight than feather peckers [108,109]. Furthermore, feather damage may impact the structural cohesiveness of the feathers and lower the aerodynamic capacity of the wings [110,111], making them less efficient in helping to maintain balance [112], which can be problematic when using perches and navigating through a complex 3D aviary environment.

Even feather removal is a strong stressor for a bird; during feather pecking, the bird being pecked often shows crouching immobility with no outward sign of pain. Gentle [113] explained this immobility as learned helplessness, which develops when an animal experiences traumatic events that are aversive and that continue to happen independently of any attempts by the animal to reduce or eliminate them. Studies have shown that during initial feather removal, the birds become agitated, with wing flapping and/or vocalisation and increased heart rate, blood pressure, and EEG arousal as clear signs of pain. Over time, the continued removal of feathers does not produce an exaggerated escape response but an immobile "helplessness" state. During this period of immobility, the EEG of the victim shows activity similar to that seen in sleep or catatonic states, such as tonic immobility. Basically, this is an anti-predator strategy following capture to prevent further damage produced by struggling and to allow escape should the occasion arise. This strategy is, however, counterproductive in production systems where hens have no possibility to escape and are, in effect, making themselves available to be pecked [113]. This type of learned helplessness or anticipation of the negative event may lead to the appearance of negative emotions in hens related to fear and anxiety [106].

Tahamtani et al. [109] suggest that feather peckers and victims experienced similar levels of negative experiences during rearing, causing stress and developmental instability, leading to either pecker or victim status. For example, it is considered that fearfulness, proactive coping, or hyperactivity may predispose chickens to develop severe feather pecking. In the study by Kops et al. [114], the severe feather-pecking problem was discussed because of the lack of monoamines (serotonin and dopamine) in certain brain areas, which affects both emotional perception and behavioural output. Due to neurochemical deficits early in life, high-feather-pecking-line chickens are prone to increased general behavioural activity. In turn, this hyperactivity seems to be a clear risk factor for the development of feather pecking.

To conclude, damaging behaviour leads to denuded overall plumage, with an increased risk of poor thermoregulation, skin damage, and possibly wounds with an increased risk of infection (infection of the skin and tissues and peritonitis). These effects act negatively on hen health and welfare and possibly lead to increased mortality [18,27,50,60].

Consequently, there is a clear need to monitor laying hen health and welfare in order to ensure early detection of damaging behaviour and/or health issues and to use corrective measures. Most modern poultry husbandry systems house thousands of animals in a single barn, leading to challenges in the assessment of individual animals. The next chapter will summarise current knowledge on monitoring systems allowing early detection of damaging behaviour and health issues in order to prevent their spread.

4. Systems for Early Monitoring of Animals in Modern Housing Facilities in Order to Limit Occurrence and Spread of Both Health Disorders and Damaging Behaviour

Monitoring of damaging behaviours and health of laying hens can be performed through monitoring of the animals themselves, e.g., behaviour or body condition, or through monitoring of resources, including feed or water consumption and egg production.

4.1. Monitoring Tools Based on Direct Observation

In order to identify the risk of compromised health and damaging behaviours at an early stage, it is essential to develop effective and efficient quantitative assessment methods that can easily be applied on commercial farms. Several methods have been developed in order to assess animal welfare in animal husbandry, consisting of the collection of different animal health or welfare parameters from a sample of birds.

The Welfare Quality® [115] method proposes an overall assessment of laying hen welfare on the farm and at the slaughterhouse. Although the evaluation is extensive, the application of the protocol in the livestock requires several hours and needs to be performed by trained assessors. In addition, part of the assessment is conducted at the slaughterhouse and consists of collecting data on indicators that are known to be related to the health and living conditions encountered by the animal on the farm, during transport, or at the slaughterhouse before being killed [116]. The disadvantage is that these are post-mortem observations, which do not allow for corrective actions to be taken on the animals or on the management of the farm, if necessary.

To assess feather damage, numeric rating scales for scoring schemes have been developed and employed in past studies. Current scoring methods [115,117–119] differ in the details they record, the type of feathers assessed, the number of body areas assessed, and whether or not birds are captured and handled during the assessment. For instance, Decina et al. [120] compared two feather scoring systems [112,119] based on user-friendliness and reliability [120]. The AssureWel scoring system is the easiest to use and achieves the most consistent outcomes among scorers for the back area of the body. The LayWel system does not provide descriptive definitions of the scores but rather provides photographs as a reference (1–4 scoring scale), while the AssureWel [121] system provides both definitions of scores (0–2 scoring scale) and photographs. AssureWel proposes an overall method of assessment based, for instance, on feather loss, bird cleanliness, observation of antagonistic behaviours, and flightiness.

Animals can be stressed by protocols that require them to be handled for close examination of their physical condition, which may affect some results [122]. To avoid this source of stress, a monitoring approach can be used based on line transects [122–128]. The transect method assesses the frequency at which animals show clear signs of impaired welfare by noting their incidence while walking along predefined paths or transects that are established among the corridors delimited by drinkers and feeder lines. A new method adapted to aviary has been developed by Vasdal et al. [128], where all the birds observed with feather loss are noted, including those on the littered floor, in the width of the space under the aviary structure, and on each tier of the structure. The scores are standardised by the estimated number of birds in the surveyed area, thus enabling comparisons of the prevalence of various welfare issues between flocks under different husbandry conditions. Several tools have recently been proposed on this principle, sometimes with the development of a smartphone app for easy collection of data and poultry welfare self-assessment by farmers, such as EBENE® for broilers and hens [129], or i-Watchturkey and i-Watchbroiler for turkeys and broilers [130]. These methods allow for shorter durations of welfare assessments. They offer producers multiple possibilities to conduct quantitative flock assessment and apply the necessary corrective actions, and multiple possibilities for the industry in the area of digitalisation and to make informed data-based decisions along the production chain.

In general, the simplicity and time efficiency of the methods are critical aspects to encourage the adoption of the protocol by farmers. The Hennovation or Featherwel projects propose recommendations to improve health and welfare and, to some extent, the consequences of damaging behaviours. For instance, Featherwel enables farmers to regularly monitor the flock via frequent inspections, observing bird behaviour and performing feather scoring to identify injurious pecking early on and to help in the implementation of strategies before the problem becomes more serious.

These methods have the advantage of relying on bird observation and reinforcing the relationships between the farmer and the layers. However, they are time-consuming and, therefore, cannot be run in a continuous manner, allowing only episodic assessment. Other automatic methods allowing continuous assessment of bird health and welfare are detailed below.

4.2. Monitoring Tools Based on Precision Livestock Farming

A wide range of sensor technologies can be used to monitor and control damaging behaviour while also minimising consequences on animal health and welfare [131]. Precision livestock farming (PLF) enables real-time and continuous monitoring and management of livestock using modern sensor technologies [132]. In this way, a problem can be identified and diagnosed during the lifetime of the animal so that appropriate corrective measures can be taken immediately if alert criteria are exceeded and before the problem worsens. PLF covers the field of sensors that carry out measurements on animals or in their environment and information and communication technologies that are used to store and transfer data.

4.2.1. Group Monitoring

According to Rowe et al. [133], most PLF strategies use image analysis to measure welfare in poultry farming (42% out of 264 publications). This is because surveillance camera systems combined with image processing techniques are inexpensive ways of providing objective measures of poultry behaviour without having to enter the barn, which involves behavioural changes in the animals. The most common video analysis method is based on counting and identifying small squares (pixels) that turn on and off for a given period of time. Specifically, these methods analyse the variation in brightness or intensity of pixels (on or off) per area of an image, both in time and space. The general idea of these methods is based on the relationship between the number of pixels that turn on and off and the activity of animals in a given unit of area. This method uses cameras to take pictures and analyse the flow. The algorithm then automatically and continuously generates four aggregated statistical values over 15 min sequences (mean, variance, skewness, and flattening) [134]. This method can quantitatively assess variations in the activity of the poultry flock (at the group level) but does not directly account for the welfare of the animals. To do this, individualised monitoring is necessary. For instance, Lee et al. [135] have used optical flow measures as indicators of bird movement, thanks to measures of disturbance using hidden Markov models. Based on these disturbance measures and age-related variables, the authors were able to predict the levels of severe feather damage in flocks in future weeks.

The use of microphones appears to be less widespread in poultry farming (14% of publications [133]). However, sound signals play an important role in animal communication, and some signals may reflect the welfare and health status of the animals. They are used to warn other animals or to communicate with each other, for example, to maintain contact or attract other animals [136]. The *Gallus gallus* species expresses at least 24 different calls to communicate. Chicks between 2 and 3 days old have a repertoire of different vocalisations, from distress calls to pleasure trills and fear trills [137,138]. Certain vocalisations can easily be seen as indicators of animal welfare status [139]. The finer characterisation of vocalisations enables the measurement of welfare indicators reflecting the emotional state of the birds (e.g., warning calls, coughing). The study of these acoustic indicators has made it possible to highlight in recent work an inverse relationship between the live weight of the animals and the peak frequency of their vocalisations. This could enable farmers to identify deterioration in poultry performance early or to predict the weight of animals at slaughter [140,141]. More specifically, acoustic studies are interesting for detecting stress or panic states or abnormal noise on the farm. For example, teams of researchers have focused on identifying rales, characteristic symptoms of respiratory infections in poultry [142,143]. A recent study has developed, under experimental conditions, an algorithm for detecting sneezing in groups of 15 to 36 broilers, with an accuracy of 88% and sensitivity of 67% [144]. Today, the digital processing of sound signals allows various digital descriptions and statistical examinations of the animals' vocalisations [136]. However, the extraction of sufficient, high-quality signals from animals remains a problem, and well-adapted procedures are required, including noise suppression to remove parasitic noises, such as ventilation noise. Like in research in imaging, artificial intelligence (AI) techniques are being developed for sound signal processing.

4.2.2. Individual Monitoring

In recent years, it has become increasingly possible to monitor individual animals, even within large groups, such as in non-cage systems [145]. A very successful example is that of the dairy cow sector, where it has become standard on many farms for every cow in the herd to be equipped with a sensor and for performance and health to be tracked continuously and fully automatically, with clear positive effects on health, welfare, and production. In the poultry industry, a range of PLF applications has been explored to track individual animals and their health. Tracking allows the recording of information at the individual level, such as the location of the animal, the distance travelled, or the speed of movement. Some solutions require the animals to be individually tagged in order to be tracked. For instance, Banerjee et al. [146] attached wireless sensors to laying hens to monitor their individual activity [146]. Zaninelli et al. [147] used radio-frequency identification (RFID) transponders that were injected into the hen's feet to collect data on individual behaviour and laying performance (the transponder was injected into the interdigital portion of each hen's right foot). Injecting the sensor technology into the animal reduces the impact of wearing a sensor, although studies have shown that this impact is minimal and that hens habituate quickly to wearing them [148,149]. Active, ultra-wideband RFID systems have proven to be promising to monitor the location and activity of individual birds, especially when combined with accelerometers, which can provide information on very specific behaviour such as feather pecking [150].

Different tagging technologies can be used, but in some cases, tagging is not suitable for young chicks [151]. So far, these individual monitoring systems are only suitable for experimental studies, and those for laying hens have been tested on rather small samples of birds in a research setting. They are not yet commercially available. Reasons for this are mainly technical, including interference of sensors with the environment, overlap of detection zones in the layer house, and short battery life of the sensors [145]. Another reason could be the cost of equipping every single bird with a sensor. However, the price of this type of technology has been dropping significantly in the last few decades as more and more researchers and producers are exploring the use of sensor technology for livestock production [152]. In the specific context of the poultry sector, individual tagging technologies can be used in a more explorative way, for instance, to assess the different housing systems and their impact on production, health, and welfare. However, it is challenging to develop this system in the field due to the very high number of animals to equip (duration for attaching and removing devices from each individual before slaughtering, data treatment, etc.).

Tagless tracking solutions are also being developed with the use of video. Several steps are required. The first stage of tracking is the detection of individuals in each frame of the video. For the detection of individuals, segmentation is a classic solution that works with low animal density but is sensitive to illumination because they are based on the intensity (brightness) of the pixels. Moreover, even though one can determine with these methods whether a pixel belongs to a chicken or to the ground, it is still complicated to determine which chicken it belongs to when two animals are close to each other.

Faced with the limitations of classic segmentation methods, for example, in the case of higher animal densities, researchers now use AI. Supervised AI allows learning by the machine by showing it thousands of labelled and categorised examples. In this way, the machine becomes capable of correctly classifying most of the images it is shown [153]. A database of characterised images is needed for learning the model, but it is most useful when the model is deemed functional. AI-based detection is much more robust and faster than conventional methods. The major limitation of AI is that more powerful and high-performance machines are needed to allow for great numbers and sometimes more complex calculations. The next step after detection is the tracking of individuals. This does not specifically require the use of AI; classic methods can be used. Recently, a team has started to carry out tracking without marking a small number of laying hens (5) in controlled conditions [154].

In summary, the recent shift to more non-cage production systems in the European Union has created the need for new ways of monitoring and managing the health and welfare of individual laying hens. At the current rate at which technology is evolving and sensor prices are dropping, a sensor for each individual laying hen is not some far-off frontier. Individual monitoring of laying hens will enable farmers to keep track of the health status and behaviour of their birds and to anticipate the spread of damaging behaviour or infections, for example, by removing birds from the flock that are showing pecking damage or symptoms of an infection, indicated by reduced activity or feeding behaviour. Finally, the data from sensors can be used to optimise breeding programmes and to breed out traits such as feather pecking in the long term.

5. Conclusions

This review shows the close relationships between damaging behaviours and health in modern husbandry systems for laying hens, which increasingly house the animals in cage-free groups of thousands of birds. These new housing conditions will offer birds more freedom to fulfil their behavioural priorities and, consequently, will reinforce interactions between animals. In case of suboptimal rearing and/or housing and management conditions, damaging behaviour or infectious diseases will be likely to spread to the whole flock. Additionally, health issues and, therefore, stimulation of the immune system may, in certain situations, lead to the development of damaging behaviours, which in turn may result in impaired body condition, leading to further health and welfare issues. This highlights the need to monitor both behaviour and health of laying hens in order to intervene as quickly as possible to preserve the health and welfare of animals, as well as farmer income and work satisfaction.

Author Contributions: Conceptualization, V.M. and M.G.; writing—original draft preparation, V.M., J.B., N.B., J.v.d.E., I.E., T.M., R.R., T.B.R., E.N.S. and M.G.; writing—review and editing, V.M. and M.G. All authors have read and agreed to the published version of the manuscript.

Funding: This research received no external funding.

Institutional Review Board Statement: Not applicable.

Informed Consent Statement: Not applicable.

Data Availability Statement: Not applicable.

Acknowledgments: This article is based upon work in the COST Action CA15134—Synergy for preventing damaging behaviour in group housed pigs and chickens (GroupHouseNet), supported by COST (European Cooperation in Science and Technology; www.cost.eu).

Conflicts of Interest: The authors declare no conflict of interest.

References

1. Michel, V.; Guinebretière, M. Technical challenges of laying hen production in cages in Sustainable animal production. In *Sustainable Animal Production: The Challenges and Potential Developments for Professional Farming*; Wageningen Academic Publishers: Wageningen, The Netherlands, 2009; pp. 329–345.
2. Relić, R.; Sossidou, E.; Dedousi, A.; Peric, L.; Božičković, I.; Đukić Stojčić, M. Behavioral and health problems of poultry related to rearing systems. *Ank. Univ. Vet. Fak. Derg.* **2019**, *66*, 423–428. [CrossRef]
3. Campbell, D.L.M.; de Haas, E.N.; Lee, C. A review of environmental enrichment for laying hens during rearing in relation to their behavioral and physiological development. *Poult. Sci.* **2019**, *98*, 9–28. [CrossRef] [PubMed]
4. European Commission Eggs Market Situation Dashboard. Available online: https://ec.europa.eu/info/sites/default/files/food-farming-fisheries/farming/documents/eggs-dashboard_en.pdf (accessed on 6 April 2022).
5. Dikmen, B.Y.; İpek, A.; Şahan, Ü.; Petek, M.; Sözcü, A. Egg production and welfare of laying hens kept in different housing systems (conventional, enriched cage, and free range). *Poult. Sci.* **2016**, *95*, 1564–1572. [CrossRef] [PubMed]
6. Bari, M.S.; Laurenson, Y.C.S.M.; Cohen-Barnhouse, A.M.; Walkden-Brown, S.W.; Campbell, D.L.M. Effects of outdoor ranging on external and internal health parameters for hens from different rearing enrichments. *PeerJ* **2020**, *8*, e8720. [CrossRef] [PubMed]
7. Rodriguez-Aurrekoetxea, A.; Estevez, I. Use of space and its impact on the welfare of laying hens in a commercial free-range system. *Poult. Sci.* **2016**, *95*, 2503–2513. [CrossRef]
8. Lin, X.; Zhang, R.; Jiang, S.; El-Mashad, H.; Xin, H. Emissions of ammonia, carbon dioxide and particulate matter from cage-free layer houses in California. *Atmos. Environ.* **2017**, *152*, 246–255. [CrossRef]
9. Zhao, Y.; Shepherd, T.A.; Swanson, J.C.; Mench, J.A.; Karcher, D.M.; Xin, H. Comparative evaluation of three egg production systems: Housing characteristics and management practices. *Poult. Sci.* **2015**, *94*, 475–484. [CrossRef]
10. Campbell, D.L.; Makagon, M.M.; Swanson, J.C.; Siegford, J.M. Litter use by laying hens in a commercial aviary: Dust bathing and piling. *Poult. Sci.* **2016**, *95*, 164–175. [CrossRef]

11. Zhao, Y.; Shepherd, T.A.; Li, H.; Xin, H. Environmental assessment of three egg production systems—Part I: Monitoring system and indoor air quality. *Poult. Sci.* **2015**, *94*, 518–533. [CrossRef]

12. Zhao, Y.; Zhao, D.; Ma, H.; Liu, K.; Atilgan, A.; Xin, H. Environmental assessment of three egg production systems—Part III: Airborne bacteria concentrations and emissions. *Poult. Sci.* **2016**, *95*, 1473–1481. [CrossRef]

13. Le Bouquin, S.; Huneau-Salaün, A.; Huonnic, D.; Balaine, L.; Martin, S.; Michel, V. Aerial dust concentration in cage-housed, floor-housed, and aviary facilities for laying hens. *Poult. Sci.* **2013**, *92*, 2827–2833. [CrossRef] [PubMed]

14. David, B.; Moe, R.O.; Michel, V.; Lund, V.; Mejdell, C. Air Quality in Alternative Housing Systems May Have an Impact on Laying Hen Welfare. Part I-Dust. *Animals* **2015**, *5*, 495–511. [CrossRef] [PubMed]

15. Rodenburg, T.B.; De Reu, K.; Tuyttens, F.A.M. Performance, welfare, health and hygiene of laying hens in non-cage systems in comparison with cage systems. In *Alternative Systems for Poultry: Health, Welfare and Productivity*; CABI: Wallingford, UK, 2012; pp. 210–224.

16. Rodenburg, T.B.; Tuyttens, F.A.M.; de Reu, K.; Herman, L.; Zoons, J.; Sonck, B. Welfare assessment of laying hens in furnished cages and non-cage systems: An on-farm comparison. *Anim. Welf.* **2008**, *17*, 363–373.

17. Riber, A.B.; Casey-Trott, T.M.; Herskin, M.S. The Influence of Keel Bone Damage on Welfare of Laying Hens. *Front. Vet. Sci.* **2018**, *5*, 6. [CrossRef]

18. Nicol, C.J.; Bestman, M.; Gilani, A.M.; De Haas, E.N.; De Jong, I.C.; Lambton, S.; Wagenaar, J.P.; Weeks, C.A.; Rodenburg, T.B. The prevention and control of feather pecking: Application to commercial systems. *Worlds Poult. Sci. J.* **2013**, *69*, 775–788. [CrossRef]

19. Bestman, M.W.P.; Koene, P.; Wagenaar, J.P. Influence of farm factors on the occurrence of feather pecking in organic reared hens and their predictability for feather pecking in the laying period. *Appl. Anim. Behav. Sci.* **2009**, *121*, 120–125. [CrossRef]

20. Coton, J.; Guinebretière, M.; Guesdon, V.; Chiron, G.; Mindus, C.; Laravoire, A.; Pauthier, G.; Balaine, L.; Descamps, M.; Bignon, L.; et al. Feather pecking in laying hens housed in free-range or furnished-cage systems on French farms. *Br. Poult. Sci.* **2019**, *60*, 617–627. [CrossRef]

21. De Haas, E.N.; Bolhuis, J.E.; de Jong, I.C.; Kemp, B.; Janczak, A.M.; Rodenburg, T.B. Predicting feather damage in laying hens during the laying period. Is it the past or is it the present? *Appl. Anim. Behav. Sci.* **2014**, *160*, 75–85. [CrossRef]

22. Leeson, S.; Morrison, W.D. Effect of Feather Cover on Feed Efficiency in Laying Birds. *Poult. Sci.* **1978**, *57*, 1094–1096. [CrossRef]

23. Tullett, S.G.; MacLeod, M.G.; Jewitt, T.R. The effects of partial defeathering on energy metabolism in the laying fowl. *Br. Poult. Sci.* **1980**, *21*, 241–245. [CrossRef]

24. El-Lethey, H.; Aerni, V.; Jungi, T.W.; Wechsler, B. Stress and feather pecking in laying hens in relation to housing conditions. *Br. Poult. Sci.* **2000**, *41*, 22–28. [CrossRef] [PubMed]

25. Heerkens, J.L.T.; Delezie, E.; Kempen, I.; Zoons, J.; Ampe, B.; Rodenburg, T.B.; Tuyttens, F.A.M. Specific characteristics of the aviary housing system affect plumage condition, mortality and production in laying hens. *Poult. Sci.* **2015**, *94*, 2008–2017. [CrossRef] [PubMed]

26. Gentle, M.J.; Hunter, L.N. Physiological and behavioural responses associated with feather removal in *Gallus gallus* var domesticus. *Res. Vet. Sci.* **1991**, *50*, 95–101. [CrossRef]

27. Green, L.E.; Lewis, K.; Kimpton, A.; Nicol, C.J. Cross-sectional study of the prevalence of feather pecking in laying hens in alternative systems and its associations with management and disease. *Vet. Rec.* **2000**, *147*, 233–238. [CrossRef]

28. Baker, J.; Greer, W. *Animal Health: A Layman's Guide to Disease Control*; IPP The Interstate Printers & Publishers Inc.: Danville, IL, USA, 1980.

29. Gunnarson, S. Definition of health and disease in text books of veterinarian medicine. Animal production in Europe: The way forward in a changing world. In Proceedings of the 13th International Society of Animal Hygien, Saint-Malo, France, 11–13 October 2004; pp. 105–109.

30. OIE. *The One Health Concept: The OIE Approach*; OIE: Paris, France, 2013.

31. Estevez, I.; Andersen, I.-L.; Nævdal, E. Group size, density and social dynamics in farm animals. *Appl. Anim. Behav. Sci.* **2007**, *103*, 185–204. [CrossRef]

32. Estevez, I.; Keeling, L.; Newberry, R. Decreasing aggression with increasing group size in young domestic fowl. *Appl. Anim. Behav. Sci.* **2003**, *84*, 213–218. [CrossRef]

33. Estevez, I.; Newberry, R.; Keeling, L. Dynamics of aggression in the domestic fowl. *Appl. Anim. Behav. Sci.* **2002**, *76*, 307–325. [CrossRef]

34. Milisits, G.; Szász, S.; Donkó, T.; Budai, Z.; Almási, A.; Pőcze, O.; Ujvári, J.; Farkas, T.P.; Garamvölgyi, E.; Horn, P.; et al. Comparison of Changes in the Plumage and Body Condition, Egg Production, and Mortality of Different Non-Beak-Trimmed Pure Line Laying Hens during the Egg-Laying Period. *Animals* **2021**, *11*, 500. [CrossRef]

35. Spindler, B.; Giersberg, M.; Andersson, R.; Kemper, N. Keeping laying hens with untrimmed beaks—A Review of the status quo in practice and science. *Züchtungskunde* **2016**, *88*, 475–493.

36. Anses. *AVIS 2016-SA-0288 de l'Anses Relatif au «Bien-être Animal: Contexte, Définition et Évaluation»*; Anses: Maisons-Alfort, France, 2008; pp. 1–34.

37. De Haas, E.N.; Newberry, R.C.; Edgar, J.; Riber, A.B.; Estevez, I.; Ferrante, V.; Hernandez, C.E.; Kjaer, J.B.; Ozkan, S.; Dimitrov, I.; et al. Prenatal and Early Postnatal Behavioural Programming in Laying Hens, with Possible Implications for the Development of Injurious Pecking. *Front. Vet. Sci.* **2021**, *8*, 693. [CrossRef]

38. Lay, D.C.; Fulton, R.M.; Hester, P.Y.; Karcher, D.M.; Kjaer, J.B.; Mench, J.A.; Mullens, B.A.; Newberry, R.C.; Nicol, C.J.; O'Sullivan, N.P.; et al. Hen welfare in different housing systems1. *Poult. Sci.* **2011**, *90*, 278–294. [CrossRef] [PubMed]
39. Jones, M.P.; Heidenreich, B. Behavior of Birds of Prey in Managed Care. *Vet. Clin. N. Am. Exot. Anim. Pract.* **2021**, *24*, 153–174. [CrossRef] [PubMed]
40. Brunberg, E.; Jensen, P.; Isaksson, A.; Keeling, L. Feather pecking behavior in laying hens: Hypothalamic gene expression in birds performing and receiving pecks. *Poult. Sci.* **2011**, *90*, 1145–1152. [CrossRef]
41. Huber-Eicher, B.; Sebö, F. The prevalence of feather pecking and development in commercial flocks of laying hens. *Appl. Anim. Behav. Sci.* **2001**, *74*, 223–231. [CrossRef]
42. Sherwin, C.M.; Richards, G.J.; Nicol, C.J. Comparison of the welfare of layer hens in 4 housing systems in the UK. *Br. Poult. Sci.* **2010**, *51*, 488–499. [CrossRef] [PubMed]
43. Campderrich, I.; Liste, G.; Estevez, I. The looks matter; aggression escalation from changes on phenotypic appearance in the domestic fowl. *PLoS ONE* **2017**, *12*, e0188931. [CrossRef] [PubMed]
44. Rodenburg, T.B.; Komen, H.; Ellen, E.D.; Uitdehaag, K.A.; van Arendonk, J.A.M. Selection method and early-life history affect behavioural development, feather pecking and cannibalism in laying hens: A review. *Appl. Anim. Behav. Sci.* **2008**, *110*, 217–228. [CrossRef]
45. Savory, C.J. Feather pecking and cannibalism. *Worlds Poult. Sci. J.* **1995**, *51*, 215–219. [CrossRef]
46. Savory, J. Nutrition, Feeding and Drinking Behaviour, and Welfare. In *The Welfare of Domestic Fowl and Other Captive Birds*; Duncan, I.J.H., Hawkins, P., Eds.; Springer: Dordrecht, The Netherlands, 2010; pp. 165–187.
47. Pötzsch, A.; Lewis, K.; Nicol, C.J.; Green, L.E. A cross-sectional study of the prevalence of vent pecking in laying hens in alternative systems and its associations with feather pecking, management and disease. *Appl. Anim. Behav. Sci.* **2001**, *74*, 259–272. [CrossRef]
48. Krause, E.T.; Petow, S.; Kjaer, J.B. A note on the physiological and behavioural consequences of cannibalistic toe pecking in laying hens (*Gallus gallus* domesticus). *Arch. Für Geflugelkd.* **2011**, *75*, 140–143.
49. Leonard, M.L.; Horn, A.G.; Fairfull, R.W. Correlates and consequences of allopecking in White Leghorn chickens. *Appl. Anim. Behav. Sci.* **1995**, *43*, 17–26. [CrossRef]
50. Rodenburg, T.B.; Van Krimpen, M.M.; De Jong, I.C.; De Haas, E.N.; Kops, M.S.; Riedstra, B.J.; Nordquist, R.E.; Wagenaar, J.P.; Bestman, M.; Nicol, C.J. The prevention and control of feather pecking in laying hens: Identifying the underlying principles. *Worlds Poult. Sci. J.* **2013**, *69*, 361–374. [CrossRef]
51. Cloutier, S.; Newberry, R.C.; Honda, K.; Alldredge, J.R. Cannibalistic behaviour spread by social learning. *Anim. Behav.* **2002**, *63*, 1153–1162. [CrossRef]
52. Tablante, N.L.; Vaillancourt, J.P.; Martin, S.W.; Shoukri, M.; Estevez, I. Spatial distribution of cannibalism mortalities in commercial laying hens. *Poult. Sci.* **2000**, *79*, 705–708. [CrossRef] [PubMed]
53. Cloutier, S.; Newberry, R.C.; Forster, C.T.; Girsberger, K.M. Does pecking at inanimate stimuli predict cannibalistic behaviour in domestic fowl? *Appl. Anim. Behav. Sci.* **2000**, *66*, 119–133. [CrossRef]
54. Kajlich, A.S.; Shivaprasad, H.L.; Trampel, D.W.; Hill, A.E.; Parsons, R.L.; Millman, S.T.; Mench, J.A. Incidence, Severity, and Welfare Implications of Lesions Observed Postmortem in Laying Hens from Commercial Noncage Farms in California and Iowa. *Avian Dis.* **2016**, *60*, 8–15. [CrossRef]
55. Craig, J.V.; Lee, H.Y. Beak trimming and genetic stock effects on behavior and mortality from cannibalism in White Leghorn-type pullets. *Appl. Anim. Behav. Sci.* **1990**, *25*, 107–123. [CrossRef]
56. Aerni, V.; El-Lethey, H.; Wechsler, B. Effect of foraging material and food form on feather pecking in laying hens. *Br. Poult. Sci.* **2000**, *41*, 16–21. [CrossRef]
57. Blokhuis, H.J.; van der Haar, J.W. Effects of pecking incentives during rearing on feather pecking of laying hens. *Br. Poult. Sci.* **1992**, *33*, 17–24. [CrossRef]
58. Dixon, L. Feather Pecking Behaviour and associated Welfare issues in Laying Hens. *Avian Biol. Res.* **2008**, *1*, 73–87. [CrossRef]
59. Gyaryahu, G.; Robinzon, B.; Snapir, N. The effect of environmental enrichment on egg-layers: Five years of research. *Poult. Sci.* **1998**, *77* (Suppl. 1), 1842.
60. Lambton, S.L.; Knowles, T.G.; Yorke, C.; Nicol, C.J. The risk factors affecting the development of gentle and severe feather pecking in loose housed laying hens. *Appl. Anim. Behav. Sci.* **2010**, *123*, 32–42. [CrossRef]
61. Guinebretière, M.; Mika, A.; Michel, V.; Balaine, L.; Thomas, R.; Keïta, A.; Pol, F. Effects of Management Strategies on Non-Beak-Trimmed Laying Hens in Furnished Cages that Were Reared in a Non-Cage System. *Animals* **2020**, *10*, 399. [CrossRef]
62. Liebers, C.J.; Schwarzer, A.; Erhard, M.; Schmidt, P.; Louton, H. The influence of environmental enrichment and stocking density on the plumage and health conditions of laying hen pullets. *Poult. Sci.* **2019**, *98*, 2474–2488. [CrossRef] [PubMed]
63. Schreiter, R.; Damme, K.; Klunker, M.; Raoult, C.; von Borell, E.; Freick, M. Effects of edible environmental enrichments during the rearing and laying periods in a littered aviary—Part 1: Integument condition in pullets and laying hens. *Poult. Sci.* **2020**, *99*, 5184–5196. [CrossRef] [PubMed]
64. Bestman, M.; Wagenaar, J.P. Health and Welfare in Dutch Organic Laying Hens. *Animals* **2014**, *4*, 374–390. [CrossRef] [PubMed]
65. Jung, L.; Brenninkmeyer, C.; Niebuhr, K.; Bestman, M.; Tuyttens, F.A.M.; Gunnarsson, S.; Sørensen, J.T.; Ferrari, P.; Knierim, U. Husbandry Conditions and Welfare Outcomes in Organic Egg Production in Eight European Countries. *Animals* **2020**, *10*, 2102. [CrossRef]

66. Falker-Gieske, C.; Iffland, H.; Preuß, S.; Bessei, W.; Drögemüller, C.; Bennewitz, J.; Tetens, J. Meta-analyses of genome wide association studies in lines of laying hens divergently selected for feather pecking using imputed sequence level genotypes. *BMC Genet.* **2020**, *21*, 114. [CrossRef] [PubMed]

67. Rodenburg, T.B.; de Haas, E.N.; Nielsen, B.L.; Buitenhuis, A.J. Fearfulness and feather damage in laying hens divergently selected for high and low feather pecking. *Appl. Anim. Behav. Sci.* **2010**, *128*, 91–96. [CrossRef]

68. Riedstra, B.; Groothuis, T.G.G. Early feather pecking as a form of social exploration: The effect of group stability on feather pecking and tonic immobility in domestic chicks. *Appl. Anim. Behav. Sci.* **2002**, *77*, 127–138. [CrossRef]

69. Kjaer, J.; Wuerbal, H.; Schrader, L. Perseveration in a guessing task by laying hens selected for high or low levels of feather pecking does not support classification of feather pecking as a stereotypy. *Appl. Anim. Behav. Sci.* **2015**, *168*, 56–60. [CrossRef]

70. Kjaer, J.B. Feather pecking in domestic fowl is genetically related to locomotor activity levels: Implications for a hyperactivity disorder model of feather pecking. *Behav. Genet.* **2009**, *39*, 564–570. [CrossRef] [PubMed]

71. Krause, E.T.; Phi-van, L.; Dudde, A.; Schrader, L.; Kjaer, J.B. Behavioural consequences of divergent selection on general locomotor activity in chickens. *Behav. Processes* **2019**, *169*, 103980. [CrossRef] [PubMed]

72. Borda-Molina, D.; Iffland, H.; Schmid, M.; Müller, R.; Schad, S.; Seifert, J.; Tetens, J.; Bessei, W.; Bennewitz, J.; Camarinha-Silva, A. Gut Microbial Composition and Predicted Functions Are Not Associated with Feather Pecking and Antagonistic Behavior in Laying Hens. *Life* **2021**, *11*, 235. [CrossRef]

73. Meyer, B.; Zentek, J.; Harlander-Matauschek, A. Differences in intestinal microbial metabolites in laying hens with high and low levels of repetitive feather-pecking behavior. *Physiol. Behav.* **2013**, *110–111*, 96–101. [CrossRef] [PubMed]

74. van der Eijk, J.A.J.; Rodenburg, T.B.; de Vries, H.; Kjaer, J.B.; Smidt, H.; Naguib, M.; Kemp, B.; Lammers, A. Early-life microbiota transplantation affects behavioural responses, serotonin and immune characteristics in chicken lines divergently selected on feather pecking. *Sci. Rep.* **2020**, *10*, 2750. [CrossRef] [PubMed]

75. Birkl, P.; Bharwani, A.; Kjaer, J.B.; Kunze, W.; McBride, P.; Forsythe, P.; Harlander-Matauschek, A. Differences in cecal microbiome of selected high and low feather-pecking laying hens. *Poult. Sci.* **2018**, *97*, 3009–3014. [CrossRef]

76. Deverman, B.E.; Patterson, P.H. Cytokines and CNS development. *Neuron* **2009**, *64*, 61–78. [CrossRef]

77. Bilbo, S.D.; Schwarz, J.M. The immune system and developmental programming of brain and behavior. *Front. Neuroendocrinol.* **2012**, *33*, 267–286. [CrossRef]

78. Borsini, A.; Zunszain, P.A.; Thuret, S.; Pariante, C.M. The role of inflammatory cytokines as key modulators of neurogenesis. *Trends Neurosci.* **2015**, *38*, 145–157. [CrossRef]

79. Capuron, L.; Miller, A.H. Immune system to brain signaling: Neuropsychopharmacological implications. *Pharmacol. Ther.* **2011**, *130*, 226–238. [CrossRef] [PubMed]

80. Dantzer, R.; Bluthé, R.M.; Gheusi, G.; Cremona, S.; Layé, S.; Parnet, P.; Kelley, K.W. Molecular basis of sickness behavior. *Ann. N. Y. Acad. Sci.* **1998**, *856*, 132–138. [CrossRef] [PubMed]

81. Dantzer, R.; Bluthé, R.M.; Layé, S.; Bret-Dibat, J.L.; Parnet, P.; Kelley, K.W. Cytokines and sickness behavior. *Ann. N. Y. Acad. Sci.* **1998**, *840*, 586–590. [CrossRef] [PubMed]

82. Miller, A.H. Norman Cousins Lecture. Mechanisms of cytokine-induced behavioral changes: Psychoneuroimmunology at the translational interface. *Brain Behav. Immun.* **2009**, *23*, 149–158. [CrossRef]

83. Miller, A.H.; Haroon, E.; Raison, C.L.; Felger, J.C. Cytokine targets in the brain: Impact on neurotransmitters and neurocircuits. *Depress. Anxiety* **2013**, *30*, 297–306. [CrossRef]

84. Silverman, J.; Garnett, N.L.; Giszter, S.F.; Heckman, C.J.; Kulpa-Eddy, J.A.; Lemay, M.A.; Perry, C.K.; Pinter, M. Decerebrate mammalian preparations: Unalleviated or fully alleviated pain? A review and opinion. *Contemp. Top. Lab. Anim. Sci.* **2005**, *44*, 34–36.

85. Dantzer, R.; O'Connor, J.C.; Freund, G.G.; Johnson, R.W.; Kelley, K.W. From inflammation to sickness and depression: When the immune system subjugates the brain. *Nat. Rev. Neurosci.* **2008**, *9*, 46–56. [CrossRef]

86. de Haas, E.N.; van der Eijk, J.A.J. Where in the serotonergic system does it go wrong? Unravelling the route by which the serotonergic system affects feather pecking in chickens. *Neurosci. Biobehav. Rev.* **2018**, *95*, 170–188. [CrossRef]

87. Koolhaas, J.M.; Korte, S.M.; De Boer, S.F.; Van Der Vegt, B.J.; Van Reenen, C.G.; Hopster, H.; De Jong, I.C.; Ruis, M.A.; Blokhuis, H.J. Coping styles in animals: Current status in behavior and stress-physiology. *Neurosci. Biobehav. Rev.* **1999**, *23*, 925–935. [CrossRef]

88. Rodenburg, T.B.; van Hierden, Y.M.; Buitenhuis, A.J.; Riedstra, B.; Koene, P.; Korte, S.M.; van der Poel, J.J.; Groothuis, T.G.G.; Blokhuis, H.J. Feather pecking in laying hens: New insights and directions for research? *Appl. Anim. Behav. Sci.* **2004**, *86*, 291–298. [CrossRef]

89. Biscarini, F.; Bovenhuis, H.; van der Poel, J.; Rodenburg, T.B.; Jungerius, A.P.; van Arendonk, J.A. Across-line SNP association study for direct and associative effects on feather damage in laying hens. *Behav. Genet.* **2010**, *40*, 715–727. [CrossRef] [PubMed]

90. Sun, Y.; Biscarini, F.; Bovenhuis, H.; Parmentier, H.K.; van der Poel, J.J. Genetic parameters and across-line SNP associations differ for natural antibody isotypes IgM and IgG in laying hens. *Anim. Genet.* **2013**, *44*, 413–424. [CrossRef] [PubMed]

91. Baumgarth, N.; Tung, J.W.; Herzenberg, L.A. Inherent specificities in natural antibodies: A key to immune defense against pathogen invasion. *Springer Semin. Immunopathol.* **2005**, *26*, 347–362. [CrossRef] [PubMed]

92. Buitenhuis, A.J.; Rodenburg, T.B.; Wissink, P.H.; Visscher, J.; Koene, P.; Bovenhuis, H.; Ducro, B.J.; van der Poel, J.J. Genetic and phenotypic correlations between feather pecking behavior, stress response, immune response, and egg quality traits in laying hens. *Poult. Sci.* **2004**, *83*, 1077–1082. [CrossRef] [PubMed]

93. Habig, C.; Geffers, R.; Distl, O. Differential Gene Expression from Genome-Wide Microarray Analyses Distinguishes Lohmann Selected Leghorn and Lohmann Brown Layers. *PLoS ONE* **2012**, *7*, e46787. [CrossRef]

94. Habig, C.; Geffers, R.; Distl, O. A Replication Study for Genome-Wide Gene Expression Levels in Two Layer Lines Elucidates Differentially Expressed Genes of Pathways Involved in Bone Remodeling and Immune Responsiveness. *PLoS ONE* **2014**, *9*, e98350. [CrossRef]

95. Brunberg, E.I.; Rodenburg, T.B.; Rydhmer, L.; Kjaer, J.B.; Jensen, P.; Keeling, L.J. Omnivores Going Astray: A Review and New Synthesis of Abnormal Behavior in Pigs and Laying Hens. *Front. Vet. Sci.* **2016**, *3*, 57. [CrossRef]

96. Buitenhuis, A.J.; Kjaer, J.B.; Labouriau, R.; Juul-Madsen, H.R. Altered circulating levels of serotonin and immunological changes in laying hens divergently selected for feather pecking behavior. *Poult. Sci.* **2006**, *85*, 1722–1728. [CrossRef]

97. van der Eijk, J.A.J.; Verwoolde, M.B.; de Vries Reilingh, G.; Jansen, C.A.; Rodenburg, T.B.; Lammers, A. Chicken lines divergently selected on feather pecking differ in immune characteristics. *Physiol. Behav.* **2019**, *212*, 112680. [CrossRef]

98. Hughes, A.L.; Buitenhuis, A.J. Reduced variance of gene expression at numerous loci in a population of chickens selected for high feather pecking. *Poult. Sci.* **2010**, *89*, 1858–1869. [CrossRef]

99. Parmentier, H.K.; Rodenburg, T.B.; De Vries Reilingh, G.; Beerda, B.; Kemp, B. Does enhancement of specific immune responses predispose laying hens for feather pecking? *Poult. Sci.* **2009**, *88*, 536–542. [CrossRef] [PubMed]

100. Nicol, C. Feather pecking. In *Poultry Feathers and Skin*; Olukosi, O.A., Olori, A.H.V., Lambton, S., Eds.; CABI International: Wallingford, UK, 2019; pp. 31–46.

101. Tarbiat, B.; Jansson, D.S.; Wall, H.; Tydén, E.; Höglund, J. Effect of a targeted treatment strategy against Ascaridia galli on egg production, egg quality and bird health in a laying hen farm. *Vet. Parasitol.* **2020**, *286*, 109238. [CrossRef] [PubMed]

102. Temple, D.; Manteca, X.; Escribano, D.; Salas, M.; Mainau, E.; Zschiesche, E.; Petersen, I.; Dolz, R.; Thomas, E. Assessment of laying-bird welfare following acaricidal treatment of a commercial flock naturally infested with the poultry red mite (*Dermanyssus gallinae*). *PLoS ONE* **2020**, *15*, e0241608. [CrossRef] [PubMed]

103. Jacobs, L.; Vezzoli, G.; Beerda, B.; Mench, J.A. Northern fowl mite infestation affects the nocturnal behavior of laying hens. *Appl. Anim. Behav. Sci.* **2019**, *216*, 33–37. [CrossRef]

104. Van Krimpen, M.M.; Kwakkel, R.P.; Reuvekamp, B.F.J.; Van der Peet-Schwering, C.M.C.; Den Hartog, L.A.; Verstegen, M.W.A. Impact of feeding management on feather pecking in laying hens. *Worlds Poult. Sci. J.* **2005**, *61*, 665–687. [CrossRef]

105. Freeman, R.E. *Strategic Management: A Stakeholder Approach*; Cambridge University Press: Boston, MA, USA, 1984.

106. Marino, L. Thinking chickens: A review of cognition, emotion, and behavior in the domestic chicken. *Anim. Cogn.* **2017**, *20*, 127–147. [CrossRef] [PubMed]

107. Rodenburg, T.B.; Koene, P. The impact of group size on damaging behaviours, aggression, fear and stress in farm animals. *Appl. Anim. Behav. Sci.* **2007**, *103*, 205–214. [CrossRef]

108. Cloutier, S.; Newberry, R. Differences in skeletal and ornamental traits between laying hen cannibals, victims and bystanders. *Appl. Anim. Behav. Sci.* **2002**, *77*, 115–126. [CrossRef]

109. Tahamtani, F.M.; Forkman, B.; Hinrichsen, L.K.; Riber, A.B. Both feather peckers and victims are more asymmetrical than control hens. *Appl. Anim. Behav. Sci.* **2017**, *195*, 67–71. [CrossRef]

110. Heers, A.M.; Tobalske, B.W.; Dial, K.P. Ontogeny of lift and drag production in ground birds. *J. Exp. Biol.* **2011**, *214*, 717–725. [CrossRef]

111. Müller, W.; Patone, G. Air transmissivity of feathers. *J. Exp. Biol.* **1998**, *201*, 2591–2599. [CrossRef] [PubMed]

112. LeBlanc, S.; Tobalske, B.; Quinton, M.; Springthorpe, D.; Szkotnicki, B.; Wuerbel, H.; Harlander-Matauschek, A. Physical Health Problems and Environmental Challenges Influence Balancing Behaviour in Laying Hens. *PLoS ONE* **2016**, *11*, e0153477. [CrossRef] [PubMed]

113. Gentle, M.J. Pain issues in poultry. *Appl. Anim. Behav. Sci.* **2011**, *135*, 252–258. [CrossRef]

114. Kops, M.S.; Kjaer, J.B.; Güntürkün, O.; Westphal, K.G.C.; Korte-Bouws, G.A.H.; Olivier, B.; Korte, S.M.; Bolhuis, J.E. Brain monoamine levels and behaviour of young and adult chickens genetically selected on feather pecking. *Behav. Brain Res.* **2017**, *327*, 11–20. [CrossRef] [PubMed]

115. Welfare Quality®. *Assessment Protocol for Poultry (Broilers, Laying Hens)*; Welfare Quality®Consortium: Lelystad, The Netherlands, 2009. Available online: https://edepot.wur.nl/233471 (accessed on 5 September 2019).

116. Averós, X.; Balderas, B.; Cameno, E.; Estevez, I. The value of a retrospective analysis of slaughter records for the welfare of broiler chickens. *Poult. Sci.* **2020**, *99*, 5222–5232. [CrossRef] [PubMed]

117. Bilčík, B.; Keeling, L.J. Changes in feather condition in relation to feather pecking and aggressive behaviour in laying hens. *Br. Poult. Sci.* **1999**, *40*, 444–451. [CrossRef] [PubMed]

118. Bright, A.; Jones, T.A.; Dawkins, M.S. A non-intrusive method of assessing plumage condition in commercial flocks of laying hens. *Anim. Welf.* **2006**, *15*, 113–118.

119. Main, D.C.J.; Mullan, S.M.; Atkinson, C.; Bond, A.; Cooper, M.; Fraser, A.; Browne, W.J. Welfare outcome assessments in laying hen farm assurance schemes. *Anim. Welf.* **2012**, *21*, 389–396. [CrossRef]

120. Decina, C.; Berke, O.; van Staaveren, N.; Baes, C.F.; Harlander-Matauscheck, A. Development of a Scoring System to Assess Feather Damage in Canadian Laying Hen Flocks. *Animals* **2019**, *9*, 436. [CrossRef]

121. AssureWel Laying Hens. Available online: http://www.assurewel.org/layinghens.html (accessed on 5 September 2019).

122. Marchewka, J.; Watanabe, T.T.N.; Ferrante, V.; Estevez, I. Welfare assessment in broiler farms: Transect walks versus individual scoring. *Poult. Sci.* **2013**, *92*, 2588–2599. [CrossRef]
123. BenSassi, N.; Averós, X.; Estevez, I. The potential of the transect method for early detection of welfare problems in broiler chickens. *Poult. Sci.* **2019**, *98*, 522–532. [CrossRef] [PubMed]
124. BenSassi, N.; Averós, X.; Estevez, I. Broiler Chickens On-Farm Welfare Assessment: Estimating the Robustness of the Transect Sampling Method. *Front. Vet. Sci.* **2019**, *6*, 236. [CrossRef] [PubMed]
125. BenSassi, N.; Vas, J.; Vasdal, G.; Averós, X.; Estévez, I.; Newberry, R.C. On-farm broiler chicken welfare assessment using transect sampling reflects environmental inputs and production outcomes. *PLoS ONE* **2019**, *14*, e0214070. [CrossRef] [PubMed]
126. Ferrante, V.; Lolli, S.; Ferrari, L.; Watanabe, T.T.N.; Tremolada, C.; Marchewka, J.; Estevez, I. Differences in prevalence of welfare indicators in male and female turkey flocks (*Meleagris gallopavo*). *Poult. Sci.* **2019**, *98*, 1568–1574. [CrossRef]
127. Marchewka, J.; Estevez, I.; Vezzoli, G.; Ferrante, V.; Makagon, M.M. The transect method: A novel approach to on-farm welfare assessment of commercial turkeys. *Poult. Sci.* **2015**, *94*, 7–16. [CrossRef]
128. Vasdal, G.; Marchewka, J.; Newberry, R.C.; Estevez, I.; Kittelsen, K. Developing a novel welfare assessment tool for loose-housed laying hens—the Aviary Transect method. *Poult. Sci.* **2022**, *101*, 101533. [CrossRef]
129. Bignon, L.; Mika, A.; Mindus, C.; Litt, J.; Souchet, C.; Bonnaud, V.; Picchiottino, C.; Warin, L.; Dennery, G.; Brame, C.; et al. *Une Méthode Pratique ET Partagée D'éValuation du Bien-êTre en Filières Avicole et Cunicole: EBENE*; 12èmes Journées de la Recherche Avicole et Palmipèdes à Foie Gras: Tours, France, 2017; pp. 1015–1019.
130. Estevez, I.; Marchewka, J.; Watanabe, T.T.N.; Ferrante, V.; Zanella, A. AWIN Welfare Assessment Protocol for Turkeys. Available online: https://www.researchgate.net/publication/279953184_AWIN_Welfare_assessment_protocol_for_Turkeys (accessed on 5 September 2019). [CrossRef]
131. BenSassi, N.; Averós, X.; Estevez, I. Technology and Poultry Welfare. *Animals* **2016**, *6*, 62. [CrossRef]
132. Berckmans, D. Automatic on-line monitoring of animals by precision livestock farming. In *Livestock Production and Society*; Geers, R., Madec, F., Eds.; Wageningen Academic Publishers: Wageningen, The Netherlands, 2006.
133. Rowe, E.; Dawkins, M.S.; Gebhardt-Henrich, S.G. A Systematic Review of Precision Livestock Farming in the Poultry Sector: Is Technology Focussed on Improving Bird Welfare? *Animals* **2019**, *9*, 614. [CrossRef]
134. Dawkins, M.S.; Lee, H.-j.; Waitt, C.D.; Roberts, S.J. Optical flow patterns in broiler chicken flocks as automated measures of behaviour and gait. *Appl. Anim. Behav. Sci.* **2009**, *119*, 203–209. [CrossRef]
135. Lee, H.J.; Roberts, S.J.; Drake, K.A.; Dawkins, M.S. Prediction of feather damage in laying hens using optical flows and Markov models. *J. R. Soc. Interface* **2011**, *8*, 489–499. [CrossRef]
136. Manteuffel, G.; Puppe, B.; Schön, P.C. Vocalization of farm animals as a measure of welfare. *Appl. Anim. Behav. Sci.* **2004**, *88*, 163–182. [CrossRef]
137. Michaud, F.; Créach, P.; Brouard, B.; Gazengel, B.; Simon, L.; Collin, A.; Métayer-Coustard, S.; Travel, A. Vocalisations du poussin: Developpement d'une méthode d'enregistrement et d'analyse. In Proceedings of the 13èmes Journées de la Recherche Avicole et Palmipèdes à Foie Gras, Tours, France, 20–21 March 2019; pp. 375–379.
138. Collias, N.E. The Vocal Repertoire of the Red Junglefowl: A Spectrographic Classification and the Code of Communication. *Condor* **1987**, *89*, 510–524. [CrossRef]
139. Dawkins, M.S. Evolution and animal welfare. *Q. Rev. Biol.* **1998**, *73*, 305–328. [CrossRef] [PubMed]
140. Carpentier, L.; Norton, T.; Berckmans, D.; Fontana, I.; Guarino, M.; Vranken, E.; Berckmans, D. Frequency Analysis of Vocalisations to Monitor Broiler Chicken Production Performance in Real-Life Farm. In Proceedings of the European Conference on Precision Livestock Farming, Nantes, France, 12–14 September 2017; pp. 138–146.
141. Fontana, I.; Tullo, E.; Butterworth, A.; Guarino, M. Broiler vocalisation analysis used to predict growth. In Proceedings of the Measuring Behavior 2014, Wageningen, The Netherlands, 27–29 August 2014.
142. Carroll, B.T.; Anderson, D.V.; Daley, W.; Harbert, S.D.; Britton, D.F.; Jackwood, M.W. Detecting symptoms of diseases in poultry through audio signal processing. In Proceedings of the 2014 IEEE Global Conference on Signal and Information Processing (GlobalSIP), Atlanta, GA, USA, 3–5 December 2014; pp. 1132–1135.
143. Rizwan, M.; Carroll, B.T.; Anderson, D.V.; Daley, W.; Harbert, S.; Britton, D.F.; Jackwood, M.W. Identifying Rale Sounds in Chickens Using Audio Signals for Early Disease Detection in Poultry. In Proceedings of the 2016 IEEE Global Conference on Signal and Information Processing (GlobalSIP), Washington, DC, USA, 7–9 December 2016; pp. 55–59.
144. Carpentier, L.; Vranken, E.; Berckmans, D.; Paeshuyse, J.; Norton, T. Development of sound-based poultry health monitoring tool for automated sneeze detection. *Comput. Electron. Agric.* **2019**, *162*, 573–581. [CrossRef]
145. Siegford, J.M.; Berezowski, J.; Biswas, S.K.; Daigle, C.L.; Gebhardt-Henrich, S.G.; Hernandez, C.E.; Thurner, S.; Toscano, M.J. Assessing Activity and Location of Individual Laying Hens in Large Groups Using Modern Technology. *Animals* **2016**, *6*, 10. [CrossRef] [PubMed]
146. Banerjee, D.; Biswas, S.K.; Daigle, C.; Siegford, J.M. Remote Activity Classification of Hens Using Wireless Body Mounted Sensors. In Proceedings of the Ninth International Conference on Wearable and Implantable Body Sensor Networks, London, UK, 9–12 May 2012; pp. 107–112.
147. Zaninelli, M.; Rossi, L.; Costa, A.; Tangorra, F.M.; Guarino, M.; Savoini, G. Performance of injected RFID transponders to collect data about laying performance and behaviour of hens. *Large Anim. Rev.* **2016**, *22*, 77–82.

148. Daigle, C.L.; Banerjee, D.; Biswas, S.; Siegford, J.M. Noncaged laying hens remain unflappable while wearing body-mounted sensors: Levels of agonistic behaviors remain unchanged and resource use is not reduced after habituation. *Poult. Sci.* **2012**, *91*, 2415–2423. [CrossRef] [PubMed]
149. Stadig, L.M.; Rodenburg, T.B.; Ampe, B.; Reubens, B.; Tuyttens, F.A.M. An automated positioning system for monitoring chickens' location: Effects of wearing a backpack on behaviour, leg health and production. *Appl. Anim. Behav. Sci.* **2018**, *198*, 83–88. [CrossRef]
150. Ellen, E.D.; van der Sluis, M.; Siegford, J.; Guzhva, O.; Toscano, M.J.; Bennewitz, J.; van der Zande, L.E.; van der Eijk, J.A.J.; de Haas, E.N.; Norton, T.; et al. Review of Sensor Technologies in Animal Breeding: Phenotyping Behaviors of Laying Hens to Select Against Feather Pecking. *Animals* **2019**, *9*, 108. [CrossRef]
151. van der Sluis, M.; de Klerk, B.; Ellen, E.D.; de Haas, Y.; Hijink, T.; Rodenburg, T.B. Validation of an Ultra-Wideband Tracking System for Recording Individual Levels of Activity in Broilers. *Animals* **2019**, *9*, 580. [CrossRef]
152. Saravanan, K.; Saraniya, S. Cloud IOT based novel livestock monitoring and identification system using UID. *Sens. Rev.* **2018**, *38*, 21–33.
153. Lecun, Y. L'apprentissage profond, une révolution en intelligence artificielle. *Lett. Collège Fr.* **2016**, *41*, 13. [CrossRef]
154. Wang, C.; Chen, H.; Zhang, X.; Meng, C. Evaluation of a laying-hen tracking algorithm based on a hybrid support vector machine. *J. Anim. Sci. Biotechnol.* **2016**, *7*, 60. [CrossRef] [PubMed]

Article

Positive Effects of Elevated Platforms and Straw Bales on the Welfare of Fast-Growing Broiler Chickens Reared at Two Different Stocking Densities

Frédérique Mocz [1,*], Virginie Michel [2], Mathilde Janvrot [1], Jean-Philippe Moysan [1], Alassane Keita [1], Anja B. Riber [3] and Maryse Guinebretière [1]

[1] Epidemiology, Health and Welfare Unit, French Agency for Food, Environmental and Occupational Health & Safety (ANSES), 22440 Ploufragan, France; mathildejanvrot@yahoo.fr (M.J.); jean-philippe.moysan@anses.fr (J.-P.M.); alassane.keita@anses.fr (A.K.); maryse.guinebretiere@anses.fr (M.G.)

[2] Direction of Strategy and Programs, French Agency for Food, Environmental and Occupational Health & Safety (ANSES), 94701 Maisons-Alfort, France; virginie.michel@anses.fr

[3] Department of Animal Science, Aarhus University, DK-8830 Tjele, Denmark; anja.riber@anis.au.dk

[*] Correspondence: frederique.mocz@anses.fr

Simple Summary: Fast-growing broiler chickens commonly experience welfare issues, such as foot and hock lesions or walking difficulties due to their genetics or the barren environment. This study assessed the impacts of elevated platforms and straw bales on the welfare of fast-growing broilers reared at two different stocking densities. The higher stocking density had negative impacts on foot and hock lesions and walking ability, whereas these welfare issues were partly positively affected by enrichments at both stocking densities.

Abstract: In conventional rearing systems, fast-growing broiler chickens commonly experience welfare issues, such as contact dermatitis, walking difficulties or a lack of expression of species-specific behaviours. Enriching their environment may be a way to improve their welfare. The objective of this study was to evaluate the benefits of elevated platforms and straw bales on the welfare of fast-growing broiler chickens reared at two different stocking densities. A total of 14,994 Ross 308 broilers were housed in 12 pens according to 4 treatments: 31 kg/m^2 with or without enrichments and 41 kg/m^2 with or without enrichments. The broilers' walking ability, footpad dermatitis (FPD), hock burns (HB), weight, mortality and litter quality were assessed. Stocking density had a negative effect on FPD and HB, whereas enrichments reduced the occurrence of FPD and HB at both densities. There was a positive enrichment effect and a negative density effect on body weight at 25 days and on walking ability, but no effect on the litter quality or mortality rate. These results confirm that an enriched environment improves animal welfare in confined chickens, regardless of the stocking density. Reducing stocking density clearly appears to be an important means of increasing animal welfare.

Keywords: broiler; enrichment; footpad dermatitis; hock burn; litter quality; stocking density; walking ability

Citation: Mocz, F.; Michel, V.; Janvrot, M.; Moysan, J.-P.; Keita, A.; Riber, A.B.; Guinebretière, M. Positive Effects of Elevated Platforms and Straw Bales on the Welfare of Fast-Growing Broiler Chickens Reared at Two Different Stocking Densities. *Animals* **2022**, *12*, 542. https://doi.org/10.3390/ani12050542

Academic Editors: Victoria Sandilands and Tina Widowski

Received: 21 January 2022
Accepted: 22 February 2022
Published: 22 February 2022

Publisher's Note: MDPI stays neutral with regard to jurisdictional claims in published maps and institutional affiliations.

1. Introduction

Rearing fast-growing broiler chickens in conventional systems is commonly associated with welfare issues, such as lameness, footpad dermatitis or a lack of expression of species-specific behaviour [1]. The impairment of welfare is generally linked to fast-growing genetics and to different elements of housing systems and management, such as a high stocking density, poor litter quality or the general barren environment. Enriching the environment could improve rearing conditions and broiler welfare. According to Newberry [2], environmental enrichment is a modification of the environment of captive animals that increases the animal's behavioural possibilities and improves biological function. There are several kinds of enrichment that can be used for broilers, e.g., elevated resting places (such as perches or platforms), panels, barriers and materials to stimulate foraging, explorative and comfort behaviours [3]. The effects of elevated platforms or perches on broiler behaviour and welfare have been the subject of recent studies that compared platforms and perches [4] or different types of platforms and configurations (number, surface, height, materials) [5–7] at different stocking densities [8] and studied their use on commercial farms [9] or under experimental conditions (with a small number of birds) [10–12]. Elevated platforms seem to be more suitable than perches for fast-growing broiler chickens due to the broilers' weight, leg weakness and difficulties in finding their balance on "traditional" perches like bars [3]. These studies assessed several parameters, such as economics [13], health [14] and animal welfare [8,15–17]. The outcomes of these studies are sometimes contradictory [3], but for the studies where only a limited improvement was found in animal welfare, it may be explained by an insufficient platform surface area [5], late provision (after 7 days old) [5] or lack of access ramps [5,6].

Similarly, straw bales may be used as elevated resting places, with the additional benefit of providing the broilers with an opportunity to express normal foraging behaviour [18]. Broilers also use them to lie against when resting [19]. Riber and colleagues [3] reviewed research on the effects of these enrichments and concluded that existing studies show either no or contradictory effects on slaughter weight, mortality and locomotion. Baxter and colleagues [20] showed no effect of adding straw bales on litter quality and ammonia levels in commercially reared fast-growing chickens and found mixed results on behaviour (decrease in locomotion and increase in sitting behaviours). For Kells and colleagues, [21] straw bales had a positive effect on resting/activity, locomotion and preening behaviours in commercial farming. Bailie and O'Connell [22] studied the difference in behaviour according to two quantities of straw bales distributed (one bale per 44 m^2 or 29 m^2) among Ross and Cobb chickens at 30 kg/m^2 but did not observe any differences in behaviour or leg health. Thus, the effect of straw bales on broiler behaviour and welfare appears to vary between studies.

The present study was designed to increase knowledge on the impact of environmental enrichment on the welfare of fast-growing broilers, especially in relation to leg health and walking ability. To this end, straw bales and elevated platforms with ramps to facilitate access for fast-growing broilers were provided in the rearing environment from the first day that the day-old chicks were placed there. Stocking densities usually varied between the reviewed studies and only one compared the impacts of enrichment (barrier perches with small groups of animals) according to stocking density [8]. The aim of this experiment was, therefore, to compare two different stocking densities to analyse the influence of space allowance in an enriched environment on the welfare of Ross 308 broilers reared in large groups: (a) at 41 kg/m^2 or 31 kg/m^2 and (b) with or without enrichment, i.e., elevated platforms with access ramps and straw bales.

2. Materials and Methods

2.1. Housing and Experimental Design

The study was conducted with 14,994 Ross 308 broiler chickens reared until 33 days of age in 6 identical rooms, each having 2 separate floor pens. The experimental design consisted of 2 × 2 modalities with 3 repetitions (3 pens) per treatment: stocking density at slaughter age of 41 kg/m^2 or 31 kg/m^2, with or without enrichments.

All the pens covered 72 m^2 (6 × 12 m) but the usable areas were considered to be 70 m^2 in non-enriched pens, and 66 m^2 in enriched pens as space under the feeders (i.e., 2 m^2) and platforms (4 m^2) were not considered as usable all the time. Indeed, we hypothesised that birds could not access the surface under the platforms during the last part of the rearing stage due to an increase in body size (the platforms being placed 30 cm off the ground). Secondly, no litter was spread on the platform surfaces, so they were not counted as usable space (in accordance with European regulations [23]). In contrast, the surface on the top of the straw bales was counted as usable since birds could access it and the straw can be considered as litter. Only the usable area was included in the calculation of the number of chicks to be placed in each pen (Table 1) in order to reach the final stocking densities (31 or 41 kg/m^2) at 33 days of age. The different treatments in the six rooms were distributed to fit with another project on the impact of enrichments on air quality.

Table 1. Distribution of broilers per treatment in the 12 pens of the 6 rooms.

Density	41 kg/m^2								31 kg/m^2			
Room	Room 1		Room 2		Room 3		Room 4		Room 5		Room 6	
Enrichment	Yes		No		Yes	No	Yes	No	No		Yes	
Pen	1	2	3	4	5	6	7	8	9	10	11	12
Number of broilers	1385	1385	1471	1471	1385	1471	1040	1103	1103	1103	1040	1040

The bedding material used in all the pens was 1 kg/m^2 of wood shavings with the addition of clean litter and the removal of dirty litter when necessary to maintain an acceptable condition. During the rearing period, dirty litter was removed from the most soiled areas of the pens (mainly under the drinkers) and clean litter was added 4 times in each pen from day 12. Each pen contained 3 lines of 29 nipple drinkers and 16 circular feeders. Artificial light was provided in addition to natural light, which birds had access to from 7 a.m. to 8 p.m. During the first week of age, chicks were exposed to a lighting programme of 23L:1D. From one week of age, artificial light was on from 5 a.m. to 11 p.m. The level of artificial lighting was managed by lux sensors (Tuffigo Rapidex$^®$, Tuffigo Rapidex, Saint-Evarzec, France) per room depending on the level of natural light so as to ensure around 100 lux on the placement day and 30 lux from 6 days on.

2.2. Enrichments

An elevated perforated platform equipped with two access ramps was placed in the middle of each enriched pen. The platform, made of plastic slatted flooring, was 30 cm high, 2 m long and 1 m wide. The access ramps on either side of the platform had a 16° slope and measured 1 m × 1 m. The total surface area of the platform plus ramps was therefore 4 m^2. We considered that 3 m^2 was potentially accessible underneath, at least during the first week of age. These platforms were available for broilers in the enriched pens from their first to their last day of life.

One straw bale was placed on each side of the barn. These straw bales were available for broilers in the enriched pens from their first day of life. The two bales were 80 cm long, 40 cm wide and 19 cm high (2 × 0.32 m^2 per pen). They weighed around 10 kg and were removed from their plastic packaging beforehand and tied up to ensure they stayed in position. They were not renewed if they disintegrated during the rearing period.

2.3. Measurements

2.3.1. Litter Quality

Litter was sampled five times throughout the rearing period (once a week) in order to assess the humidity level. A handful of litter (around 10 cm diameter on the ground) was collected from four areas (between feeders) in every pen. For each pen, the samples of bedding from the four areas were manually mixed to ensure a representative sample. A subsample of approximately 20 cL was then weighed, dried for 24 h at 70 °C and reweighed to measure the dry matter [24].

2.3.2. Walking Ability

Walking ability was assessed at 26 and 32 days of age on 20 randomly chosen birds in each pen. The observer walked towards one bird at a time. Birds either moved of their own volition or were stimulated vocally or by a gentle touch with the foot or hand to encourage them to walk. Scores were assigned using a 0–3 scale adapted from Meyer et al. [25] where 0 = ability to walk with no signs of lameness, 1 = unevenness in steps or stopped and sat down but able to walk 1.5 m, 2 = severe disability, birds can walk a few steps but not 1.5 m and 3 = birds unable to walk.

2.3.3. Body Weight, Mortality and Contact Dermatitis

Every day during the rearing period, the number of birds that had to be culled or were found dead was recorded.

At 25 days of age, 50 sexed birds per pen (25 males and 25 females) were randomly selected for weighing (Signum 3 from Minebea®, Minebea Intec, Hamburg, Germany) and an evaluation of footpad dermatitis and hock burns. Contact dermatitis was assessed quite early, at 25 days, because we started to observe a high prevalence of lesions on the birds' feet during regular inspections. To assess footpad dermatitis/hock burns, both the feet and hocks were inspected, and the worst was scored. When feet/hocks were dirty, they were gently brushed with a toothbrush and soapy water. The scoring systems were adapted from the Welfare Quality Protocol® [26]: a = no evidence of footpad dermatitis/hock burns, b = minimal evidence of footpad dermatitis/hock burns (mild lesions), c = evidence of footpad dermatitis/hock burn (severe lesions). The distinction between mild and severe lesions depended on the size and depth of the lesions, according to a photographic reference [26].

2.3.4. Welfare Indicators Obtained Post-Mortem

At the slaughterhouse, footpad dermatitis was evaluated on the whole batch for each treatment with an automatic camera system (Meyn® footpad inspection system, Meyn, Oostzaan, Amsterdam) providing three scores, depending on the size and colour of lesions: no lesions (score a), medium/minor lesions (score b) and severe footpad dermatitis (score c). Due to the incorrect positioning of feet and other errors, only 75–95% of the pads in each batch were examined. In addition, for each treatment, carcasses were visually observed for 15 min on the slaughter line after bleeding to score hock burns with the same scoring system as used at 25 days (a = no evidence of hock burn, b = minimal evidence of hock burn, c = evidence of hock burn). A total of 1850 carcasses, i.e., hocks (both hocks were inspected, and the worst was scored), were observed per treatment, corresponding to 42–63% of the total carcasses per treatment (speed of the line: 7400 chickens per hour).

2.4. Statistical Analysis

The results were analysed using R (version 4.0.3) [27] and RStudio. For each of the five ages, litter humidity values were analysed using an ANOVA, with the main effects being enrichment and stocking density as well as the interaction between the two. Body weights (at 25 days and from automatic weighing scales) were analysed with the geeglm function. Pen repetition was taken into account in the analysis of manual weighing data, as was time repetition for the automatic weighing scale data. The daily cumulative mortality was analysed with a survival analysis and a Cox mixed-effects model on the number of broilers found dead during the rearing period. The walking ability scores were analysed with a generalised linear model (GLM) distinguishing birds free of lameness (score 0) from all others. To go further in the analysis, a pairwise comparison was made using the estimated marginal means model. The footpad dermatitis scores assessed on the farm were analysed with two GLMs: one distinguished score c from scores a + b to evaluate the severity of footpad dermatitis, and the other distinguished score a from scores b + c to evaluate the prevalence of lesions, whatever their severity. The hock burn scores assessed on the farm were analysed with a GLM that distinguished score a from score b (there being no or very few c scores observed). As only one data point was available per treatment (pens were not distinguished at slaughter), footpad dermatitis and hock burn at the slaughterhouse were analysed with a chi-square test for each severity score between treatments.

3. Results

3.1. Litter Quality

In the lower density, the mean levels of litter humidity varied from $25.1 \pm 5.1\%$ to $48.1 \pm 7.7\%$ in the enriched pens and from $19.4 \pm 3.4\%$ to $45.1 \pm 9.9\%$ in the unenriched pens. In the higher density, they varied from $26.8 \pm 9.8\%$ to $51.1 \pm 5.2\%$ in the enriched pens and from 22.1 ± 2.7 to $53.1 \pm 8.8\%$ in the unenriched pens. There was no effect of density ($p = 0.55$) or enrichment ($p = 0.12$) on litter humidity at any age.

3.2. Weight and Mortality

There was an effect of enrichment ($p = 0.01$) and of density ($p = 0.05$) on the body weight assessed at 25 days of age. Broilers from the enriched pens were heavier than those from unenriched pens, and broilers from the lower density pens were heavier than those from the higher density pens (mean body weight: 31 kg/m^2-1376 ± 149 g with enrichment and 1357 ± 142 g without enrichment; 41 kg/m^2-1350 ± 146 g with enrichment and 1314 ± 136 g without enrichment). No effect was found for the interaction of density and enrichment ($p = 0.86$).

The cumulative mortality rates (found dead and culled) never exceeded 5.8%. Mortality was neither affected by stocking density ($p = 0.58$), enrichment ($p = 0.91$), nor the interaction of both ($p = 0.70$).

3.3. Walking Ability

Broilers reared at the lower stocking density of 31 kg/m^2 were able to walk better than those from the pens with a stocking density of 41 kg/m^2 at 26 days ($p = 0.001$) and at 32 days of age ($p = 0.0002$) (Figure 1). Pairwise comparisons showed a significant effect of density in unenriched groups ($p < 0.0001$ and 0.004 at 26 and 32 days of age, respectively), whereas differences were not significant in enriched groups ($p = 0.98$ and 0.17 at 26 and 32 days of age, respectively).

Enrichment had an effect on walking ability at 26 days of age but only in the higher density groups. In groups of broilers reared at 41 kg/m^2, there were more birds walking normally in the enriched group (83%) than in the unenriched group (63%) ($p = 0.03$) at 26 days of age. This effect disappeared at 32 days, however, though a statistical tendency remained ($p = 0.08$). This enrichment effect was not present in the lower density groups at either 26 ($p = 0.79$) or 32 days of age ($p = 1$).

3.4. Welfare Indicators Assessed on the Farm

3.4.1. Footpad Dermatitis (FPD) at 25 Days of Age

An effect of stocking density on footpad dermatitis was found (Figure 2). Broilers reared at the lower density of 31 kg/m^2 had less severe footpad dermatitis (score c) ($p = 0.0001$) and the prevalence of birds with signs of lesions (score b + c) ($p = 0.008$) was lower than those raised at the higher density of 41 kg/m^2. There was no effect of enrichment on the percentages of severe footpad dermatitis ($p = 0.56$) or on the prevalence of lesions ($p = 0.16$).

No effect of sex was found on the level of footpad dermatitis (scores b + c: 93.5% of females and 93.8% of males) ($p = 0.92$).

3.4.2. Hock Burns at 25 Days of Age

Birds raised at the lower stocking density had fewer hock burns (scores b + c) than those raised at the higher density ($p = 0.0009$) (Figure 3). No impact of enrichment was found on the occurrence of hock burns ($p = 0.62$). There were so few c scores (one bird at 41 kg/m^2 with and one bird at 41 kg/m^2 without enrichment) that we could not compare the lesions' severity between groups.

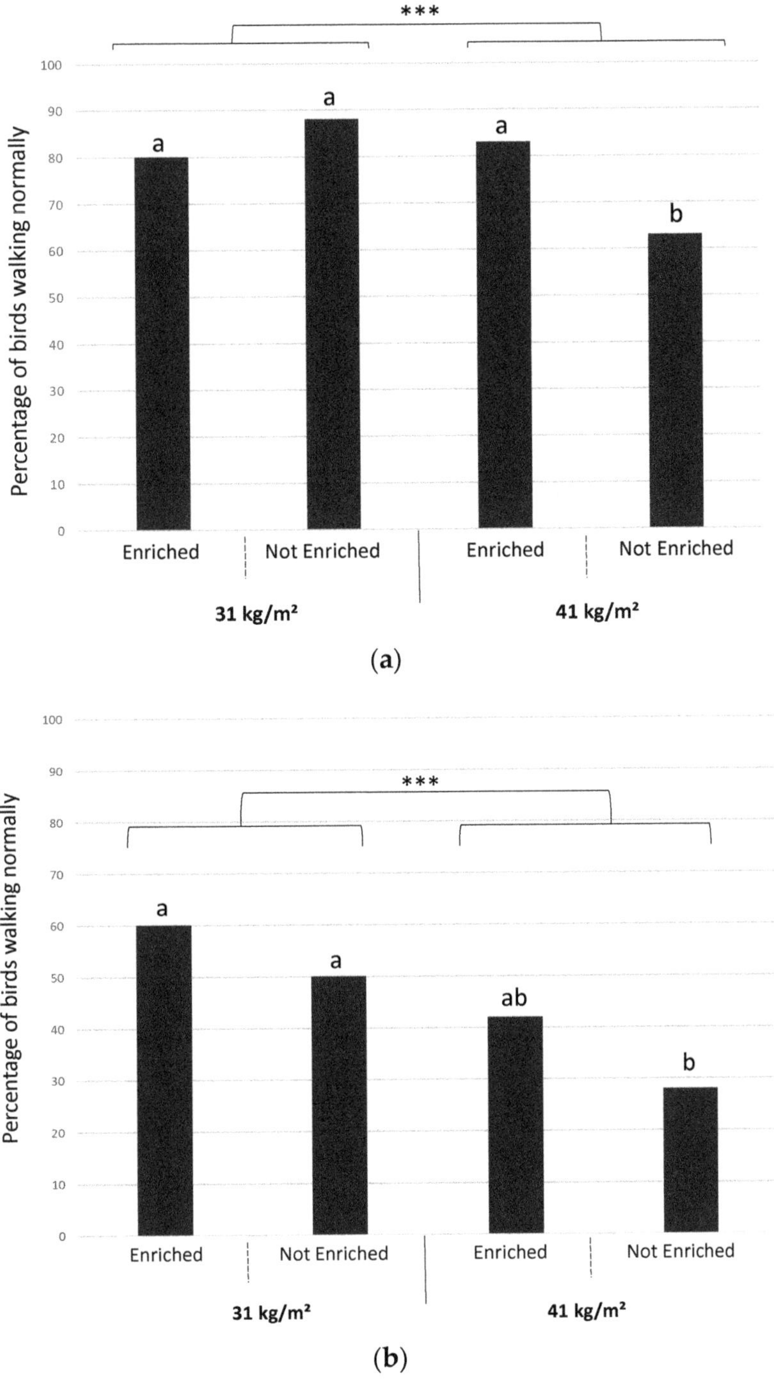

(**a**)

(**b**)

Figure 1. Percentage of broilers walking normally (score 0) per treatment at 26 days (**a**) and 32 days of age (**b**) (n = 60 per treatment). *** $p \leq 0.001$. Different letters (a or b) above the columns indicate a significant difference between the groups ($p < 0.05$).

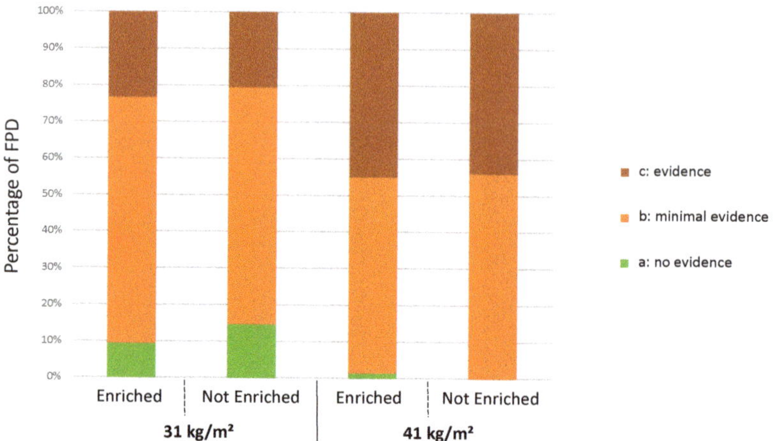

Figure 2. Distribution of broilers following the same treatment according to the level of footpad dermatitis (n = 150 per treatment) at 25 days of age.

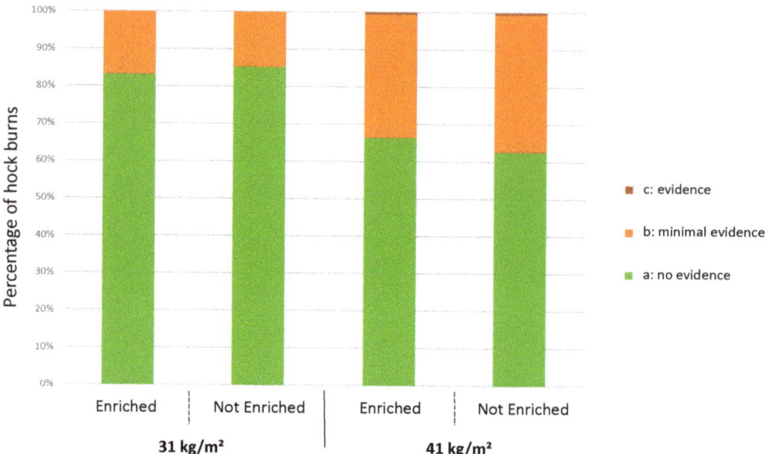

Figure 3. Distribution of broilers following the same treatment according to the level of hock burns (n = 150 per treatment) at 25 days of age.

Males had more hock burns (scores b + c) than females at 25 days (males: 31.6%; females: 20%; $p = 0.0009$).

3.5. Welfare Indicators Assessed Post-Mortem

3.5.1. Footpad Dermatitis

As observed at 25 days of age, stocking density negatively impacted the levels of footpad dermatitis assessed post-mortem (Figure 4). Broiler chickens raised in pens with the higher stocking density of 41 kg/m^2 had more severe foot lesions (score c) ($p < 0.0001$) than those at 31 kg/m^2. There was also a lower prevalence of broilers with signs of lesions (score b + c) ($p < 0.0001$) when raised at a stocking density of 31 kg/m^2. Minor footpad dermatitis (score b) was more common in birds raised at 31 kg/m^2 than at 41 kg/m^2 ($p < 0.0001$).

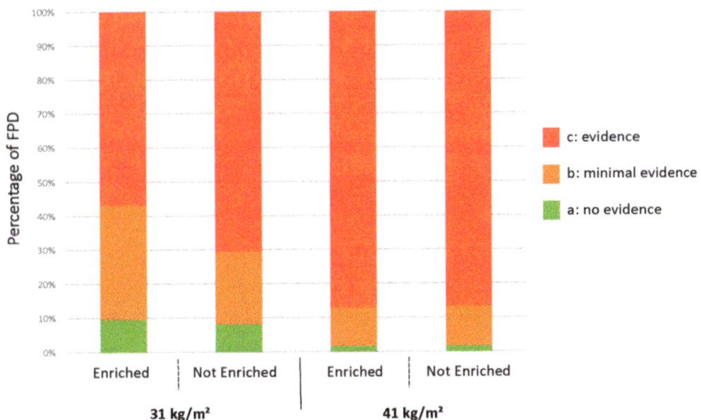

Figure 4. Distribution of broilers following the same treatment according to the level of footpad dermatitis observed during a post-mortem examination by an automatic camera system at the slaughterhouse.

There was an effect of enrichments in broilers housed at 31 kg/m^2 but not in birds reared at 41 kg/m^2. At 31 kg/m^2 without enrichments, more birds had severe footpad lesions (score c) ($p < 0.0001$) and fewer had minor footpad dermatitis (score b) ($p < 0.0001$) than birds raised at the same density (31 kg/m^2) but with enrichments. There was no effect of enrichment at 31 kg/m^2 on the absence of footpad dermatitis (score a) ($p = 0.23$). No effect of enrichment on FPD scores was observed in birds raised at 41 kg/m^2 (score a: $p = 1$; score b: $p = 0.58$; score c: $p = 0.43$).

3.5.2. Hock Burns

Stocking density and enrichment impacted the occurrence of hock burns scored on the slaughter line (Figure 5). Broiler chickens raised at the higher stocking density had more severe (score c) ($p < 0.0001$) and minor hock burns (score b) ($p < 0.0001$) than those raised at a lower density, whereas more birds raised at 31 kg/m^2 had absolutely no sign of hock burns (score a) than those reared at 41 kg/m^2 ($p < 0.0001$).

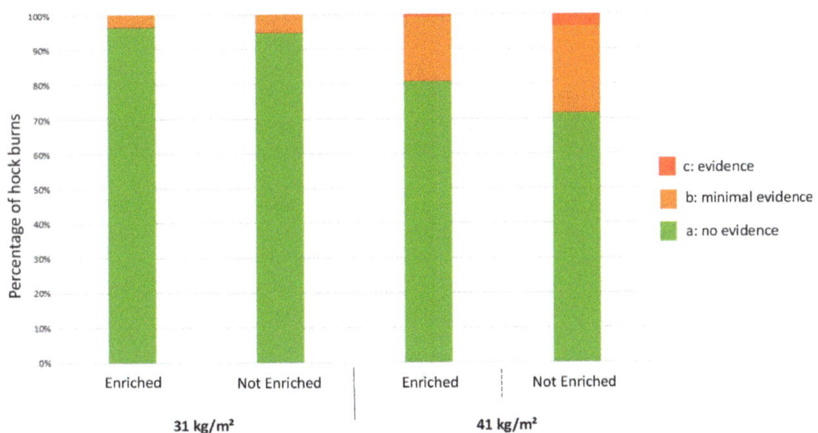

Figure 5. Hock burns of broilers following each treatment at post-mortem.

There was also an effect of enrichment at both 31 kg/m^2 and 41 kg/m^2 densities. Birds housed at 31 kg/m^2 with enrichments had fewer minor lesions (score b) ($p = 0.02$) than those raised at the same stocking density without any enrichments. They also had no sign of hock burns (score a) ($p = 0.02$) more often than those at the same stocking density without enrichments. At 31 kg/m^2, no severe hock burns (score c) were observed. At 41 kg/m^2, there were more severe hock burns (score c) ($p < 0.0001$) and minor lesions (score b) ($p < 0.0001$) without enrichments than with. The opposite was observed with score a, i.e., there were more birds with no hock burns in enriched pens at 41 kg/m^2 than in unenriched pens ($p < 0.0001$).

4. Discussions

The present study found that stocking density negatively impacted every measured indicator of broiler welfare except mortality, whereas enrichments had a positive effect on some of the welfare indicators, whether in one stocking density or both.

4.1. Contact Dermatitis and Litter Humidity

Stocking density had a negative effect on FPD and hock burns at 25 days of age. This effect was also visible in the post-mortem examination. Stocking density is often linked to welfare issues like dermatitis (e.g., [1,8]), so this result was expected. The present study shows a positive enrichment effect on FPD only in the 31 kg/m^2 group, visible at the post-mortem examination. Nevertheless, the high level of FPD in the 41 kg/m^2 group could have masked a potential effect of the enriched environment. Indeed, in the present study, FPD levels were high, probably due to litter management issues. Our litter was quite damp in all the pens, with moisture levels between 41.7% and 53.1% during the last week, despite the regular additions of litter. Other studies that analysed litter humidity in enriched environments measured a maximum humidity of 33% [5,12,20] (though with different experimental designs). Nevertheless, no effect of enrichment or stocking density was found on litter humidity, so the differences in FPD and hock burns between the groups could not be explained by this humidity. In the majority of previous studies, enrichments (e.g., straw bales, perches, elevated platforms and dustbathing areas) were found not to impact FPD levels [5,6,8,12,20,22,28,29]. However, in most of these studies (e.g., [6,12,29]), the great majority of birds had no signs or very low levels of FPD (i.e., between 3.88% to 7.89% of broilers observed had FPD in the study by Baxter et al. [6]). Two previous studies showed an effect of enrichment on FPD [16,18]. In the first one, Tahamtani et al. [16] demonstrated a positive effect of platforms on FPD in comparison with straw bales (groups having access to platforms 30 cm or 5 cm off the ground had better FPD scores than groups having access to straw bales only). In the second study, Ohara et al. [18] also found a positive effect of enrichments, i.e., straw bales and perches, but only on the foot lesions of females. This difference in FPD between males and females was not observed in the present study but the assessment was conducted earlier (25 days) and on another strain (Ross 308) than in Ohara's study [18] (Japanese broilers, assessment of FPD at 60 days old, at slaughter). However, we found more hock burns in males at 25 days old. The cause of hock burns is multifactorial, i.e., hock burns may be related to inactivity [30], growth rate [31], litter moisture [32], genetics [33] or body weight [1,34,35]. The difference between the levels of hock burns in males and females could be explained by the body weight, males being heavier than females. Beyond the sex effect, a positive enrichment impact was noted on hock burns observed at slaughter. In contrast to FPD, where the enrichment only affected the broilers from the lower density of 31 kg/m^2, there was a positive enrichment effect on hock burns at both densities, with broilers from the enriched groups having fewer hock burns than those from the unenriched groups. To the authors' knowledge, no previous study has shown an effect of enrichment (i.e., barrier perches, elevated platforms, ramps, straw bales, dustbathing area) on levels of hock burns in broiler chickens (with [8] or without comparing stocking densities [5,6,22,28,29]). Thus, the present study shows that enrichment, such as elevated platforms and straw bales, may improve hock health as well as footpad health. Future research could be carried out to see whether elevated platforms and straw bales have the same positive impact on leg health or whether differences are also observed between straw bales and perches [18].

4.2. Walking Ability

Our finding that stocking density negatively impacts walking ability at 26 and 32 days of age is consistent with the literature showing evidence of a decrease in walking ability when density is increased (e.g., [1]). However, in pairwise comparisons, this negative effect of stocking density was found only in the groups of broilers without access to enrichments. Thus, the presence of enrichments seems to mitigate the negative consequences of stocking density on walking ability. As there was no general significant effect of enrichments, this result nonetheless needs to be further examined with more repetitions and more or different enrichment materials. Generally, in previous studies on platforms and straw bales, no effect of these enrichment materials was found on walking ability [5,6,9,12,20,22]. However, in these studies, walking difficulties were rarely observed, unlike in the present study where the number of broilers with walking difficulties (scores 1, 2 and 3) was quite high, which may explain why an effect of enrichment was observed. The exception is the study by Kaukonen et al. [9], whose results are in agreement with ours in that they show a positive effect of elevated platforms on the mean gait score of broilers. Finally, it is possible that the poor litter quality observed in our study gave us the opportunity to demonstrate both enrichment and density impacts on contact dermatitis and walking ability, whereas other studies rarely observed such welfare issues.

4.3. Mortality and Weight

Our results showing no effect of either enrichment or stocking density on mortality are consistent with the literature. In previous studies on the same genotype (Ross 308) with different types of enrichment (straw bales, various shapes and the height of perches and platforms), whether on commercial farms or in experimental facilities, enrichments did not impact mortality rates [5–7,20]. No effect of stocking density on mortality is commonly found in the literature (e.g., [8,36,37]).

In contrast, body weights were slightly impacted by both enrichment and stocking density at 25 days of age. Broilers reared at a density of 31 kg/m^2 were heavier than those from the higher density of 41 kg/m^2. This result is consistent with previous studies showing a negative impact of stocking density on body weight (e.g., [36,37]). Furthermore, broilers from enriched pens were heavier than those from unenriched ones at 25 days. We can hypothesise that increased activity due to the use of enrichments leads to more muscle mass and to heavier birds. To compare this finding with the literature, the effect of enrichment is in agreement with Ohara et al. [18], who found a greater final body weight among Tatsuno slow-growing broilers in enriched pens (straw bales and perches) than among controls. In contrast, De Jong et al. [29] found that broilers (males from two strains: Ross 308 slaughtered at 38 days and JA757 slaughtered at 53 days) reared without any enrichment were heavier from day 17 onwards than birds reared with enrichments (barrier perches, ramps, platforms and a dustbathing area). De Jong et al. [29] concluded that enrichments increased the activity of birds, which then had an adverse effect on performance (average body weight and other parameters). This conclusion differs from that of Ohara et al. [18], who also observed increased activity in an enriched environment but deduced that enhancing broilers' activity with enrichments may not have adverse effects on productivity.

4.4. Platform Use and Impact on Stocking Density

In the present study, the surfaces below and above the platforms were, for different reasons, not counted as usable areas in the enriched pens. However, broilers perched on the platform throughout the rearing period, at times covering the entire surface of platforms and ramps (personal observations). Moreover, the area underneath the platforms was fully occupied by the broilers throughout the rearing period, mostly for the purpose of resting. Thus, the effective stocking density, if including the platform surfaces (below and above), was around 27–28 kg/m^2 as compared to the 31 kg/m^2 in lower density pens and 37–38 kg/m^2 as compared to the 41 kg/m^2 in higher density pens. Thus, the positive enrichments effects observed (on weights, walking ability, FPD, and hock burns) cannot be completely differentiated from the lower stocking density impact. The addition of enrichments in the rearing environment can then be considered as positive, intrinsically due to the increased possibilities for the expression of natural behaviours like perching and foraging, but also due to the increase in space allowance that it comes with.

5. Conclusions

Our results suggest that providing elevated platforms and straw bales helps to improve broiler welfare by reducing footpad dermatitis, hock burns and walking difficulties even at a high stocking density. However, reducing stocking density remains the key to improving broiler welfare. Further investigations are needed to deepen the knowledge of the effect of enrichments on birds' walking ability and to distinguish between the effects of different types of enrichments, examined separately, using a variety of stocking densities.

Author Contributions: Conceptualisation and methodology, M.G., F.M. and A.B.R.; statistical analysis, F.M. and M.J.; experimentation, F.M., M.G., J.-P.M. and A.K.; experiment supervision, A.K.; data curation, writing—original draft preparation, F.M.; review and editing, M.G., V.M., A.B.R., A.K., J.-P.M. and F.M.; project administration, M.G. and V.M.; funding acquisition, V.M. All authors have read and agreed to the published version of the manuscript.

Funding: This research was funded by the EURCAW-Poultry-SFA. European Commission Grant number: SANTE/EURC/2020/SI.824038.

Institutional Review Board Statement: Ethical review and approval were waived for this study due to the absence of invasive acts susceptible to cause harm or suffering to the animals.

Informed Consent Statement: Not applicable.

Data Availability Statement: Data is contained within the article.

Acknowledgments: The authors would like to thank the staff of the Avian Experimental Unit (ANSES SELEAC, 22440 Ploufragan, France) for rearing and taking care of the poultry and for their assistance in experiments. Our thanks to Mathieu Andraud, Stéphanie Bougeard, Pachka Hammami and Camille Lucas for their assistance in statistical analysis. Additionally, the authors would like to thank Delphine Libby-Claybrough for the English language review.

Conflicts of Interest: The authors declare no conflict of interest.

References

1. de Jong, I.C.; Berg, C.; Butterworth, A.; Estevez, I. Scientific report updating the EFSA opinions on the welfare of broilers and broiler breeders. *Support. Publ.* **2012**, *9*, 295E. [CrossRef]
2. Newberry, R.C. Environmental enrichment: Increasing the biological relevance of captive environments. *Appl. Anim. Behav.* **1995**, *44*, 229–243. [CrossRef]
3. Riber, A.B.; van de Weerd, H.A.; de Jong, I.C.; Steenfeldt, S. Review of environmental enrichment for broiler chickens. *Poult. Sci.* **2018**, *97*, 378–396. [CrossRef] [PubMed]
4. Malchow, J.; Berk, J.; Puppe, B.; Schrader, L. Perches or grids? What do rearing chickens differing in growth performance prefer for roosting? *Poult. Sci.* **2019**, *98*, 29–38. [CrossRef] [PubMed]
5. Bailie, C.L.; Baxter, M.; O'Connell, N.E. Exploring perch provision options for commercial broiler chickens. *Appl. Anim. Behav. Sci.* **2018**, *200*, 114–122. [CrossRef]
6. Baxter, M.; Richmond, A.; Lavery, U.; O'Connell, N.E. Investigating optimal levels of platform perch provision for windowed broiler housing. *Appl. Anim. Behav. Sci.* **2020**, *225*, 104967. [CrossRef]
7. Malchow, J.; Puppe, B.; Berk, J.; Schrader, L. Effects of elevated grids on growing male chickens differing in growth performance. *Front. Vet. Sci.* **2019**, *6*, 203. [CrossRef]
8. Ventura, B.A.; Siewerdt, F.; Estevez, I. Effects of barrier perches and density on broiler leg health, fear, and performance. *Poult. Sci.* **2010**, *89*, 1574–1583. [CrossRef]
9. Kaukonen, E.; Norring, M.; Valros, A. Perches & elevated platforms in commercial broiler farms: Use & effect on walking ability, incidence of tibial dyschondroplasia & bone mineral content. *Animal* **2017**, *11*, 864–871. [CrossRef]
10. Chuppava, B.; Visscher, C.; Kamphues, J. Effect of different flooring designs on the performance and foot pad health in broilers and Turkeys. *Animals* **2018**, *8*, 70. [CrossRef]

11. Liu, Z.; Torrey, S.; Newberry, R.C.; Widowski, T. Play behaviour reduced by environmental enrichment in fast-growing broiler chickens. *Appl. Anim. Behav. Sci.* **2020**, *232*, 105098. [CrossRef]

12. Yang, X.; Huo, X.; Li, G.; Purswell, J.L.; Tabler, T.; Chesser, D.; Zhao, Y. Application of elevated perching platform and robotic vehicle in broiler production. In Proceedings of the 2019 ASABE Annual International Meeting, Boston, MA, USA, 7–10 July 2019.

13. Jones, P.J.; Tahamtani, F.M.; Pedersen, I.J.; Niemi, J.K.; Riber, A.B. The productivity and financial impacts of eight types of environmental enrichment for broiler chickens. *Animals* **2020**, *10*, 378. [CrossRef] [PubMed]

14. Pedersen, I.J.; Tahamtani, F.M.; Forkman, B.; Young, J.F.; Poulsen, H.D.; Riber, A.B. Effects of environmental enrichment on health and bone characteristics of fast growing broiler chickens. *Poult. Sci.* **2020**, *99*, 1946–1955. [CrossRef]

15. Bach, M.H.; Tahamtani, F.M.; Pedersen, I.J.; Riber, A.B. Effects of environmental complexity on behaviour in fast-growing broiler chickens. *Appl. Anim. Behav. Sci.* **2019**, *219*, 104840. [CrossRef]

16. Tahamtani, F.M.; Pedersen, I.J.; Riber, A.B. Effects of environmental complexity on welfare indicators of fast-growing broiler chickens. *Poult. Sci.* **2020**, *99*, 21–29. [CrossRef] [PubMed]

17. Tahamtani, F.M.; Pedersen, I.J.; Toinon, C.; Riber, A.B. Effects of environmental complexity on fearfulness and learning ability in fast growing broiler chickens. *Appl. Anim. Behav. Sci.* **2018**, *207*, 49–56. [CrossRef]

18. Ohara, A.; Oyakawa, C.; Yoshihara, Y.; Ninomiya, S.; Sato, S. Effect of environmental enrichment on the behavior and welfare of Japanese broilers at a commercial farm. *J. Poult. Sci.* **2015**, *52*, 323–330. [CrossRef]

19. Bergmann, S.; Schwarzer, A.; Wilutzky, K.; Louton, H.; Bachmeier, J.; Schmidt, P.; Erhard, M.; Rauch, E. Behavior as welfare indicator for the rearing of broilers in an enriched husbandry environment—A field study. *J. Vet. Behav. Clin. Appl. Res.* **2017**, *19*, 90–101. [CrossRef]

20. Baxter, M.; Bailie, C.L.; O'Connell, N.E. Evaluation of a dustbathing substrate and straw bales as environmental enrichments in commercial broiler housing. *Appl. Anim. Behav. Sci.* **2017**, *200*, 78–85. [CrossRef]

21. Kells, A.; Dawkins, M.S.; Cortina Borja, M. The effect of a 'freedom food' enrichment on the behaviour of broilers on commercial farms. *Anim. Welf.* **2001**, *10*, 347–356.

22. Bailie, C.L.; O'Connell, N.E. The effect of level of straw bale provision on the behaviour and leg health of commercial broiler chickens. *Animal* **2014**, *8*, 1715–1721. [CrossRef] [PubMed]

23. European Commission. Council Directive 2007/43/EC of 28 June 2007 laying down minimum rules for the protection of chickens kept for meat production. *Off. J. Eur. Union* **2007**, *182*, 19–28.

24. McLean, J.A.; Savory, C.J.; Sparks, N.H.C. Welfare of male and female broiler chickens in relation to stocking density, as indicated by performance, health and behaviour. *Anim. Welf.* **2002**, *11*, 55–73.

25. Meyer, M.M.; Johnson, A.K.; Bobeck, E.A. Development and Validation of Broiler Welfare Assessment Methods for Research and On-farm Audits. *J. Appl. Anim. Welf. Sci.* **2020**, *23*, 433–446. [CrossRef]

26. Welfare Quality®. *Assessment Protocol for Poultry (Broilers, Laying Hens)*; Welfare Quality®Consortium: Lelystad, The Netherlands, 2009.

27. R Development Core Team. *R: A Language and Environment for Statistical Computing*; R Foundation for Statistical Computing: Vienna, Austria, 2020.

28. Kaukonen, E.; Norring, M.; Valros, A. Evaluating the effects of bedding materials and elevated platforms on contact dermatitis and plumage cleanliness of commercial broilers and on litter condition in broiler houses. *Br. Poult. Sci.* **2017**, *58*, 480–489. [CrossRef] [PubMed]

29. de Jong, I.C.; Blaauw, X.E.; van der Eijk, J.A.J.; Souza da Silva, C.; van Krimpen, M.M.; Molenaar, R.; van den Brand, H. Providing environmental enrichments affects activity and performance, but not leg health in fast- and slower-growing broiler chickens. *Appl. Anim. Behav. Sci.* **2021**, *241*, 105375. [CrossRef]

30. Dawkins, M.S.; Cain, R.; Merelie, K.; Roberts, S.J. In search of the behavioural correlates of optical flow patterns in the automated assessment of broiler chicken welfare. *Appl. Anim. Behav. Sci.* **2013**, *145*, 44–50. [CrossRef]

31. Hepworth, P.J.; Nefedov, A.V.; Muchnik, I.B.; Morgan, K.L. Hock burn: An indicator of broiler flock health. *Vet. Rec.* **2011**, *168*, 303. [CrossRef]

32. Haslam, S.M.; Knowles, T.G.; Brown, S.N.; Wilkins, L.J.; Kestin, S.C.; Warriss, P.D.; Nicol, C.J. Factors affecting the prevalence of foot pad dermatitis, hock burn and breast burn in broiler chicken. *Br. Poult. Sci.* **2007**, *48*, 264–275. [CrossRef]

33. Ask, B. Genetic variation of contact dermatitis in broilers. *Poult. Sci.* **2010**, *89*, 866–875. [CrossRef]

34. Kjaer, J.B.; Su, G.; Nielsen, B.L.; Sorensen, P. Foot pad dermatitis and hock burn in broiler chickens and degree of inheritance. *Poult. Sci.* **2006**, *85*, 1342–1348. [CrossRef] [PubMed]

35. Hepworth, P.J.; Nefedov, A.V.; Muchnik, I.B.; Morgan, K.L. Early warning indicators for hock burn in broiler flocks. *Avian Pathol.* **2010**, *39*, 405–409. [CrossRef] [PubMed]

36. Şekeroğlu, A.; Sarica, M.; Gulay, M.; Duman, M. Effect of Stocking Density on Chick Performance, Internal Organ Weights and Blood Parameters in Broilers. *J. Anim. Vet. Adv.* **2011**, *10*, 246–250. [CrossRef]

37. Zuowei, S.; Yan, L.; Yuan, L.; Jiao, H.; Song, Z.; Guo, Y.; Lin, H. Stocking density affects the growth performance of broilers in a sex-dependent fashion. *Poult. Sci.* **2011**, *90*, 1406–1415. [CrossRef] [PubMed]

 **_animals_**

Article

The Utility of Scatter Feeding as Enrichment: Do Broiler Chickens Engage with Scatter–Fed Items?

Brittany Wood, Christina Rufener, Maja M. Makagon and Richard A. Blatchford *

Center for Animal Welfare, Department of Animal Science, University of California, Davis, 1 Shields Avenue, Davis, CA 95616-8531, USA; bwood@ucdavis.edu (B.W.); cbrufener@ucdavis.edu (C.R.); mmakagon@ucdavis.edu (M.M.M.)
* Correspondence: rablatchford@ucdavis.edu; Tel.: +1-530-752-8763

Simple Summary: In recent years, there has been increasing interest in providing an enriched environment to broiler chickens. Indeed, many welfare certification companies encourage or require enrichment to be provided. Most of these companies suggest the use of scatter feeding as enrichment material, though there is little scientific evidence to support the implementation of a scatter feeding program. One of the potential benefits of scatter feeding programs may be an observed increase in foraging behavior, and hence overall activity of the birds. This study aimed to understand the impact of scatter feeding on the foraging behavior of broilers. Six groups of broilers were provided with either dried mealworms, whole wheat, shredded cabbage, alfalfa pellets, wood shavings, or no scatter feeding. To maintain the birds' interest in the enrichment, feed items were only scattered on the first three days of each week. Foraging and feeding behavior were observed via video for one-hour periods immediately after scattering, 2 h later, and 6 h later. Immediately following the scattering of feed items, broilers in all groups showed an increase in foraging, though this was most pronounced in the dried mealworm group. Foraging behavior decreased with age for all groups. The mealworm group also fed less during hour one compared to the later hours. These results did not provide evidence that scatter feeding encourages foraging behavior, except for a short-term effect of a high value food item. Therefore, future studies should examine the feed item and delivery in more detail.

Abstract: In recent years, welfare certification companies have encouraged the use of scatter feeding as enrichment material, though there is little scientific evidence to support a scatter feeding program. This study aimed to understand the impact of scatter feeding on the foraging behavior of broilers. One hundred eighty Ross 308 chicks were allocated into six treatment groups (six replicates/treatment). Broilers were scatter fed dried mealworms, whole wheat, shredded cabbage, alfalfa pellets, wood shavings, or no scatter feeding, respectively. Enrichment was provided on the first three days of each week. Total foraging, active foraging, and feeding were observed for one-hour periods immediately after scattering, 2 h later, and 6 h later. In all groups, broilers increased both total ($p = 0.001$) and active ($p = 0.001$) foraging, though this was most pronounced in the dried mealworm group. Across all groups, active foraging decreased with age ($p = 0.001$). The mealworm group also showed a corresponding decrease in feeding during hour one compared to the later hours ($p = 0.001$). These results did not provide evidence that scatter feeding encourages foraging behavior, except for a short-term effect of a high value feed item. This finding suggests that the item scattered and the delivery method should be studied further.

Keywords: broiler; environmental enrichment; scatter feed; foraging; engagements

Citation: Wood, B.; Rufener, C.; Makagon, M.M.; Blatchford, R.A. The Utility of Scatter Feeding as Enrichment: Do Broiler Chickens Engage with Scatter–Fed Items? . *Animals* **2021**, *11*, 3478. https://doi.org/10.3390/ani11123478

Academic Editor: Harry J. Blokhuis

Received: 3 November 2021
Accepted: 23 November 2021
Published: 7 December 2021

Publisher's Note: MDPI stays neutral with regard to jurisdictional claims in published maps and institutional affiliations.

1. Introduction

Environmental enrichments are modifications of the environment that aim to promote the performance of normal behaviors and/or animal health. In broiler chicken production, environmental enrichment is often provided with the goal of increasing overall activity, and preventing the development of lameness, hock burns, and breast blisters [1]. In the US, third-party animal welfare certification programs typically require that broiler chickens are housed with enrichment materials ex. [2,3]. Scatter feeding is often listed among the recommended forms of enrichment. Its application is based on the hypothesis that this practice will promote foraging behavior and overall activity. However, studies investigating foraging in broilers have failed to find a definitive relationship between scattering of feed stuff and broiler activity levels. The scattering of whole wheat, for example, did not increase broiler activity as reported in several recent studies [4–6]. Other feed items, such as mealworms, have resulted in only short-term increases in foraging activity [6]. The current study aims to add to the existing body of literature by evaluating how feeding and foraging behavior change immediately following the scattering of feed and on days and at times when feed stuff is not scattered, as well as assessing which feed items have the most pronounced impact on broiler behavior.

The idea that the scattering of whole grains or other feed items can be used as a form of environmental enrichment for broilers is grounded in the assumption that foraging behavior is an important part of the normal behavioral repertoire of broilers. For example, a welfare certification organization [2] states that "if chickens are provided with edible material contained in their litter, they will be actively engaged in foraging behavior for extended periods," while another one [3] lists foraging as an example of a natural behavior that should be encouraged. However, what constitutes a normal behavior (i.e., behavior that is important to the animal in a particular environmental context) can be modified by selective breeding and environmental conditions. It has been suggested that, along with selection for fast growth, the broiler behavioral repertoire has shifted towards the performance of behaviors that allow the birds to conserve their energy [7]. For example, while red junglefowl allocate a large proportion of their day to foraging-related activities such as pecking the ground or scratching the litter [8], modern-day broilers have been shown to spend over 60% of the day inactive and less than 4% of daylight hours foraging [9,10]. Moreover, while red junglefowl will search for feed even when they are provided with a "free" feed option [11], broilers do not show this tendency [7]. If scattering of feed is to be recommended as a form of environmental enrichment for broilers, further investigation into whether scattering of feed effectively promotes foraging behavior is warranted. The overall goal of this study is to evaluate whether scatter feeding is a viable form of enrichment for broiler chickens. Specifically, we assessed whether broiler chickens would forage for scattered feed items when provided, and whether increased activity would be observed on days when scattered feed was not offered. We further investigated whether broilers would engage with some feed items compared to others.

2. Materials and Methods

The study was conducted over six weeks between June and July 2018 at the Hopkins Avian Facility, University of California, Davis, with approval from the UC Davis Institutional Animal Care and Use Committee (Protocol #20212, approved November 2017).

2.1. Animals and Housing

One hundred ninety-five mixed-sex day-old Ross 308 chicks were reared on wood shavings. The chicks were acquired from a commercial hatchery, and individually marked with food coloring on day one of age. Chicks were colored with 5 colors, grouped into 5 pens by color and brooded for five days under ceramic heating bulbs in six identical pens (3.05 m × 1.52 m). A sixth group of 15 extra chicks were not colored. During brooding, chicks had ad libitum access to water and commercially supplied starter feed delivered in a standard 3.5-gal (13.25 L) waterer and 30 lb (13.61 kg) round feeder, respectively. For the first three days, chicks received 23 L:1 D hours light: dark. Daylight hours were subsequently reduced to 20 L:4 D. On day six of age, chicks were placed into 36 pens (3.05 m × 1.52 m; in groups of 5–6. The groups were composed of one chick from each of the colored brooding pens. Unmarked (extra) chicks were distributed as evenly as possible across the pens. An opaque tarp was hung across pen partitions to prevent chicks in one group from seeing those in adjacent pens. Ad libitum access to water and commercially supplied starter (1 week), grower (2 weeks), and finisher (2 weeks) diets continued to be provided. Researchers and staff entered the barn only to conduct daily wellness and equipment checks, and to clean and refill feeders/waterers.

2.2. Research Treatments

Each of the thirty-six pens was assigned to one of six treatments (six replicates/treatment): (1) dried mealworms (MW), (2) whole wheat (WW), (3) cabbage (CA), (4) alfalfa pellets (AP), (5) wood shavings (SH), and (6) feed-only/no-scattering control (Control). Treatments were assigned in blocks to ensure that they were uniformly distributed across the barn. Enrichments were scattered evenly across the pen floor on the first three days of each week between 10:00 and 11:00 h. Birds in the MW, CA, AP, and SH treatment pens received a half cup (118.3 mL) of enrichment. Due to the difference in grain size (grain units), only $\frac{1}{4}$ cup (59.15 mL) of WW was used. The Control group served as a non-scatter-feeding control. Specifically, we tested how the scattering of whole wheat, cabbage, alfalfa, dried mealworms, broiler feed, and shavings impacted the feeding and foraging activity of broiler chickens. Due to the fact that the proposed benefits of scatter feeding are linked to increased activity and load placed on the legs, we further differentiated between active and inactive foraging, where active foraging was performed while the bird was standing up or walking. The whole wheat and cabbage treatment were selected for evaluation as they are currently listed as effective forms of enrichment by one or more broiler welfare assurance programs ex. [2,3]. Alfalfa pellets were included because they are easily accessible to producers; therefore, they were selected due to their potential to serve as a practical enrichment. Wood shavings were used to test the impact of the act of scattering (a non-nutritive resource) on broiler behavior. The dried mealworm treatment was included as a positive control. Mealworms are considered a high-value feed item, are commonly used as a feed reward in research [12–14], and have been shown to have some impact on foraging behavior in previous studies [6]. The feed-only treatment served as the negative control.

2.3. Behavioral Observations

A DVR furnished with GeoVision-1480 video surveillance system software and connected to 36 video cameras (Clinton, Model CE-VF540, Clinton Electronics, Loves Park, IL, USA; 1 video camera per pen) recording chick behavior within the entire floor area of each pen. Video recorded on the first and fourth day of weeks 2 and 4, and the first day of week 6, was subsequently analyzed. The first day of each week (ON day) represented the first day of enrichment delivery, while day four (OFF day) represented the first day within the week when enrichment was not scattered. Three one-hour observation periods were monitored on each focal day. The exact start time for the observation was established independently for each pen each week. Scattering occurred between 10:00 and 11:00. During ON days, behavioral observations commenced immediately after enrichment was scattered and the researcher moved completely out of the video frame (H1). The same observation start time was used for the OFF day observations within a given pen each week. Control was not provided with enrichment. The observations for those 5 pens began 30 min after the adjacent pens received enrichment. The remaining daily observation periods took place two hours after the last set of H1 observations, at approximately 13:00–14:00 (H2) and 17:00–18:00 (H3).

Behavioral data were collected using the 1-0 scan sampling strategy. Observers reviewed the first of every five minutes of video and recorded whether each of the color-marked birds participated in feeding from the feeder or in foraging behavior during that minute. Chicks were assumed to be feeding from the feeder if they were observed pecking within the feeder trough. Chicks observed pecking at the ground or raking their beaks across or scratching the wood shavings were assumed to be foraging. We considered chicks to be "foraging active" if they were standing or walking while foraging, and "foraging inactive" if they were sitting or lying down. In total, 12 observations were recorded per hour, per chick, and per pen.

Five observers assisted with data collection. Before engaging in data collection, the observers were trained on the data collection protocol, and their inter-rater reliability was evaluated against that of the lead researcher (B.W.; at least 90% agreement was required). Inter-rater reliability was assessed based on a review of two hours of video footage (one hour recorded in the morning and one in the evening). Additional two-hour video clips taken from a variety of cameras and representing a variety of chick ages were assigned to all of the observers over the course of the study to ensure that reliability remained consistently high.

2.4. Data Processing

For each focal hour of behavioral observations, we calculated the percentage of observations (out of 12 possible) during which each individual broiler chicken engaged in "foraging active", "foraging inactive", and "feeding". The proportion of observations during which each individual engaged in any type of foraging ("foraging active" + "foraging inactive") was also calculated ("foraging total"). An initial visual comparison of means revealed that means were similar across the observations when no scattering was provided, i.e., during H2 and H3 on ON days, and all observations on OFF days. Therefore, the analysis included data from ON days only. Observations during H2 and H3 were combined to allow for comparison between observations immediately after scattering (H1) vs. later in the day (H2 and H3). Based on the visual comparison of means, we combined treatments to Control, Other (WW, CA, AP, SH; all treatments where scattering was provided except for MW), and Mealworms (MW).

2.5. Statistical Analysis

Statistical analysis was completed in R 3.4.2 (R Core Team, 2017) using linear mixed-effect models (LMER) with package 'lme4' [15]. A graphical analysis was used to confirm homoscedasticity of explanatory variables and normality of residuals. Outcome variables were transformed as needed. The final model was obtained with a stepwise backward reduction and a p-value > 0.05 as a criterion of exclusion using parametric bootstrap tests (package 'pbkrtest' [16]). The 'effects' package [17] was used to calculate model estimates.

Outcome variables were the percent of observations spent foraging total, the percent of observations spent foraging actively (square root transformed), and the percent of observations spent feeding. Fixed effects were age (factor with 3 levels: 2, 4, 6 weeks of age), treatment (factor with 3 levels: Control, Other, Mealworms), hour (factor with 2 levels: H1, H2 and H3), and the interaction of hour and treatment. To account for repeated measures and pseudo-replication, as well as pen-to-pen and individual-to-pen variation, hour nested in week nested in individual nested in pen was included as a random effect.

3. Results

3.1. Foraging Total

The percentage of observations spent foraging was higher immediately after the scattering was provided (H1) than later in the day (H2 and H3) in all treatments, but this pattern was most pronounced in the MW treatment (Figure 1, $p = 0.001$). More specifically, the estimated means [95% confidence interval] were similar during H2 and H3 irrespective of treatments (Control: 22.7 [17.2, 28.2] %, Other: 21.8 [19.1, 24.6] %, MW: 20.8 [15.3, 26.3] %). Whereas the percent of observations spent foraging during H1 was increased by only few percent in Control and Other birds (Control: 28.9 [23.0, 34.7] %, Other: 29.0 [26.1, 32.0] %), MW birds were foraging during 52.8 [47.0, 58.7] % of the observations in H1.

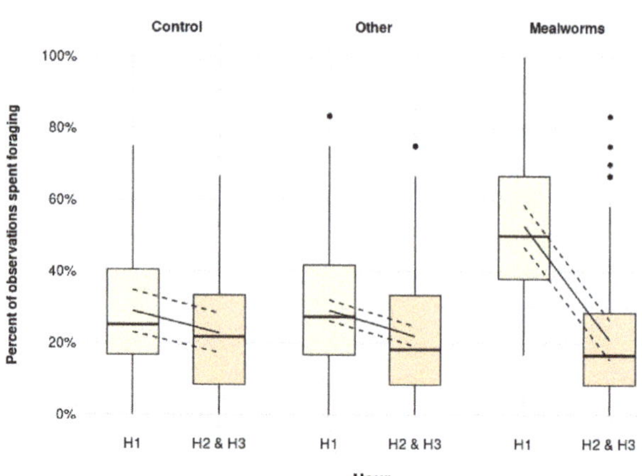

Figure 1. The percentage of observations spent foraging (including foraging active and foraging inactive) immediately after scattering was provided (H1) and later in the day (H2 and H3) for Control (no scattering), Other (scattering of shavings, whole wheat, alfalfa, or cabbage), and Mealworms ($p = 0.001$). Boxplots show medians, lower, and upper interquartile range of raw data. Whiskers indicate 1.5 times the interquartile range. Solid lines represent estimated means and dashed lines represent 95 % confidence intervals.

The percent of observations spent foraging was further affected by age (Figure 2, $p = 0.001$), though the effect was small (2 weeks of age: 25.2 [22.7, 27.7] %, 4 weeks of age: 28.8 [26.3, 31.3] %, 6 weeks of age: 22.6 [20.1, 25.1] %).

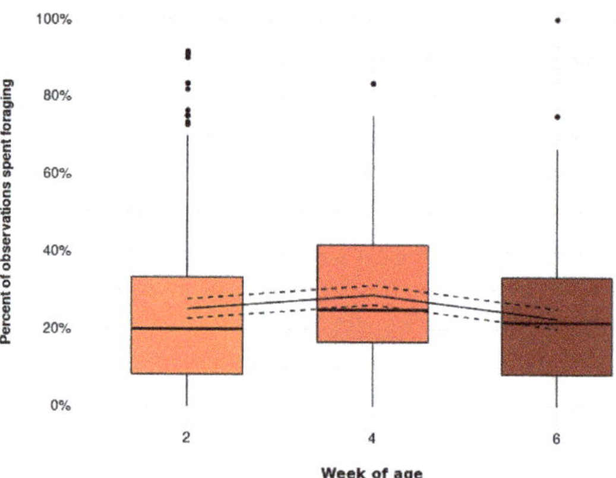

Figure 2. The percent of observations spent foraging (including foraging active and foraging inactive) at 2, 4, and 6 weeks of age ($p = 0.001$). Boxplots show medians, lower, and upper interquartile range of raw data. Whiskers indicate 1.5 times the interquartile range. Solid lines represent estimated means and dashed lines represent 95% confidence intervals.

3.2. Foraging Active

Similar to the percentage of observations spent foraging in total, the percentage of observations spent foraging actively was higher immediately after the scattering was provided (H1) than later in the day (H2 and H3) in all treatments. Again, this pattern was most pronounced in the MW treatment (Figure 3, $p = 0.001$). Whereas birds were foraging actively to a low percent irrespective of treatment during H2 and H3 (Control: 4.6 [2.6, 7.1] %, Other: 3.5 [2.6, 4.6] %, MW: 3.2 [1.6, 5.4] %), the percent of observations spent foraging actively was increased by a few percent during H1 in Control (10.2 [6.9, 14.2] %)

and Other (10.8 [9.0, 12.8] %). MW birds, on the other hand, spent 36.6 [30.0, 43.9] % of the observations during H1 foraging actively.

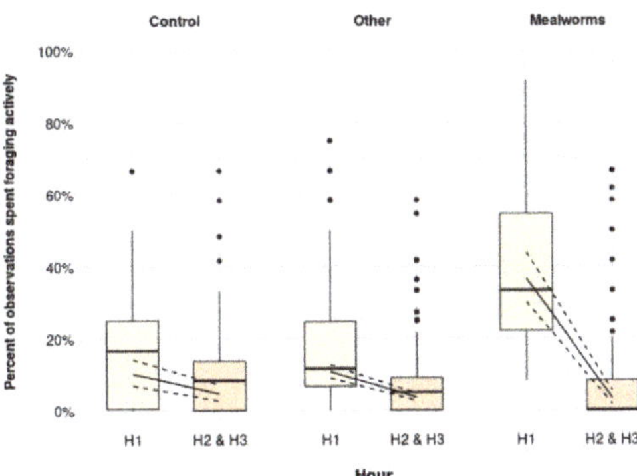

Figure 3. The percentage of observations spent foraging actively immediately after scattering was provided (H1) and later in the day (H2 and H3) for Control (no scattering), Other (scattering of shavings, whole wheat, alfalfa, or cabbage), and Mealworms ($p = 0.001$). Boxplots show medians, lower, and upper interquartile range of raw data. Whiskers indicate 1.5 times the interquartile range. Solid lines represent estimated means and dashed lines represent 95% confidence intervals.

In addition, the percentage of observations spent foraging actively decreased with increasing age (Figure 4, $p = 0.001$). Birds were foraging actively in 11.5 [9.9, 13.2] % of the observations at 2 weeks of age, in 6.6 [5.4, 8.0] % of the observations at 4 weeks of age, and in 2.5 [1.8, 3.4] %) of the observations at 6 weeks of age.

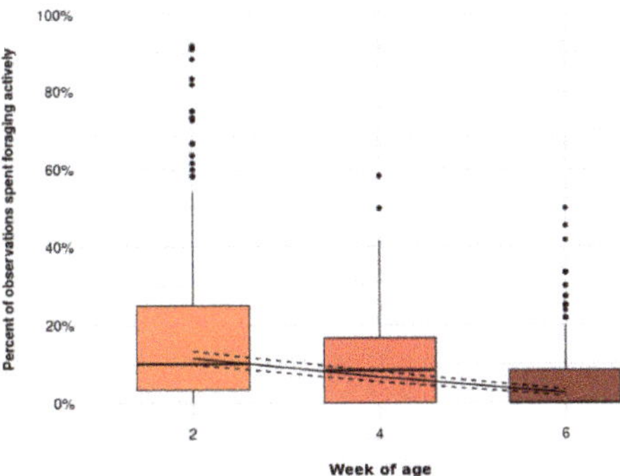

Figure 4. The percentage of observations spent foraging actively at 2, 4, and 6 weeks of age ($p = 0.001$). Boxplots show medians, lower, and upper interquartile range of raw data. Whiskers indicate 1.5 times the interquartile range. Solid lines represent estimated means and dashed lines represent 95% confidence intervals.

3.3. Feeding

The percentage of observations spent feeding at the feeder was similar across all observations, but birds in the Mealworms treatment were observed feeding less during H1 (Figure 5, $p = 0.001$). More specifically, Mealworm birds were feeding in 8 [4.7, 11.3] % of the observations during H1, which was lower compared to all other observations (H1 Control: 16.7 [13.4, 20.0] %, H1 Other: 19.2 [17.5, 20.8], H2 and H3 control: 16.9 [13.8, 19.8] %, H2 and H3 Other: 15.7 [14.2, 17.3], H2 and H3 Mealworms: 13.1 [10.0, 16.1] %).

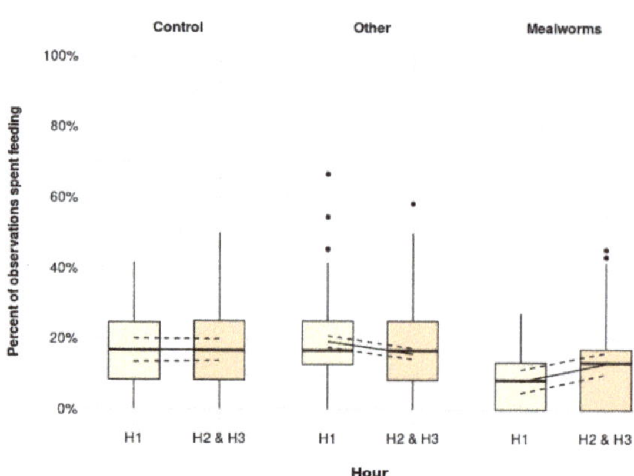

Figure 5. The percentage of observations spent feeding at the feeder immediately after scattering was provided (H1) and later in the day (H2 and H3) for Control (no scattering), Other (scattering of shavings, whole wheat, alfalfa, or cabbage), and Mealworms (p = 0.001). Boxplots show medians, lower, and upper interquartile range of raw data. Whiskers indicate 1.5 times the interquartile range. Solid lines represent estimated means and dashed lines represent 95% confidence intervals.

The percentage of observations spent feeding at the feeder was further affected by age (Figure 6, p = 0.001), though the effect was small (2 weeks of age: 18.4 [17.0, 19.8] %, 4 weeks of age: 16.8 [15.4, 18.3] %, 6 weeks of age: 12.6 [11.2, 14.0] %).

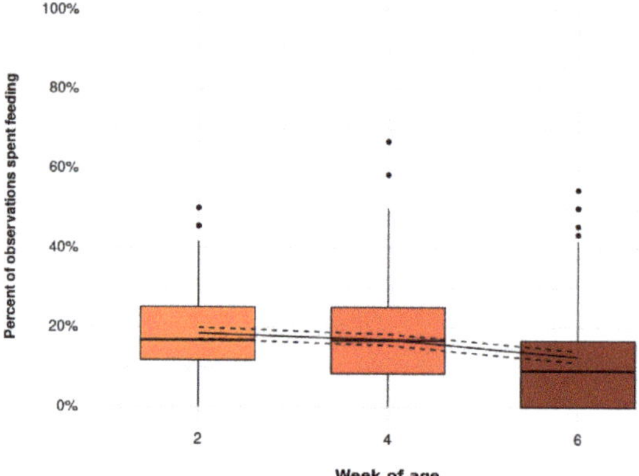

Figure 6. The percentage of observations spent feeding at the feeder at 2, 4, and 6 weeks of age (p = 0.001). Boxplots show medians, lower, and upper interquartile range of raw data. Whiskers indicate 1.5 times the interquartile range. Solid lines represent estimated means and dashed lines represent 95% confidence intervals.

4. Discussion

Scattering of feed with the goal of promoting foraging behavior and overall activity is a recommended form of broiler enrichment. Several previous studies have, however, failed to find a relationship between the scattering of feed and foraging and/or overall activity [4–6]. When noted, impacts of scatter feeding on broiler behavior have been associated with the scattering of high-value feed items, such as mealworms, but have had short-term impacts on broiler behavior [6].

Across all treatments, foraging activity was higher during H1 than subsequently in the day. In line with previous work, different feed items had different impacts on broiler behavior [4–6]. Along with type, the presence of the feed item also contributes to observed increases in foraging activity, as indicated by the fact that foraging activity increased only during H1 and only on days when scatter feed was delivered. However, as is consistent with other studies [4,9,10], broiler activity (in the current study, foraging) still suffered a reduction as age increased, even in treatments that initially stimulated foraging. This is a reoccurring issue with birds selected for fast growth, as it is difficult to bypass the confounding effect of age (and size) on overall locomotive activity levels.

In the current study, mealworms stimulated the most total and active foraging activity. However, the increase in foraging activity in the mealworm treatment only increased during H1 of the observations. Pichova et al. [6] investigated the effects of whole wheat, wood shavings, and mealworms on activity levels in broilers. They also found that scattering mealworms on the litter once per day encouraged activity such as litter pecking and scratching, and that the change in behavior only occurred immediately after mealworm delivery. In both studies, enrichment items were only scattered once per day, and this may have had an impact on the amount of foraging behavior observed. It could be that the significance of the items scattered decreased over time, as the items were consumed, and their presence in the environment decreased. Pichova et al. [6] also observed motivational differences in litter directed behavior for scattered feed items on litter. This may be attributable to the fact that, as the visual stimulation of the items decreased, the motivation of the birds to forage also decreased.

The increase in foraging among mealworm-fed birds during H1 was associated with a reduction in the percentage of observations of birds feeding. This is not surprising, as these behaviors are mutually exclusive. Overall, the percentage of observations that the broilers spent feeding was similar across treatments, suggesting that the amount of scattered feed items did not interfere with feed intake. Although there was a very small effect, observed feeding behavior decreased with age, as also shown by Alvino et al. [10]. The small change in feeding behavior observed may be due to sampling time. In the current study, behavior was observed over three one-hour periods. Feeding behavior has been shown to have a strong circadian rhythm [9,18,19], and it could be stated that the chosen observation times were not optimal for measuring feeding behavior.

The results from the current study suggest that there is little effect of scatter feeding enrichments on stimulating broilers to perform foraging activity. However, Pichova et al. [6] suggested future research should focus on developing feed items that are highly attractive and distributed in ways that would increase broiler activity. For instance, the current study used whole wheat, which is a frequently recommended scatter enrichment item, and showed no effect on activity, which is in line with several studies. However, there is often little consideration given to the environments in which studies are conducted. It may be that experimental scale studies, small numbers of animals, small spaces, etc., may have an effect on the behavior of the broilers, which may differ under large scale commercial conditions. In the current study, scattering of items was carried out in close proximity to the feeders, where feed was available ad libitum. Therefore, the broilers may have been less motivated to seek out scattered items through foraging. Jordan et al. [5] found that when feeders were removed, broilers were more likely to be active when pelleted feed was scattered. Although they reported no increased activity when whole wheat was scattered, they did not remove feeders for this treatment. Therefore, future scatter feeding studies should test items under commercial conditions where broilers are likely to have times that they are not near feeder lines [20], and this may increase the likelihood of foraging even for items shown experimentally not to be as effective. Recently, Ferreira et al. [14] also found individual differences in the display of foraging behavior by broiler chickens, which may have implications on measuring this behavior when comparing treatments across the group level. Future studies should consider investigating and comparing results between individuals and groups.

5. Conclusions

As indicated, this study did not provide evidence that scatter feeding, as currently recommended in welfare certification guidelines, promotes long term foraging. In this study, foraging behavior was only encouraged in broilers scatter fed with mealworms, a well-known, high-value feed item. These results coincide with previous studies using mealworms as a source of environmental enrichment. Although broilers were stimulated to forage, this occurred only on the days when enrichment was provided (ON days) and only directly after distribution. Despite not being able to modify foraging behavior in the long term, there is a chance that, with the appropriate enrichment, delivery method, and schedule, broiler activity can be increased. It may also be of interest to look at the entire behavioral repertoire, instead of only foraging, as scatter feeding may have effects outside of the measures in the current study.

It is recommended that broilers are provided with high-value feed items that will motivate foraging behavior and general activity. Suggested environmental enrichment that promotes locomotor activity, such as mealworms, bales of straw, feed pellets, and various light intensities, may be used in order to increase activity. Enrichment has been found to effectively induce broiler activity when used in combinations. For example, straw bales and light intensity show promising results in promoting activity, which can lead to a reduction in lameness [21]. As enrichment stimulated foraging primarily within the first hour, in the future, broilers should probably receive enrichment more frequently and at various times of the day to initiate and prolong foraging activity. The use of various or a combination of enrichments throughout the development period may be worth exploring to keep enrichment novel and engaging, encourage locomotion, increase foraging behavior, and possibly contribute to improved leg health.

Author Contributions: All authors contributed significantly to the content of this manuscript: Conceptualization, M.M.M. and R.A.B.; methodology, B.W., M.M.M. and R.A.B.; validation B.W., C.R., M.M.M. and R.A.B.; formal analysis, B.W. and C.R.; investigation, B.W., M.M.M. and R.A.B.; resources, M.M.M. and R.A.B.; data curation, B.W. and C.R.; writing—original draft preparation, B.W.; writing—review and editing, B.W., C.R., M.M.M. and R.A.B.; visualization, B.W., C.R., M.M.M. and R.A.B.; supervision, M.M.M. and R.A.B.; project administration, B.W., M.M.M. and R.A.B.; funding acquisition, M.M.M. and R.A.B. All authors have read and agreed to the published version of the manuscript.

Funding: This project was supported in part by USDA Multistate Research Project NE 1942 as well as the Arthur W. Perdue Graduate Fellowship and a generous gift from the Arthur W. Perdue Foundation.

Institutional Review Board Statement: The study was conducted according to the guidelines of the Declaration of Helsinki, and approved by the Institutional Animal Care and Use Committee of the University of California, Davis (protocol number 20212, approved November 2017).

Data Availability Statement: The raw data supporting the conclusions of this article will be made available by the authors, without undue reservation.

Acknowledgments: We thank the staff at the Hopkins Avian Facility, particularly Kristy Portillo and Kevin Bellido, and all of the student assistants and interns for excellent animal care.

Conflicts of Interest: The authors declare no conflict of interest.

References

1. Riber, A.B.; van de Weerd, H.A.; de Jong, J.C.; Steenfeldt, S. Review of environmental enrichment for broiler chickens. *Poult. Sci.* **2018**, *97*, 378–396. [CrossRef] [PubMed]
2. Certified Humane Raised and Handled. Available online: https://certifiedhumane.org/our-standards/ (accessed on 2 November 2021).
3. Global Animal Partnership. Available online: https://globalanimalpartnership.org/standards/chicken/ (accessed on 2 November 2021).
4. Bizeray, D.; Estevez, I.; Leterrier, C.; Faure, J.M. Effects of increasing environmental complexity on the physical activity of broiler chickens. *Appl. Anim. Behav. Sci.* **2002**, *79*, 27–41. [CrossRef]
5. Jordan, D.; Štuhec, I.; Bessei, W. Effect of whole wheat and feed pellets distribution in the litter on broilers' activity and performance. *Arch. Geflügelk.* **2011**, *75*, 98–103.

6.	Pichova, K.; Nordgreen, J.; Leterrier, C.; Kostal, L.; Moe, R.O. The effects of food-related environmental complexity on litter directed behaviour, fear and exploration of novel stimuli in young broiler chickens. *Appl. Anim. Behav. Sci.* **2016**, *174*, 83–89. [CrossRef]
7.	Lindqvist, C.; Zimmerman, P.; Jensen, P. A note on contrafeeloading in broilers compared to layer chicks. *Appl. Anim. Behav. Sci.* **2006**, *101*, 161–166. [CrossRef]
8.	Dawkins, M.S. Time budgets in Red Junglefowl as a baseline for the assessment of welfare in domestic fowl. *Appl. Anim. Behav. Sci.* **1989**, *24*, 77–80. [CrossRef]
9.	Deep, A.; Schwean-Lardner, K.; Crowe, T.G.; Fancher, B.I.; Classen, H.L. Effect of light intensity on broiler behavior and diurnal rhythms. *Appl. Anim. Behav. Sci.* **2012**, *136*, 50–56. [CrossRef]
10.	Alvino, G.M.; Blatchford, R.A.; Archer, G.S.; Mench, J.A. Light intensity during rearing affects the behavioural synchrony and resting patterns of broiler chickens. *Br. Poult. Sci.* **2009**, *50*, 275–283. [CrossRef] [PubMed]
11.	Lindqvist, C.; Schutz, K.E.; Jensen, P. Red jungle fowl have more contrafreeloading than White Leghorn layers: Effect of food deprivation and consequences for information gain. *Behaviour* **2002**, *139*, 1195–1209. [CrossRef]
12.	Bokkers, E.A.M.; Koene, P. Sex and type of feed effects on motivation and ability to walk for a food reward in fast growing broilers. *Appl. Anim. Behav. Sci.* **2002**, *79*, 247–261. [CrossRef]
13.	Taylor, P.S.; Hamlin, A.S.; Crowley, T.M. Anticipatory behavior for a mealworm reward in laying hens is reduced by opiod receptor antagonism but not standard feed intake. *Front. Behav. Neurosci.* **2020**, *13*, 290. [CrossRef] [PubMed]
14.	Ferreira, V.H.B.; Germain, K.; Calandreau, L.; Guesdon, V. Range use is related to free-range broiler chickens behavioral responses during food and social conditioned place preference tests. *Appl. Anim. Behav. Sci.* **2020**, *230*, 105083. [CrossRef]
15.	Bates, D.; Mächler, M.; Bolker, B.; Walker, S. Fitting linear mixed-effects models using lme4. *J. Stat. Soft.* **2015**, *67*, 1–48. [CrossRef]
16.	Halekoh, U.; Højsgaard, S. A Kenward-Roger approximation and parametric bootstrap methods for tests in linear mixed models—The R package pbkrtest. *J. Stat. Soft.* **2014**, *59*, 1–32. [CrossRef]
17.	Fox, J. Effect displays in R for generalized linear models. *J. Stat. Soft.* **2003**, *8*, 1–27. [CrossRef]
18.	Blatchford, R.A.; Klasing, K.C.; Shivaprasad, H.L.; Wakenell, P.S.; Archer, G.S.; Mench, J.A. The effect of light intensity on the behavior, eye and leg health, and immune function of broiler chickens. *Poult. Sci.* **2009**, *88*, 20–28. [CrossRef] [PubMed]
19.	Blatchford, R.A.; Archer, G.S.; Mench, J.A. Contrast in light intensity, rather than daylength, influences the behavior and health of broiler chickens. *Poult. Sci.* **2012**, *91*, 1768–1774. [CrossRef] [PubMed]
20.	Febrer, K.; Jones, T.A.; Donnelly, C.A.; Dawkins, M.S. Forced to crowd or choosing to cluster? Spatial distribution indicates social attraction in broiler chickens. *Anim. Behav.* **2006**, *72*, 1291–1300. [CrossRef]
21.	Bailie, C.; Ball, M.E.E.; O'Connell, N.E. Influence of the provision of natural light and straw bales on activity levels and leg health in commercial broiler chickens. *Animal* **2013**, *7*, 618–626. [CrossRef] [PubMed]

Article

Broiler Chicken Behavior and Activity Are Affected by Novel Flooring Treatments

Leonie Jacobs [1,*], Shawnna Melick [1], Nathan Freeman [1], An Garmyn [2] and Frank A. M. Tuyttens [2,3]

[1] Department of Animal and Poultry Sciences, Virginia Polytechnic Institute and State University, 175 West Campus Drive, Blacksburg, VA 24061, USA; shawnna5@vt.edu (S.M.); nathanf1@vt.edu (N.F.)
[2] Faculty of Veterinary Medicine, Ghent University, 9820 Merelbeke, Belgium; An.Garmyn@UGent.be (A.G.); frank.tuyttens@ilvo.vlaanderen.be (F.A.M.T.)
[3] Flanders Research Institute for Agriculture, Fisheries and Food (ILVO), 9090 Melle, Belgium
* Correspondence: jacobsl@vt.edu; Tel.: +1-540-231-4735

Simple Summary: Broiler chickens should be able to express highly motivated behaviors, such as foraging and dustbathing. Health status and housing conditions impact the expression of these behaviors. This study compared the impact of novel flooring treatments on broiler chicken behavioral repertoire. We found that broilers' behavior was impacted by novel flooring treatments at 5 and 6 weeks of age. Differences were found in prevalences of drinking, foraging, preening, locomoting, and in generally being active. Generally, broilers with access to clean friable litter spent more time drinking, foraging, locomoting, preening and being active compared to when housed with a partially slatted floor and/or a disinfectant mat. Thus, access to clean, regularly replaced litter is beneficial for broiler chicken welfare, especially for their ability to perform normal behaviors.

Abstract: The objective was to determine broiler chicken behavioral differences in response to novel flooring treatments. Broilers ($n = 182$) were housed in 14 pens (a random subset from a larger-scale study including 42 pens), with 13 birds/pen. One of seven flooring treatments were randomly allocated to 14 pens (2 pens per treatment). The flooring treatments (provided from day 1 {1} or day 29 {29}) included regularly replaced shavings (POS), a mat with 1% povidone-iodine solution (MAT), and the iodine mat placed on a partially slatted floor (SLAT). In addition, a negative control treatment was included with birds kept on used litter from day 1 (NEG). Behavior was recorded in weeks 1, 2, 5, and 6. In week 5, treatments affected the behavioral repertoire ($p \leq 0.035$). Birds in POS-1 showed more locomoting, preening and activity overall compared to MAT and/or SLAT treatments. Birds in POS-29 showed more drinking, foraging, preening and overall activity than birds in MAT and/or SLAT treatments. In week 6, birds in the POS-1 treatment spent more time foraging compared to birds in all MAT and SLAT treatments ($p \leq 0.030$). In addition, birds in the POS-1 treatment spent more time preening than birds in the MAT-1 treatment ($p = 0.046$). Our results indicate that access to partially slatted flooring and/or disinfectant mats does not benefit broiler chicken welfare in terms of their ability to express highly motivated behaviors. Access to clean, regularly replaced litter is beneficial for broiler chicken welfare in terms of their ability to express their normal behavioral repertoire.

Keywords: animal welfare; animal behavior; normal behavior; flooring; meat birds; ethology

Citation: Jacobs, L.; Melick, S.; Freeman, N.; Garmyn, A.; Tuyttens, F.A.M. Broiler Chicken Behavior and Activity Are Affected by Novel Flooring Treatments. *Animals* **2021**, *11*, 2841. https://doi.org/10.3390/ani11102841

Academic Editor: Victoria Sandilands

Received: 2 July 2021
Accepted: 23 September 2021
Published: 29 September 2021

Publisher's Note: MDPI stays neutral with regard to jurisdictional claims in published maps and institutional affiliations.

1. Introduction

To ensure good animal welfare it is important to allow animals the ability to express highly motivated behaviors [1,2]. For broiler chickens, species-specific, highly motivated behaviors include foraging (or scratching) and dustbathing [2–4]. Broiler chickens were observed spending approximately 1–10% of their time foraging [5,6] and between 0.2 and 3.7% of their time dustbathing when housed on a range of litter types [6]. These behaviors and behaviors such as preening, stretching and play are generally considered to be positive indicators for animal welfare as they are deemed to be indicative of positive affective states [7–9]. Inducing positive affective states is considered a key component to ensure good animal welfare [10].

Previous work has shown that accessible resources such as a substrate can impact the expression of these behaviors. For instance, access to sand resulted in more frequent dustbathing compared to pine wood shavings, rice hulls, or a paper bedding product [3]. Broilers showed more foraging when housed with access to maize roughage compared to the control group, yet access to straw bales did not impact foraging behavior [11]. More play, ground scratching and ground pecking was observed when broilers had access to enrichments (peat, hay bales and elevated platforms) compared to a control [12]. Broilers housed on partially or fully slatted floors grew faster than broilers housed on litter, and this was attributed to birds on slatted floors directing their foraging behavior to the feeder, rather than the litter [13]. However, direct observations of behavior were not performed, thus the impact of slatted flooring on broilers' behavioral repertoire was not investigated. Other studies did not find an impact of accessible resources on foraging [14] or play [15,16].

Flooring substrate can also impact broilers' physical condition. Contact dermatitis on feet, hocks, or breast due to prolonged contact with moisture and irritants in litter is a prevalent health and welfare concern in broiler chickens [17–22]. In turn, these skin conditions can affect the birds' behavioral repertoire due to discomfort and pain [23,24]. Birds with contact dermatitis in a commercial setting are generally not treated to allow healing, thus lesions will likely worsen over time. We propose to investigate topical application of an antiseptic whether or not combined with a slatted floor to reduce time spent in contact with litter as potential flock-level approaches to prevent or remedy contact dermatitis. Early (day 1) or late (day 29) flock-level treatment approaches may prove beneficial to reduce dermatitis prevalence and severity, yet these treatments could affect the birds' ability to perform highly motivated behaviors. We previously determined the impact of novel pen-level preventative and remedial approaches on contact dermatitis (including footpad dermatitis), cleanliness, gait and body weight [25]. Those results demonstrated that flooring treatments and timing of treatments affected contact dermatitis severity, with access to regularly replaced clean litter showing the lowest prevalence and severity of the welfare issues [25]. With our previous work indicating worsened footpad dermatitis when birds had access to partially slatted floors and/or mats [25], it is likely the behavioral repertoire would be impacted. Therefore, the objective of this study was to assess behavioral differences in response to novel flooring treatments at the flock level, provided at different ages (starting on day 1 or day 29 of age). We hypothesized that flooring treatments would affect the behavioral repertoire, with more highly motivated behaviors (foraging and dustbathing) shown when birds had access to clean, regularly replaced pine shavings, and a potential detrimental effect of slatted flooring and mats on the birds' ability to show highly motivated behaviors.

2. Materials and Methods

This experiment was carried out between March and May 2019 and was approved by the Institutional Animal Care and Use Committee (IACUC) of Virginia Tech (protocol 18-246). For the behavioral data reported in this manuscript, a random selection of birds (182 out of 546) and pens (14 out of 42) were used from those reported in [25].

For this study, 182 commercial strain (Ross × Hubbard) male broiler chicks were housed at the poultry facility of Virginia Tech from day 1 until day 49 of age, with 13 birds per pen (1.25 m^2). Upon arrival at the facility, birds were randomly allocated to 1 of 7 treatment groups (3 × 2 factorial + 1 industry control group). Pens contained pine shavings, a drinker line with three nipples, a feeder, and in the first week, a heat lamp and a feed flat with feed. After the first week lighting was provided for 18 h, followed by 6 h of uninterrupted darkness. Feed was provided ad libitum following commercial standards with a starter, grower and finisher phase. More details on housing conditions were reported in [25].

The experiment consisted of an incomplete factorial design, with four flooring treatments and two timing treatments. The flooring treatments included a negative control (NEG), a positive control (POS), and two novel flooring treatments with disinfectant mats containing a povidone-iodine solution (MAT and SLAT). The two timing treatments consisted of access to the flooring treatments from day 1 of age or day 29 of age onwards. Pens that received the flooring treatments from day 29 of age were kept under identical conditions as the negative control up until day 29. Treatments were randomly allocated over blocks, resulting in 14 pens and 7 treatment groups (2 replicates per treatment).

Flooring treatments (see [25] for more details) consisted of the following:

1. The NEG flooring treatment consisted of pens with used litter (19.1% moisture content as measured prior to bird placement) that was collected from a previous broiler flock, to model an industry standard in the United States and other countries [26,27]. Litter was collected from experimental pens in the same facility and piled in the center hallway. The used litter was mixed manually and returned to the pens the next day to ensure an equal distribution of litter at a depth of approximately 6 cm. The NEG treatment was solely provided from day 1, no timing treatment at day 29 was included (as that would mean NEG until day 29 followed by NEG until day 49).

2. The MAT flooring treatment (Figure 1a) consisted of a disinfectant mat (60 × 70 cm mat; product 802010, Agri-Pro Enterprises of Iowa Inc., Iowa Falls, IA, USA) placed in the back middle of the pen under the drinker line. The mats covered 34% of the pen floor surface, and were filled with 3 L of a 1% povidone-iodine solution (diluted with tap water; 050AB Povidone Iodine Solution 10%, Vi-Jon Inc., Breckenridge Hills, MO, USA). Mats were provided on day 1 (MAT-1) or day 29 (MAT-29). Prior to day 29, conditions were identical to the NEG treatment. Every four days, the mats were removed from the pen, rinsed, and refilled with the disinfectant solution. The remainder of the pen contained used litter (66% of floor surface) as in the NEG treatment.

3. The SLAT treatment (Figure 1b) consisted of the mat with the disinfectant solution, placed on top of a black plastic slatted floor (60 × 120 cm, DURA-SLAT® Black Poultry and Kennel Flooring, Southwest Agri-Plastics Inc., Addison, TX, USA). The slatted floor covered 58% of floor surface, but only 24% was accessible to birds as the mat was placed on top of the slatted flooring. The slat and mat were placed on top of the litter but not elevated from the ground. The mat was placed on top of the slatted floor, and both were placed in the back of the pen under the drinker, provided on day 1 (SLAT-1) or day 29 (SLAT-29). The remainder of the pen contained used litter (42% of floor surface) as in the NEG treatment. Prior to day 29, conditions were identical to the NEG treatment. The slatted flooring was removed as needed to eliminate excess litter and fecal content, but was not rinsed. The mat was rinsed and refilled every four days.

4. The POS flooring treatment was provided from day 1 (POS-1) or day 29 (POS-29), with new pine shavings (10.7% moisture content prior to bird placement) at a depth of 6 cm, and shavings completely replaced every four days. Prior to day 29, conditions were identical to the NEG treatment.

Fourteen video cameras (IP Bullet camera FLPB133F, FLIR Systems Inc., Wilsonville, OR, USA) were installed to record behavior in each pen (total of 14 pens). Videos were recorded on Sundays to limit human disturbance during recording. For 2 pens (MAT-1 and MAT-29 treatments) in week 5, the recording from Saturday (no human disturbance during recording) was used instead of the Sunday recording, which was missing.

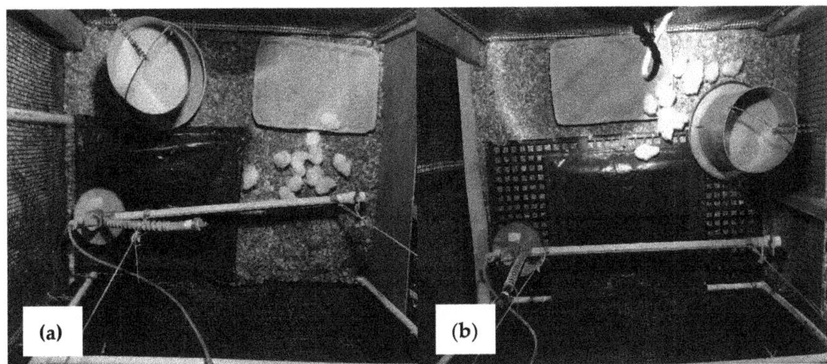

Figure 1. Top view photos of pen-based flooring treatments, with (**a**) a mat filled with 1% povidone-iodine solution (MAT) and (**b**) the iodine mat placed on a slatted floor (SLAT). The pens contained a hanging drinker line, a metal feeder, used litter, a disinfection mat, and for the SLAT treatment, a plastic slatted floor. In the first week, a feed flat with feed was provided.

Behavior was recorded at individual bird-level using scan sampling with a 1 min inter-sampling interval during two 15 min time periods (starting at 10 AM and 6 PM) for all birds in a pen (13 birds per pen), for 4 weeks (week 1, 2, 5, and 6 of age). Birds were not marked for identification, yet the observer was able to ensure all birds were observed based on the bird location at the start of the scan. This resulted in 15 scans per time period ($n = 195$ observations per time period per pen), and a targeted total of 21,840 behavioral entries (13 birds × 15 scans × 4 weeks × 14 pens × 2 time points). These individual behavioral observations were used to calculate pen-level percentages of time spent on each behavior. Due to normal mortality, fewer birds were observed resulting in a total of 21,350 behavioral entries available for analysis. A single observer used Solomon coder version 17.03.22 (Andras Peter, https://solomon.andraspeter.com/ (accessed on 17 May 2019)) to quantify broiler chicken behaviors (Table 1).

Table 1. Ethogram of recorded broiler chicken behaviors.

Behavior	Description
Eat	Beak inside or above feeder, may include extension of the neck
Drink	Beak near or in contact with the drinker, may include extension of the neck
Forage [1]	Pecking/scratching at the flooring substrate
Stretch	Extension of the wing or leg, may include fluffing of the feathers
Preen [2]	Feathers are raised, cleaned and realigned with the beak
Locomotion [3]	Moving using legs in a continuous forward motion (walking or running)
Dustbathe [1]	Vertical wing shakes, interacting with flooring substrate, performing side-rubs, and intermittent ground pecking with beak
Play [4]	Spontaneous motor behavior that occurs without apparent purpose. Includes frolicking (sudden running with no apparent stimulus, flapping wings) and food running (object in beak and locomotion at high speed)
Passive [1,3]	Bird sits resting its abdomen on the flooring substrate or stands with feet in contact with any flooring. Bird may have head tucked under the wing or have head at or below body level. Bird may stand without showing other behaviors.
Other	Other behaviors or behavior cannot be identified
Out of View	Bird is out of camera view
Active	Sum of all active behaviors, including eat, drink, forage, stretch, preen, locomotion, dustbathe, and play

[1] Adapted from [28]. [2] [29]. [3] Adapted from [30]. [4] Adapted from [16].

Behavioral entries for all birds of the group were converted to percentage (%) of total observations (n = 150–195) at pen level within a 15 min time period. Thus, we calculated the percentage of time spent on a specific behavior per pen, per time point, per week. Frequencies were analyzed in JMP® Pro 15.1 (SAS institute Inc., Cary NC, USA). In addition to the analysis of individual behaviors, we grouped behaviors into an "Active" category (Table 1), including all behaviors besides "Passive" (Table 1). The behavioral categories "Out of View" and "Other" (Table 1) were not analyzed, but were included in the total n of observations within a specific time period. Visual inspection of data residuals using normal quantile plots showed normal distribution for all but the residuals of "Dustbathing" and "Play". Normally distributed data were analyzed by age (1, 2, 5, 6 weeks) to account for the decrease in activity (more sitting) as birds age [31]. Analyses with week as fixed factor did show an age effect for all behavioral categories at $p < 0.05$. For these behaviors, mixed models were used for each age category, with treatment (flooring × timing treatment combination; n = 7), time (10 AM and 6 PM; n = 2) as fixed factors, with pen nested in block as random factors. Dustbathing and play were analyzed with non-parametric Wilcoxon chi-square test, assessing the effect of treatment (flooring × timing treatment combination; n = 7) for each sampling week. Post hoc analyses were performed using a nonparametric comparison for all pairs using the Dunn method for joint ranking, which includes a Bonferroni correction.

3. Results

Effect of flooring treatments on individual behaviors varied during different weeks (Figure 2). In week 1, the percentage of observations spent eating, drinking, foraging, locomoting, preening, stretching, playing and being passive did not differ between the 7 treatment groups (all $p \geq 0.13$). In week 2, treatments affected the percentage of time spent playing ($p = 0.047$), but pairwise comparisons were non-significant ($p > 0.45$).

In week 5, treatments affected the percentage of time spent drinking, foraging, locomoting, preening, and being active ($p \leq 0.035$; Figure 2). Percentage of time spent drinking was greater for birds in the POS-29 treatment, compared to the MAT-29 treatment ($p = 0.025$), and the SLAT-1 treatment ($p = 0.032$; Figure 2). Percentage of time spent foraging was greater for broilers in the POS-29 treatment than broilers in the SLAT-29 treatment ($p = 0.027$; Figure 2). Birds in the POS-1 treatment spent more time locomoting than birds in the SLAT-29 treatment ($p = 0.035$; Figure 2). Birds in the POS-1 and POS-29 treatments spent more time preening than birds in the SLAT-1 or MAT-29 treatments (all $p \leq 0.042$). The sum of all active behaviors in week 5 was greater in the POS-1 treatment than in the SLAT-29 ($p = 0.005$) and MAT-1 treatments ($p = 0.025$). Similarly, the time spent on active behaviors in the POS-29 treatment was greater compared to the SLAT-1 and SLAT-29 treatments, and the MAT-1 treatment (all $p \leq 0.01$; Figure 2). In week 5, birds in the NEG treatment were more active than birds in the SLAT-29 treatment ($p = 0.018$).

In week 6, treatments affected percentage of time spent foraging ($p = 0.009$) and preening ($p = 0.022$), while other behavior was unaffected ($p > 0.172$). Birds in the POS-1 treatment spent more time foraging compared to birds in all MAT and SLAT treatments (all $p \leq 0.030$; Figure 2). In week 6, birds in the POS-1 treatment spent more time preening than birds in the MAT-1 treatment ($p = 0.046$; Figure 2).

The behavioral repertoire of broiler chickens changed as they aged (all $p < 0.05$; Table 2), with more time spent passive ($p < 0.001$), and less time spent active ($p < 0.001$; Figure 3).

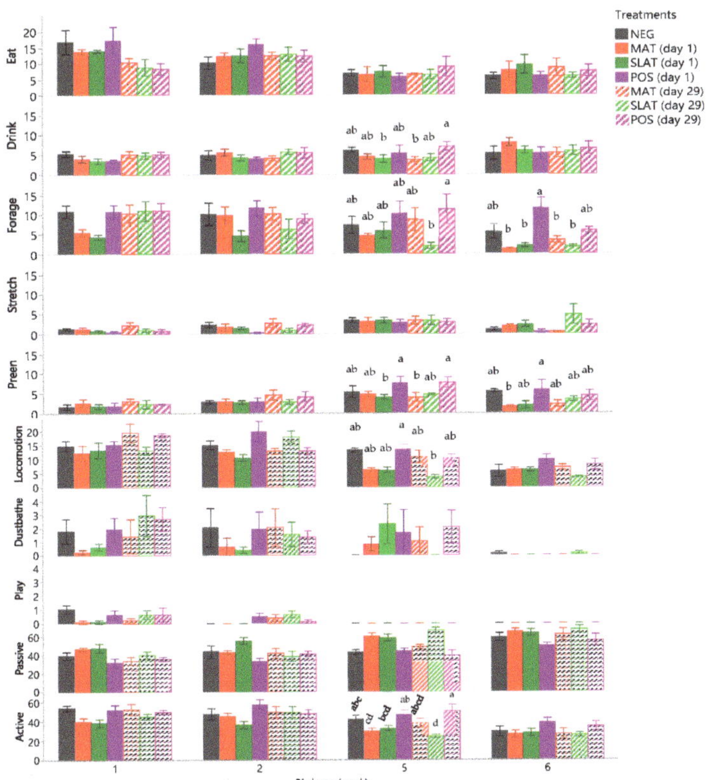

Figure 2. Percentage of observations (mean % ± SEM) spent on eating, drinking, foraging, stretching, preening, locomotion, dustbathing, play, passive, and active behaviors by bird age (in weeks) and by treatments group, with negative control (NEG), iodine mat (MAT), iodine mat with slatted floor (SLAT), and clean litter (POS) provided either from day 1 of age or 29 of age. Within week means without a common superscript ([a–d]) differed at $p < 0.05$.

Table 2. Behaviors as percentage of observations at pen-level (mean % ± SE) during week 1, 2, 5 and 6 of life.

Behavior	Week				Age Effect (*p*-Value)
	1	2	5	6	
Eat	12.80 ± 1.07 [a]	12.61 ± 0.69 [a]	6.88 ± 0.63 [b]	7.29 ± 0.73 [b]	<0.001
Drink	4.47 ± 0.29 [b]	4.87 ± 0.34 [ab]	4.97 ± 0.43 [ab]	6.05 ± 0.47 [a]	0.034
Forage	9.14 ± 0.78 [a]	8.85 ± 0.81 [a]	7.17 ± 1.00 [a]	4.56 ± 0.77 [b]	<0.001
Stretch	1.12 ± 0.18 [b]	1.64 ± 0.26 [b]	3.15 ± 0.30 [a]	1.86 ± 0.45 [b]	<0.001
Preen	2.22 ± 0.28 [c]	3.26 ± 0.34 [bc]	5.47 ± 0.47 [a]	3.57 ± 0.49 [b]	<0.001
Locomotion	15.26 ± 0.89 [a]	14.49 ± 0.87 [a]	9.06 ± 0.81 [b]	6.57 ± 0.56 [c]	<0.001
Dustbathe [1]	1.67 ± 0.36 [a]	1.42 ± 0.35 [a]	1.13 ± 0.39 [ab]	0.04 ± 0.03 [b]	<0.001
Play [1]	0.50 ± 0.11 [a]	0.24 ± 0.07 [ab]	0.00 ± 0.00 [c]	0.00 ± 0.00 [c]	<0.001
Passive	38.92 ± 1.68 [c]	41.93 ± 1.96 [c]	51.28 ± 2.20 [b]	60.43 ± 1.86 [a]	<0.001
Active	47.16 ± 1.70	47.37 ± 1.97	37.81 ± 2.13	29.92 ± 1.65	<0.001
Other	4.56 ± 0.35	4.06 ± 0.28	2.96 ± 0.24	2.64 ± 0.37	-
Out of view	9.36 ± 0.78	6.63 ± 0.73	7.95 ± 0.70	7.01 ± 0.69	-

Means within a row without a common superscript ([a–c]) differed at $p < 0.05$; [1] Non-parametric analysis; - Not assessed.

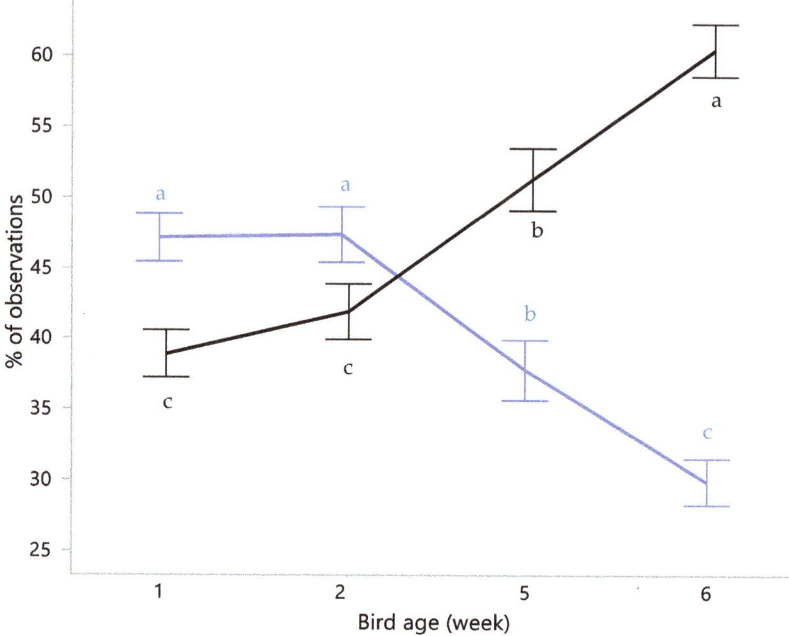

Figure 3. Percentage (means ± SEM) of observations that birds were active (sum of all active behaviors; in blue) or passive (in black) in week 1, 2, 5 and 6. Percentages of active and passive behaviors do not sum to 100% as "out of view" and "other" were excluded from the calculations. Means within a behavioral category without common superscripts differed at $p < 0.05$.

4. Discussion

This study evaluated the impact of pen-level flooring treatments on the behavioral repertoire of broiler chickens. In our study this impact was found when birds were 5 and 6 weeks old. Birds housed in the MAT and SLAT treatments generally showed less frequent foraging, preening, locomotion and overall active behaviors than in the POS treatment in week 5 of age, as well as less frequent foraging and preening than in the POS treatment in week 6 of age. Some of these differences in behavioral repertoire could be due to more severe (and possibly painful [23,24]) footpad dermatitis in all MAT and SLAT treatments compared to the NEG and POS treatments. In week 5 and 6, mean footpad dermatitis scores (0–4 categorical scale [32]) were between 1.2 and 2.3 for both MAT treatments and between 1.1 and 2.6 for both SLAT treatments, compared to between 0 and 0.6 for birds in the NEG and POS treatments [25].

Foraging and dustbathing are part of a chicken's natural behavioral repertoire [2,33]. Birds housed with access to clean, regularly replaced pine wood shavings foraged more than birds housed in some of the MAT and SLAT treatment groups, but similarly to the used litter (NEG) treatment in those weeks. Birds in MAT and SLAT treatment groups may have redirected their foraging behavior towards the feeder because of the limited litter access, as was theorized for broilers housed on a slatted or partially slatted floor [13]. However, time spent eating did not differ between treatments [25]. Access to clean litter promoted foraging at 5 and 6 weeks of age. This result is comparable to findings by [5], who found that clean wood shavings were an attractive foraging substrate. They assessed substrate preferences of commercially housed broilers by providing five different substrates within metal rings. Birds preferred to forage in clean wood shavings compared to straw pellets and the control litter (new shavings from day 1 but not replaced or replenished [5]). However, their birds also preferred to sit inactive in clean shavings compared to some of the other substrates, whilst in our study birds tended to be equally or more active on clean litter. Although not formally analyzed, birds in our MAT and SLAT treatments seemed to prefer spending time in litter rather than on the mat or the slatted flooring, with least behaviors observed on the mat (34% of floor surface; 21% of observations in MAT and 23% in SLAT) or slatted flooring (24% of floor surface; 17% of observations in SLAT) compared to on litter (79% of observations in MAT (66% of floor surface), 60% in SLAT (42% of floor surface); results not presented). This possible aversion to the novel flooring treatments and possible overcrowding in litter could have contributed to the detrimental effect of those treatments on behavior.

As foraging frequency decreased with age in our study and in previous research [5], it is an especially important finding that foraging behavior can be stimulated in older broilers by providing an appropriate substrate. Flooring treatments did not impact the observed prevalence of eating, stretching, dustbathing, play, or general passivity in this study. Prevalences of eating and passivity were comparable as reported for conventionally raised Ross 308 broilers (eat: 10–16% of observations; lie/rest: 52–77% of observations [34]), while dustbathing was observed more frequently in the current study (less than 1% in [34]). Somewhat comparable to our study, enrichments in a commercial setting or in a pilot study also did not impact play behavior in broilers [15,16].

Most active behaviors were observed less frequently with increasing age, including eating, foraging, locomotion, dustbathing, and play. Dustbathing is performed in short periods of time, with peak frequency around midday in laying hens and broilers [31,34]. With observations in our study performed during morning and evening hours, our findings align with expectations and previous research, in that the proportion of time spent dustbathing is small [31,35]. Dustbathing was reduced in week 6 compared to week 1 and 2, but was not significantly impacted by treatments. This suggests that age rather than available substrate affected the ability or motivation to perform dustbathing. As expected, passivity was observed more frequently with increasing age, which is in line with previous research on broilers [5,30,34,35], with 50–60% of observations of fast-growing broilers spent sitting on the floor or resting [31,36]. Drinking increased as birds aged (in line with [35]), whilst stretching and preening peaked at week 5. Early work showed that preening was more frequent and showed different patterns in feed-deprived and thwarted hens compared to hens under normal conditions, suggesting preening as a redirected behavior indicative of frustration [37]. Bokkers and Koene [31] observed broiler behaviors until 12 weeks of age, and found an increase in preening with age. The authors theorized that increased preening may be due to frustration related to poor mobility while birds had an equal motivation to walk compared to at an earlier age. Our results somewhat support the theory that poor mobility at a later age affected the birds' behavior. Foraging, dustbathing, stretching and locomotion require energetic movements and exercise of the legs [38], and these behaviors were decreased in week 6 compared to week 5 of age. Preening, the behavior potentially indicative of frustration at a later age, was more frequent in week 6 than week 1, but similar to week 2, and less frequent than in week 5. This suggests that as birds age their ability to express certain behaviors is inhibited, likely by their body weight [36,39]. A further investigation of broilers' ability to perform active behaviors at a later age could focus on the use of analgesia to reduce the pain experience and the hypothesized increase in active behaviors thereafter. In addition, to determine the impact of this inhibited ability, an assessment of affective states could be valuable, for instance using a cognitive bias test [40].

Although treatments did not impact passivity, they did impact active behaviors in week 5, with birds spending most time on active behaviors in the POS-29 treatment. Thus, the novelty of clean litter after 4 weeks of access to used litter stimulated broilers' activity, although only in week 5 and no longer in week 6. This implies that providing commercially housed broilers with fresh litter later in life could boost their activity even at relatively high body weights, but only for a short period of time.

With two replicates per treatment this study had limited statistical power. Therefore, further research on broiler chickens' behavioral repertoire and the impact of flooring treatments is recommended. Especially for behaviors that were not impacted by treatments in the current study as low statistical power could lead to type II error (not identifying a significant difference between treatment groups, thus a false negative).

5. Conclusions

Our results indicate that access to partially slatted flooring and/or disinfectant mats does not benefit broiler chicken welfare in terms of their ability to express highly motivated behaviors. We identified a detrimental effect of these novel flooring types (slats and/or mats) on the expression of foraging, preening, locomotion and overall active behaviors in week 5 of age, as well on foraging and preening in week 6 of age. Thus, our results suggest that access to clean, regularly replaced litter, is beneficial for broiler chicken welfare in terms of their ability to express their normal behavioral repertoire.

Author Contributions: Conceptualization, L.J., F.A.M.T. and A.G.; methodology, L.J., F.A.M.T., and A.G.; validation, L.J.; formal analysis, L.J.; investigation, S.M., N.F.; resources, L.J.; data curation, L.J.; writing—original draft preparation, L.J.; writing—review and editing, L.J., F.A.M.T. and A.G.; visualization, L.J.; supervision, L.J.; project administration, L.J.; funding acquisition, L.J. and F.A.M.T.; All authors have read and agreed to the published version of the manuscript.

Funding: This research was supported by the Flanders Research Institute for Agriculture, Fisheries and Food (ILVO).

Institutional Review Board Statement: This experiment was approved by the Institutional Animal Care and Use Committee (IACUC) of Virginia Tech (protocol 18-246).

Informed Consent Statement: Not Applicable.

Data Availability Statement: The data presented in this study are available upon request from the corresponding author.

Acknowledgments: We would like to thank Gracie Anderson, Carley Elliot, Caroline Flood, Kylie Hericks, and Sarina Williams for their valuable help during the experiment. We owe many thanks to the late Dale Shumate for his help as the farm manager.

Conflicts of Interest: The authors declare no conflict of interest. The funders had no role in the design of the study; in the collection, analyses, or interpretation of data; in the writing of the manuscript, or in the decision to publish the results.

References

1. Newberry, R.C. Environmental enrichment: Increasing the biological relevance of captive environments. *Appl. Anim. Behav. Sci.* **1995**, *44*, 229–243. [CrossRef]
2. Bracke, M.B.M.; Hopster, H. Assessing the importance of natural behavior for animal welfare. *J. Agric. Environ. Ethics* **2006**, *19*, 77–89. [CrossRef]
3. Shields, S.J.; Garner, J.P.; Mench, J.A. Dustbathing by broiler chickens: A comparison of preference for four different substrates. *Appl. Anim. Behav. Sci.* **2004**, *87*, 69–82. [CrossRef]
4. Lindberg, A.C.; Nicol, C.J. Dustbathing in modified battery cages: Is sham dustbathing an adequate substitute? *Appl. Anim. Behav. Sci.* **1997**, *55*, 113–128. [CrossRef]
5. Baxter, M.; Bailie, C.L.; O'Connell, N.E. An evaluation of potential dustbathing substrates for commercial broiler chickens. *Animal* **2018**, *12*, 1933–1941. [CrossRef]
6. Toghyani, M.; Gheisari, A.; Modaresi, M.; Tabeidian, S.A.; Toghyani, M. Effect of different litter material on performance and behavior of broiler chickens. *Appl. Anim. Behav. Sci.* **2010**, *122*, 48–52. [CrossRef]
7. Boissy, A.; Manteuffel, G.; Jensen, M.B.; Moe, R.O.; Spruijt, B.; Keeling, L.J.; Winckler, C.; Forkman, B.; Dimitrov, I.; Langbein, J.; et al. Assessment of positive emotions in animals to improve their welfare. *Physiol. Behav.* **2007**, *92*, 375–397. [CrossRef]
8. Fraser, D.; Duncan, I. 'Pleasures', 'Pains' and Animal Welfare: Toward a Natural History of Affect. *Anim. Welf. Collect.* **1998**, *7*, 383–396.
9. Dawkins, M.S. From an animal's point of view: Motivation, fitness, and animal welfare. *Behav. Brain Sci.* **1990**, *13*, 1–9. [CrossRef]
10. Fraser, D. Understanding animal welfare. *Acta Vet. Scand.* **2008**, *50*, 61–78. [CrossRef]
11. Bach, M.H.; Tahamtani, F.M.; Pedersen, I.J.; Riber, A.B. Effects of environmental complexity on behaviour in fast-growing broiler chickens. *Appl. Anim. Behav. Sci.* **2019**, *219*, 104840. [CrossRef]
12. Vasdal, G.; Vas, J.; Newberry, R.C.; Moe, R.O. Effects of environmental enrichment on activity and lameness in commercial broiler production. *J. Appl. Anim. Welf. Sci.* **2019**, *22*, 197–205. [CrossRef] [PubMed]

13. Chuppava, B.; Visscher, C.; Kamphues, J. Effect of Different Flooring Designs on the Performance and Foot Pad Health in Broilers and Turkeys. *Animals* **2018**, *8*, 70. [CrossRef] [PubMed]
14. Bizeray, D.; Estevez, I.; Leterrier, C.; Faure, J.M. Effects of increasing environmental complexity on the physical activity of broiler chickens. *Appl. Anim. Behav. Sci.* **2002**, *79*, 27–41. [CrossRef]
15. Liu, Z.; Torrey, S.; Newberry, R.C.; Widowski, T. Play behaviour reduced by environmental enrichment in fast-growing broiler chickens. *Appl. Anim. Behav. Sci.* **2020**, *232*, 105098. [CrossRef]
16. Baxter, M.; Bailie, C.L.; O'Connell, N.E. Play behaviour, fear responses and activity levels in commercial broiler chickens provided with preferred environmental enrichments. *Animal* **2019**, *13*, 171–179. [CrossRef]
17. Opengart, K.; Bilgili, S.F.; Warren, G.L.; Baker, K.T.; Moore, J.D.; Dougherty, S. Incidence, severity, and relationship of broiler footpad lesions and gait scores of market-age broilers raised under commercial conditions in the southeastern United States. *J. Appl. Poult. Res.* **2018**, *27*, 424–432. [CrossRef]
18. Lund, V.P.; Nielsen, L.R.; Oliveira, A.R.S.; Christensen, J.P. Evaluation of the Danish footpad lesion surveillance in conventional and organic broilers: Misclassification of scoring. *Poult. Sci.* **2017**, *96*, 2018–2028. [CrossRef] [PubMed]
19. Martland, M.F. Ulcerative dermatitis in broiler chickens: The effects of wet litter. *Avian Pathol.* **1985**, *14*, 353–364. [CrossRef] [PubMed]
20. Grimes, J.L.; Smith, J.; Williams, C.M. Some alternative litter materials used for growing broilers and turkeys. *Worlds. Poult. Sci. J.* **2002**, *58*, 515–526. [CrossRef]
21. de Jong, I.C.; Gunnink, H.; van Harn, J. Wet litter not only induces footpad dermatitis but also reduces overall welfare, technical performance, and carcass yield in broiler chickens. *J. Appl. Poult. Res.* **2014**, *23*, 51–58. [CrossRef]
22. Bassler, A.W.; Arnould, C.; Butterworth, A.; Colin, L.; De Jong, I.C.; Ferrante, V.; Ferrari, P.; Haslam, S.; Wemelsfelder, F.; Blokhuis, H.J. Potential risk factors associated with contact dermatitis, lameness, negative emotional state, and fear of humans in broiler chicken flocks. *Poult. Sci.* **2013**, *92*, 2811–2826. [CrossRef]
23. Gentle, M.J.; Tilston, V.; McKeegan, D.E.F. Mechanothermal nociceptors in the scaly skin of the chicken leg. *Neuroscience* **2001**, *106*, 643–652. [CrossRef]
24. Michel, V.; Prampart, E.; Mirabito, L.; Allain, V.; Arnould, C.; Huonnic, D.; Le Bouquin, S.; Albaric, O. Histologically-validated footpad dermatitis scoring system for use in chicken processing plants. *Br. Poult. Sci.* **2012**, *53*, 275–281. [CrossRef]
25. Freeman, N.; Tuyttens, F.A.M.; Johnson, A.; Marshall, V.; Garmyn, A.; Jacobs, L. Remedying Contact Dermatitis in Broiler Chickens with Novel Flooring Treatments. *Animals* **2020**, *10*, 761. [CrossRef]
26. Xavier, D.B.; Broom, D.M.; Mcmanus, C.M.P.; Torres, C.; Bernal, F.E.M. Number of flocks on the same litter and carcase condemnations due to cellulitis, arthritis and contact foot-pad dermatitis in broilers. *Br. Poult. Sci.* **2010**, *51*, 586–591. [CrossRef] [PubMed]
27. Bilgili, S.F.; Hess, J.B.; Blake, J.P.; Macklin, K.S.; Saenmahayak, B.; Sibley, J.L. Influence of bedding material on footpad dermatitis in broiler chickens. *J. Appl. Poult. Res.* **2009**, *18*, 583–589. [CrossRef]
28. Baxter, M.; Bailie, C.L.; O'Connell, N.E. Evaluation of a dustbathing substrate and straw bales as environmental enrichments in commercial broiler housing. *Appl. Anim. Behav. Sci.* **2018**, *200*, 78–85. [CrossRef]
29. Jacobs, L.; Vezzoli, G.; Beerda, B.; Mench, J.A. Northern fowl mite infestation affects the nocturnal behavior of laying hens. *Appl. Anim. Behav. Sci.* **2019**, *216*, 33–37. [CrossRef]
30. Meyer, M.M.; Johnson, A.K.; Bobeck, E.A. Development and Validation of Broiler Welfare Assessment Methods for Research and On-farm Audits. *J. Appl. Anim. Welf. Sci.* **2020**, *23*, 433–446. [CrossRef]
31. Bokkers, E.A.M.; Koene, P. Behaviour of fast- and slow growing broilers to 12 weeks of age and the physical consequences. *Appl. Anim. Behav. Sci.* **2003**, *81*, 59–72. [CrossRef]
32. Welfare Quality®Network. *Welfare Quality®Assessment Protocol for Poultry (Broilers, Laying Hens)*; Welfare Quality®Consortium: Lelystad, The Netherlands, 2009.
33. Vestergaard, K. Dust-bathing in the domestic fowl—diurnal rhythm and dust deprivation. *Appl. Anim. Ethol.* **1982**, *8*, 487–495. [CrossRef]
34. Bergmann, S.; Schwarzer, A.; Wilutzky, K.; Louton, H.; Bachmeier, J.; Schmidt, P.; Erhard, M.; Rauch, E. Behavior as welfare indicator for the rearing of broilers in an enriched husbandry environment—A field study. *J. Vet. Behav.* **2017**, *19*, 90–101. [CrossRef]
35. Cornetto, T.; Estevez, I. Behavior of the Domestic Fowl in the Presence of Vertical Panels. *Poult. Sci.* **2001**, *80*, 1455–1462. [CrossRef] [PubMed]
36. Castellini, C.; Mugnai, C.; Moscati, L.; Mattioli, S.; Guarino Amato, M.; Cartoni Mancinelli, A.; Dal Bosco, A. Adaptation to organic rearing system of eight different chicken genotypes: Behaviour, welfare and performance. *Ital. J. Anim. Sci.* **2016**, *15*, 37–46. [CrossRef]
37. Duncan, I.J.H.; Wood-Gush, D.G.M. An Analysis of Displacement Preening in the Domestic Fowl. *Anim. Behav.* **1972**, *20*, 68–71. [CrossRef]
38. Arnould, C.; Bizeray, D.; Faure, J.; Leterrier, C. Effects of the addition of sand and string to pens on use of space, activity, tarsal angulations and bone composition in broiler chickens. *Anim. Welf.* **2004**, *8*, 87–94.

39. Bokkers, E.A. *Behavioural Motivations and Abilities in Broilers*; Wageningen University: Wageningen, The Netherlands, 2004.
40. Anderson, M.G.; Campbell, A.M.; Crump, A.; Arnott, G.; Newberry, R.C.; Jacobs, L. Effect of Environmental Complexity and Stocking Density on Fear and Anxiety in Broiler Chickens. *Animals* **2021**, *11*, 2383. [CrossRef] [PubMed]

MDPI

St. Alban-Anlage 66

4052 Basel

Switzerland

www.mdpi.com

Animals Editorial Office

E-mail: animals@mdpi.com

www.mdpi.com/journal/animals

9 783725 811045